Aquatic Microbial Ecology and Biogeochemistry: A Dual Perspective

Patricia M. Glibert • Todd M. Kana
Editors

Aquatic Microbial Ecology and Biogeochemistry: A Dual Perspective

Springer

Editors
Patricia M. Glibert
Horn Point Laboratory
University of Maryland Center for Environmental Science
Cambridge, MD, USA

Todd M. Kana
Horn Point Laboratory
University of Maryland Center for Environmental Science
Cambridge, MD, USA

ISBN 978-3-319-30257-7 ISBN 978-3-319-30259-1 (eBook)
DOI 10.1007/978-3-319-30259-1

Library of Congress Control Number: 2016942786

Printed on acid-free paper

This Springer imprint is published by Springer Nature
The registered company is Springer International Publishing AG Switzerland

Preface: Building on a History of Dual Careers in the Sciences

Aquatic microbial ecology and the processes of biogeochemistry are coupled and synergistic. The aim of this book is to highlight some perspectives, insights, and data in the coupled fields of aquatic microbial ecology and biogeochemistry when viewed through the lens of collaborative duos—dual career couples. Their synergy and collaborative interactions have contributed substantially to our contemporary understanding of pattern, process, and dynamics. This is thus a book *by* dual career couples, not *about* them.

A great deal has been written about dual career couples—the challenges they face in job searches, spousal hiring negotiations as well as the daily challenges and rewards of working in like-minded professions, often in shared offices or work environments (e.g., Ferber and Loeb 1997). Little has been written to highlight the important contributions many dual career couples have made to contemporary aquatic sciences. This book aims to do just that.

Science has long attracted bright minds, and bright minds often think alike. Most notable are the spousal collaborations of the Curies and the Einsteins, but, in fact, there is a long history of creative couples working together—especially in fields such as ecology and plant sciences. As noted by Slack (1996), "Collaborative marriages were common in botany even in the first half of the nineteenth century in England. However, many of these partnerships were highly imbalanced, as was the case, for example, of the collaborative marriage of Elizabeth G. Knight and Nathaniel Lord Britton. He, a professor of Botany at Columbia by 1891, and she a Hunter College graduate. Their collaboration was one in which he had the sole paid position, and she conducted her research unpaid. Nevertheless, she went on to publish over 300 articles and reviews, edited several journals and even supervised at least one graduate student" (Slack 1996). Theirs was a "personal and botanical partnership." While their partnership was scientifically fulfilling, it highlights the frustrations of antinepotism regulations, sexism, and bias that have pervaded academia for decades. The Comstocks of Cornell, Anna Botsford Comstock and John Henry Comstock, had a similar professional and personal partnership, with Anna only embarking on a teaching career late in life, having been the "power behind the throne" for decades. Because Cornell, where she had been an instructor of nature-study, did not permit full professorships for women except in the Department of Home Economics (Henson 1996), only 2 years before her retirement, in 1922, was she finally advanced to full professor.

These vignettes are interesting reading, and the collection of essays edited by Pycior et al. (1996) describes many more, but many of the experiences are not so dissimilar from experiences today, where many dual career couples have to make do in order to advance their collective personal and professional careers. Fortunately, pathways for dual career couples are improving, with many institutions recognizing that spousal collaborative teams often bring a unique and special strength to a collaboration. Whether they work closely on a daily basis, or whether they share only a similar passion for their work but maintain distinct disciplines, couples share "in the joys and the sorrows of their life-companions, but they also have a part in their thoughts, their studies, their labors, their achievements" (Mozens 1913, cited in Pycior et al. 1996). This book hopes to celebrate those thoughts, studies, labors, and achievements.

In the field of aquatic microbial ecology, as is most likely the case in all sciences, dual career couples are more and more common. As recently reported in an article in *Nature* highlighting dual career couples, and based on a National Science Foundation survey, over 25 % of "married people with doctorates had a spouse working in science or engineering" (Smith 2014). The goal of this book was to assemble a series of chapters written by dual career couples—and only by them. The couples selected were invited because we aimed to represent established scientists as well as up-and-coming scientists, those working in both marine and freshwater, across a spectrum of topics, and who represented a diverse geographic area. The end result is the nearly two dozen papers herein, representing contributions from across the globe, from USA to Australia and Europe (France, Germany), and the UK (Wales). More than that, several of the couples originally hail from outside the USA, including Africa, India and Israel. In fact, based on the couples we knew and came to know during this process, we could have filled three volumes rather than just one.

The chapters presented here represent a wide-ranging collection of topics. This is not a structured book, wherein a set suite of topics was outlined and authors chosen to each write their piece of the assignment. Rather, this collection of chapters flowed organically. The first set of chapters begins to unravel microbial diversity, from a discussion of the history of ideas of the microbial web (Sherr and Sherr) to an overview of the drivers of biodiversity (Rynearson and Menden-Deuer) and selective pressures of phytoplankton shape (Karp Boss and Boss), followed by a phylogenetic analysis of bacteria across the freshwater–saltwater continuum (Bižić-Ionescu and Ionescu), and an analysis of the challenges linking biogeochemical models with 'omics data (Coles and Hood). These chapters are followed by two chapters that address the history of ideas related to stoichiometry (Kilham and Kilham) and their implications related to plankton predation (Mitra and Flynn). Understanding the mysteries of light and nitrogen and their regulation is the topic of the next set of chapters. This part begins with a perspective on the dynamics of saturation responses (Kana and Glibert) and moves to the metabolism of nitrogen in terms of nitrate reductase (Young and Berges), and ammonium as a paradoxical nutrient (Wilkerson and Dugdale), and the mysteries of nitrogen fixation (Marino and Howarth), and an analysis of the global distribution of subsurface chlorophyll maximum layers (Silsbe and Malkin). The chapters then transition to a series of analyses in which the dynamics of changing ecosystems is the focus, generally using longer-term records or observations, including the Arabian Sea (Goes and Gomes), the Arctic Ocean (Grebmeier and Cooper), the coastal North Sea (Wiltershire and Boersma), and Pensacola Bay (Caffrey and Murrell), to the even fresher waters of the Seine River (Garnier and Billen) and Maine's Mount Desert Island lakes (Roesler and Culbertson). Finally, the last suite of chapters highlights unique systems, processes, and dynamics, from the Sargasso Sea (Pinckney and Richardson) to Australia's Gulf of Carpentaria (Rothlisberg and Burford), the Chesapeake Bay (Sellner and Sellner), and the freshwater hydrothermal vents of the Yellowstone National Park (Aguilar and Cuhel). In all, these chapters take us from the Arctic to Africa, from the Arabian Sea to Australia, from small lakes and Yellowstone hot vents to the Sargasso Sea, and in the process provide analyses that make us think about the structure and function of all of these systems in the aquatic realm.

While the chapters speak for themselves, each chapter is told with a little bit of backstory—a small snapshot into the lives of the couple, who they are, how they met, their children (they are couples, after all), or what else they may enjoy outside of science. It is something special to learn about friendships turned more serious on station, underwater marriage proposals, and at-sea weddings on oceanographic ships. Aquatic science is in the veins of the couples whose work is presented here. The more traditional bios of the authors can be found elsewhere; we are interested to peek into the human dimension of these great teams. This is a book about dual perspectives, and a snapshot into work–life balance helps to put it all in "perspective."

Our hope is that this book is useful not only for the depth and breadth of knowledge that is conveyed in the chapters, but that this book also serves to be useful for future dual career couples faced with the challenges only dual career couples face. Great teams do make great science. Of the couples highlighted here, 78 % have positions at the same institution. Another interesting statistic here is that 75 % of the chapters herein were first-authored by the distaff side; this just shows who really does the work in these collaborations! This book should give new meaning to the concept of coupled biogeochemical cycles.

We wish to thank our authors for their contributions; we thank all those who wanted to contribute but could not for varying reasons. We thank our editor, Janet Slobodien at Springer, for welcoming this book into the portfolio of Springer science books. Finally, we thank Kate, Zander and Coe, Austin and Hannah, and Patrick and Emily for filling the nonscientific parts of our lives with joy—and for allowing just a little bit of the science in their blood to shine through.

Cambridge, MD, USA

Patricia M. Glibert
Todd M. Kana

References

1. Ferber MA, Loeb JA (eds) Academic couples: problems and promises. University of Illinois Press, Urbana, IL, p 323
2. Henson PM (1996) The Comstocks of Cornell: a marriage of interests. In: Pycior HM, Slack NG, Abir-Am PG (eds) Creative couples in the sciences. Rutgers University Press, New Brunswick, NJ, pp 112–125
3. Pycior HM, Slack NG, Abir-Am PG (eds) Creative couples in the sciences. Rutgers University Press, New Brunswick, NJ, p 369
4. Slack NG. Botanical and ecological couples: a continuum of relationships. In: Pycior HM, Slack NG, Abir-Am PG (eds) Creative couples in the Sciences. Rutgers University Press, New Brunswick, NJ, pp 235–253
5. Smith K (2014) Love in the lab: close collaborations. Nature 510:460

Contents

Part III Understanding the Mysteries of Light and Nitrogen

Contributors

Carmen Aguilar School of Freshwater Sciences, University of Wisconsin-Milwaukee, Milwaukee, WI, USA

John A. Berges Department of Biological Sciences and School of Freshwater Sciences, University of Wisconsin-Milwaukee, Milwaukee, WI, USA

Gilles Billen Université Pierre et Marie Curie, CNRS UMR 7619 Metis, Paris, France

Mina Bižić-Ionescu Department of Experimental Limnology, Leibniz Institute for Freshwater Ecology and Inland Fisheries, IGB, Neuglobsow, Stechlin, Germany

Maarten Boersma Alfred-Wegener-Institut Helmholtz-Zentrum für Polar- und Meeresforschung, Biologische Anstalt Helgoland, Helgoland, Germany

University of Bremen, Bremen, Germany

Emmanuel Boss School of Marine Science, University of Maine, Orono, ME, USA

Michele A. Burford Australian Rivers Institute, Griffith University, Nathan, QLD, Australia

Jane M. Caffrey Center for Environmental Diagnostics and Bioremediation, University of West Florida, Pensacola, FL, USA

Victoria J. Coles Horn Point Laboratory, University of Maryland Center for Environmental Science, Cambridge, MD, USA

Lee W. Cooper Chesapeake Biological Laboratory, University of Maryland Center for Environmental Science, Solomons, MD, USA

Russell Cuhel School of Freshwater Sciences, University of Wisconsin-Milwaukee, Milwaukee, WI, USA

Charles Culbertson U.S. Geological Survey, New England Water Science Center, Augusta, ME, USA

William C. Dennison University of Maryland Center for Environmental Science, Cambridge, MD, USA

Richard Dugdale Romberg Tiburon Center of Environmental Studies, San Francisco State University, Tiburon, CA, USA

Kevin J. Flynn Swansea University, Swansea, UK

Josette Garnier Université Pierre et Marie Curie, CNRS UMR 7619 Metis, Paris, France

Patricia M. Glibert Horn Point Laboratory, University of Maryland Center for Environmental Science, Cambridge, MD, USA

Joaquim I. Goes Marine Biology and Paleoenvironment, Lamont Doherty Earth Observatory at Columbia University, New York, NY, USA

Helga do R. Gomes Marine Biology and Paleoenvironment, Lamont Doherty Earth Observatory at Columbia University, New York, NY, USA

Jacqueline M. Grebmeier Chesapeake Biological Laboratory, University of Maryland Center for Environmental Science, Solomons, MD, USA

Raleigh R. Hood Horn Point Laboratory, University of Maryland Center for Environmental Science, Cambridge, MD, USA

Robert W. Howarth Department of Ecology and Evolutionary Biology, Cornell University, Ithaca, NY, USA

Danny Ionescu Department of Experimental Limnology, Leibniz Institute for Freshwater Ecology and Inland Fisheries, IGB, Neuglobsow, Stechlin, Germany

Todd M. Kana Horn Point Laboratory, University of Maryland Center for Environmental Science, Cambridge, MD, USA

Lee Karp-Boss School of Marine Science, University of Maine, Orono, ME, USA

Susan S. Kilham Department of Biodiversity, Earth and Environmental Science, Drexel University, Philadelphia, PA, USA

Peter Kilham (deceased)

Sairah Y. Malkin Horn Point Laboratory, University of Maryland Center for Environmental Science, Cambridge, MD, USA

Department of Marine Sciences, University of Georgia (UGA), Athens, GA, USA

Roxanne Marino Department of Ecology and Evolutionary Biology, Cornell University, Ithaca, NY, USA

Susanne Menden-Deuer Graduate School of Oceanography, University of Rhode Island, Narragansett, RI, USA

Aditee Mitra Swansea University, Swansea, UK

Michael C. Murrell Gulf Ecology Division, US Environmental Protection Agency, Office of Research and Development, National Health and Environmental Effects Research Laboratory, Gulf Breeze, FL, USA

Judith M. O'Neil Horn Point Laboratory, University of Maryland Center for Environmental Science, Cambridge, MD, USA

James L. Pinckney Marine Science Program and Department of Biological Sciences, University of South Carolina, Columbia, SC, USA

Tammi L. Richardson Marine Science Program and Department of Biological Sciences, University of South Carolina, Columbia, SC, USA

Collin Roesler Department of Earth and Oceanographic Science, Bowdoin College, Brunswick, ME, USA

Peter C. Rothlisberg CSIRO Oceans and Atmosphere Flagship, Brisbane, QLD, Australia

Tatiana A. Rynearson Graduate School of Oceanography, University of Rhode Island, Narragansett, RI, USA

Kevin G. Sellner Center for Coastal and Watershed Studies, Hood College, Frederick, MD, USA

Stella G. Sellner Patuxent Environmental and Aquatic Research Laboratory, Morgan State University, St. Leonard, MD, USA

Evelyn B. Sherr College of Earth, Atmospheric and Ocean Sciences, Oregon State University, Corvallis, OR, USA

Barry F. Sherr College of Earth, Atmospheric and Ocean Sciences, Oregon State University, Corvallis, OR, USA

Greg M. Silsbe Department of Botany and Plant Pathology, Oregon State University, Corvallis, OR, USA

Horn Point Laboratory, University of Maryland Center for Environmental Science, Cambridge, MD, USA

Frances Wilkerson Romberg Tiburon Center of Environmental Studies, San Francisco State University, Tiburon, CA, USA

Karen H. Wiltshire Alfred-Wegener-Institut Helmholtz-Zentrum für Polar- und Meeresforschung, Wattenmeerstation Sylt, List/Sylt, Germany

Jacobs University Bremen, Bremen, Germany

Erica B. Young Department of Biological Sciences and School of Freshwater Sciences, University of Wisconsin-Milwaukee, Milwaukee, WI, USA

Part I
Unraveling Microbial Diversity and Their Processes

Phagotrophic Protists: Central Roles in Microbial Food Webs

Evelyn B. Sherr and Barry F. Sherr

Overview

Protists, single-celled eukaryotes, the vast majority of which are between about 2 and 200 μm in size, are ubiquitous in marine and freshwater ecosystems. Initial grouping of microbial eukaryotes into phototrophs (algae), and heterotrophs (protozoa), has morphed into categorization of protists based on molecular genetic lineage trees (Caron et al. 2012). Many non-pigmented flagellates are closely related to photosynthetic species, photosynthetic flagellates have been demonstrated to ingest bacteria or other prey (Zubkov and Tarran 2008), and acquisition of functional chloroplasts from ingested algae is common among ciliates and dinoflagellates (Flynn et al. 2012; Stoecker et al. 2009). Commonly studied aquatic protists are algae in the phytoplankton, non-pigmented, phagotrophic flagellates in the nanoplankton (2–20 μm) size class, and phagotrophic protists (microzooplankton), mainly ciliates and non-pigmented dinoflagellates, in the microplankton (20–200 μm) size class (Sieburth et al. 1978). Microzooplankton have been shown to be significant consumers of photosynthetic microbes in the sea (Calbet and Landry 2004). Our own research has focused primarily on phagotrophic nanoflagellates and microzooplankton (examples shown in Fig. 1). Here we review method development and research findings that underpin the current understanding of the roles of protists in aquatic food webs, highlighting our own work in this field. Discovering the importance of phagotrophic protists in planktonic food webs was a crucial part of the history of marine microbial ecology (Sherr and Sherr 2008).

E.B. Sherr (✉) • B.F. Sherr
College of Earth, Atmospheric and Ocean Sciences, Oregon State University, Corvallis, OR, USA
e-mail: sherre@coas.oregonstate.edu

P.M. Glibert, T.M. Kana (eds.), *Aquatic Microbial Ecology and Biogeochemistry: A Dual Perspective*, DOI 10.1007/978-3-319-30259-1_1

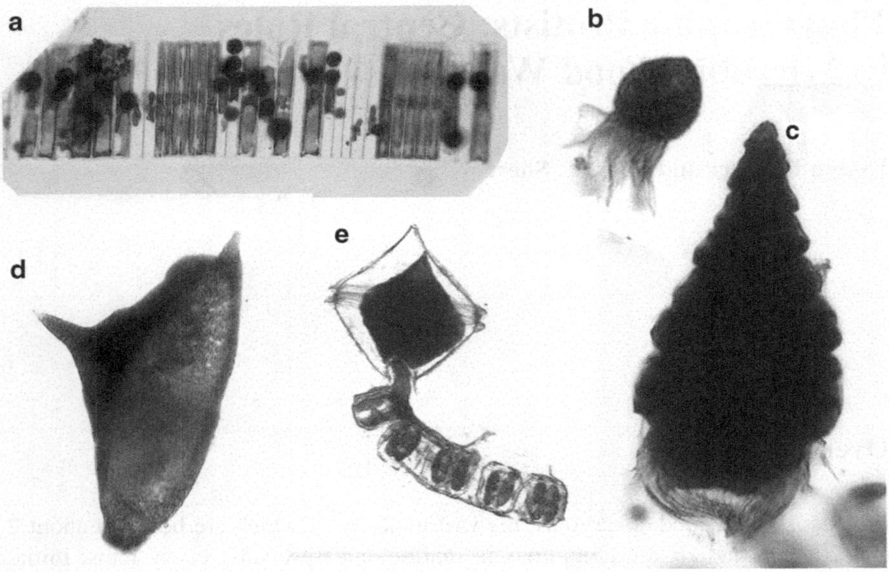

Fig. 1 Examples of protists observed in various marine systems during our career: (**a**) Parasitoid flagellates (4–6 μm) feeding on cells of a pennate diatom chain in the Bering Sea; (**b**) small spirotrichous ciliate (20×25 μm) in the plankton of the Bering Sea; (**c**) large mixotrophic ciliate, *Laboea spiralis* (50×110 μm) in the plankton of the Bering Sea; (**d**) non-thecate heterotrophic dinoflagellate, *Gyrodinium* sp. (60×90 μm), cell shape greatly distorted with an ingested pennate diatom cell, in the plankton of the coastal Oregon upwelling system; (**e**) thecate dinoflagellate, *Protoperidinium* sp. (50×60 μm) feeding extracellularly on a centric diatom chain with an extruded pseudopodial pallium in the plankton of the Bering Sea

Protists as Elemental Recyclers

The reviews of Pomeroy (1974), Azam et al (1983), and Ducklow (1983) advanced the idea that in the sea (and by implication in freshwaters as well) bacteria and their protist grazers were responsible for the bulk of ecosystem respiration and thus recycling of organic matter. Investigations of respiration and nutrient regeneration in marine habitats demonstrated that a multistep microbial food web was necessary for complete remineralization of nitrogen and phosphorus from the organic compounds in bacterial biomass.

While rates of respiration could, with caveats, be determined in seawater samples by measuring the rates of decrease of dissolved oxygen, rates of nutrient release in such experiments were too low to be accurately quantified using standard chemical analyses. Phosphate regeneration could be sensitively assessed by short-term release of radioactive P into the dissolved fraction from organic matter produced by bacteria and phytoplankton grown with added ^{32}P. However, the longest-lived radioactive form of nitrogen, ^{13}N, with a half-life of about 10 min, could not be easily used to evaluate rates of nitrogen remineralization.

Harrison (1978) and Glibert et al. (1982) pioneered the ^{15}N isotope dilution technique, in which ^{15}N labeled ammonium was added to aliquots of seawater. Seawater samples were either whole (unfiltered) or size fractioned by passing water through mesh netting or filters of progressively smaller pore size. Experimental seawater treatments with added ^{15}N ammonium were incubated under in situ conditions of light and temperature for 12–24 h. During that time, a portion of the added ^{15}N ammonium would be incorporated into phytoplankton and bacterial biomass, while the unenriched nitrogen present in organic matter would be remineralized into the pool of ammonium in the seawater. Quantitative extraction of ammonium from experimental water samples at initial and final times of incubation, followed by measurement of the ^{15}N:^{14}N isotope ratios of the extracted ammonium, allowed estimation of the extent to which the time zero ^{15}N ammonium was diluted with unenriched recycled ammonium. Harrison's initial results documented that organisms less than 35 μm in size, including bacteria and heterotrophic protists, were responsible for about 90 % of the rate of ammonium regeneration in the coastal systems that he studied. This clever, although challenging, approach was used by investigators to establish the impact of bacteria and their grazers in nutrient recycling in other aquatic systems (Glibert et al. 1992). Additional investigations focused on the amount of regeneration of nutrients by protists feeding on bacterial and phytoplankton prey with a range of C:N and C:P ratios (Caron and Goldman 1988).

Our own contribution to protist regeneration of N and P nutrients occurred at the beginning of our work as a team, during our collaboration with Tom Berman on microbial activity in Lake Kinneret, Israel from 1979 to 1981. Research on heterotrophic protists, especially bacterivorous flagellates, was cutting edge at that time, stimulated by the work on significance of protists in nutrient regeneration and by studies of Tom Fenchel on bacterivorous flagellates in Danish coastal waters (Fenchel 1982). Berman introduced us to another Israeli ecologist, Utsa Pollinger, who told us that during the decline phase of an annual dinoflagellate bloom in the lake, there was a secondary burst of tiny non-pigmented flagellates. She wondered what caused these heterotrophic flagellates to grow up. We experimented with lake water containing both bacteria and flagellates to which we added dried dinoflagellates collected from a previous year's bloom. The dinoflagellate armor plating, the theca, was composed of sugary carbohydrate that the lake bacteria readily used as a substrate. Bacterivorous flagellates then grew up, consuming the bacteria and releasing ammonium and phosphate into the water that the bacteria could use for further growth (Sherr et al. 1982). We also evaluated ammonium excretion rates by a heterotrophic flagellate isolated from lake water fed different species of bacteria (Sherr et al. 1983).

Later, while at Oregon State University, we supervised the M.S. thesis research of Marcelino Suzuki in which he used a ^{15}N tracer approach to sensitively measure the rate of regeneration of ammonium by grazing of marine flagellates on ^{15}N labeled bacteria (Suzuki et al. 1996). Suzuki's results corroborated the earlier findings based on the ^{15}N isotope dilution method; in his study nitrogen regeneration efficiencies from bacterial-N were 30–35 % for a one-step trophic link (bacterivorous flagellates), 60 % for 5 μm filtered seawater (bacterivorous flagellates plus small flagellate predators), and 90 % for unfiltered seawater with an intact microbial food web.

Phagotrophic nanoflagellates that can participate in nutrient recycling also include parasitoid flagellates that either internally infect host cells (Guillou et al. 2008; Siano et al. 2011) or feed externally on diatoms (Fig. 1a) (Tillmann et al. 1999; Sherr et al. 2013).

Protists as Consumers of Bacteria

The discovery in the 1980s of the true abundance of heterotrophic bacteria in marine and freshwaters by using fluorescent dyes to visualize bacteria with epifluorescence microscopy, coupled with measurement of rates of bacterial biomass production via incorporation of radiolabeled thymidine and leucine (Sherr and Sherr 2008), turned attention to the processes controlling bacteria growth in aquatic systems. It was obvious that heterotrophic nanoflagellates were effective grazers of the small-sized bacterial cells present in seawater (Sieburth 1981, Fenchel 1982; Andersen and Fenchel 1985). However, approaches were needed both to accurately enumerate the in situ abundances of heterotrophic flagellates, and to determine their rates of bacterial consumption. Fluorescent dyes, notably acridine orange (AO) and DAPI, already used to enumerate aquatic bacterioplankton, were applied to determining abundance of non-pigmented flagellates (Sherr et al. 1993). UV-excited, blue fluorescing DAPI was judged to be superior to blue light-excited, green fluorescing dyes such as AO, as blue-fluorescing cells made it easier to distinguish heterotrophic flagellates from chloroplast-bearing phytoflagellates.

Development of standard methods to quantify rates of flagellate bacterivory in aquatic systems proved more difficult. A variety of approaches were tried, including comparative growth of bacteria in whole water and in water filtered to remove most bacterivores, dilution of whole water with particle free water to decrease abundance of bacterivores, ingestion of radiolabeled bacteria, and use of inhibitors of eukaryotic cell growth to eliminate, or at least decrease, protist grazing on bacteria (reviewed in Strom 2000). All of these methods had serious artifactual problems and/or were too time-consuming for routine measurements.

Visualizing direct ingestion of added fluorescent particles by heterotrophic flagellates was among the methods proposed. Initial efforts used plastic, bacterial-sized beads, with or without pre-treatment with an organic compound such as bovine serum albumin to minimize bead clumping and to make the beads more palatable to flagellates (Sherr and Sherr 1993; Strom 2000). Concern remained that the plastic beads were not adequate analogs for in situ bacterioplankton, and subsequent studies showed that some protist species did show selective preference for ingestion of bacterial cells compared to bacterial-sized beads.

With help from colleagues, including our Israeli mentor Tom Berman, we developed a method to concentrate and then stain estuarine bacteria with the green fluorescing dye DTAF (Sherr et al. 1987). Known abundances of fluorescently labeled bacteria (FLB) were added to subsamples of coastal water. Flagellates in the samples fed on bacteria, including FLB, and after a short incubation time (15–60 min) aliquots of the incubated water were preserved and processed for microscopy. By switching epifluorescence filter sets between UV light for DAPI-stained protists, and blue light for DTAF-stained FLB, ingested FLB could be visualized inside the flagellates. After determining the average number of FLB per flagellate, and the abundances of flagellates, FLB, and total bacteria in the sample, an estimate of the rate of clearance of bacteria by flagellates could be made. This approach proved to be fairly robust, and was subsequently applied in other aquatic systems. The main consumers of bacterioplankton proved to be flagellates less than 5 μm in size (Sherr and Sherr 1991). Results of FLB ingestion and other methods of estimating rates of protist bacterivory demonstrated that this source of mortality could crop a large fraction of bacterial production in both oligotrophic as well as coastal and estuarine waters (Strom 2000). Selective ingestion by flagellates of larger-sized and metabolically active bacteria was also documented (Sherr et al. 1992; Gonzalez et al. 1993; Strom 2000).

In our bacterivory experiments, we discovered that ciliates in coastal water also ingested FLB (Sherr and Sherr 1987; Sherr et al. 1989). Marine ciliates in the plankton were thought to feed on algal cells, not bacteria. But most of the bacteria-ingesting ciliates were tiny, some only 10–12 μm in diameter (Sherr et al. 1986a). In a review of our 1986 manuscript, Tom Fenchel stated that the existence of a 10 μm-sized ciliate was as incredible as that of a 1 cm-sized mammal. We rebutted that a newborn pigmy shrew was about 1 cm long; our paper was published.

Protists as Consumers of Phytoplankton

The role of aquatic protists in marine food webs as herbivores, consumers of photosynthetic cells, is quantitatively more significant than is protistan bacterivory. This fact is a consequence of the greater biomass of photosynthetic bacteria and algae compared to the biomass of heterotrophic bacteria in both coastal and open ocean systems, and of the capacity of protists, from nanoflagellates to microzooplankton, to consume a wide range of sizes of photosynthetic cells (Sherr and Sherr 1992, 1994, 2002, 2007, 2009).

Landry and Hassett (1982) developed the dilution assay to evaluate grazing impact of microzooplankton on phytoplankton. In this approach, whole water samples are diluted with freshly prepared particle free water to yield subsamples that range from 100 % whole water to 5 % or 10 % whole water, in order to proportionally reduce protist grazing on individual phytoplankton cells. Nutrients are added to ensure equivalent cell-specific phytoplankton growth across the dilution series, and initial chlorophyll concentration is determined. After a 24 h incubation under in situ conditions of light and temperature, each experimental dilution is sampled for final chlorophyll concentration, and phytoplankton growth rates are calculated based on change in chlorophyll from initial to final time in each treatment. Typically, there is a linear decrease of phytoplankton growth rate with amount of whole water in the samples. A plot of growth rate versus proportional amount of whole water yields a regression in which the y-axis intercept estimates phytoplankton growth rate in the absence of grazing (i.e. at infinite dilution), and the negative slope is the microzooplankton grazing rate. Although various concerns about the method have been raised, and dilution experiments are time-consuming to perform, the dilution assay has been adopted as the standard approach to evaluating protistan herbivory in the sea. The results have demonstrated that microzooplankton herbivory is a major source of phytoplankton mortality in all marine regions (Sherr and Sherr 2002; Calbet and Landry 2004; Sherr et al. 2009, 2013).

Protists in High Latitude Food Webs

In 1990, the year that we moved to Oregon State University, there was still surprisingly little research on microbial plankton in the Arctic Ocean, an important and climate-sensitive region of the earth. That was about to change, and we were fortunate to get onboard with developing arctic research programs.

Together with a fellow OSU oceanographer, Pat Wheeler, we were funded to participate in the Arctic Ocean Section (AOS). Two icebreakers, the U.S. Coast Guard's Polar Sea and the Canadian Coast Guard's Louis S. St. Laurent, carried an international group of marine scientists to the North Pole, the first time that North American surface ships had reached the Pole. Our part involved assessment of the abundances of microbial eukaryotes, both photosynthetic and heterotrophic, in the central Arctic Ocean (Sherr et al. 1997). In 1997–1998, again in collaboration with Wheeler, we participated in another multi-national expedition, the Surface Heat Budget of the Arctic Ocean (SHEBA), in which we were able to follow the seasonal cycle of abundances and activity of marine microbes in the central Arctic Ocean from a Canadian icebreaker, Des Groseilliers, that drifted with the ice pack for a full year (Sherr and Sherr 2003; Sherr et al. 2003). Results from both projects demonstrated that there is a diverse and active microbial food web in the Arctic Ocean, with heterotrophic flagellates feeding on both bacteria and small algae.

In subsequent high latitude projects, we shifted our focus to microzooplankton, both as herbivores and as a food resource for copepods and krill. We had long considered this latter trophic link to be a fundamental part of aquatic food webs (Sherr et al. 1986b; Sherr and Sherr 1988). We collaborated with zooplankton ecologists: Carin Ashjian at the Woods Hole Oceanographic Institution and Robert Campbell at the Graduate School of Oceanography at the University of Rhode Island, as part of the Shelf-Basin Interactions (SBI) project in the western Arctic Ocean. During cruises on the new U.S. Coast Guard icebreaker Healy in spring and summer of 2002 and 2004, our team carried out dilution assays and mesozooplankton grazing assays using on-deck incubators. Sea ice was extensive over the shelf regions during the summer of 2002, but had melted back to the central Arctic in 2004, providing a good contrast in system conditions. There was significant microzooplankton herbivory in both years (Sherr et al. 2009), and protists in the microzooplankton were avidly consumed by zooplankton, especially smaller copepods (Campbell et al. 2009).

We worked with Ashjian and Campbell on similar research in the Bering Sea in 2008–2010. The Bering Sea Ecosystem Study (BEST) was a multi-investigator program in which fisheries scientists and oceanographers collaborated in examining the impact of extent of spring sea ice on fish and marine mammal stocks in this important region. About half of the total U.S. fish catch (including pollock, halibut and salmon as well as crabs) comes from the Bering Sea, due to the extraordinary diatom blooms that start each spring at the edge of the sea ice pack, and then follow the ice as it retreats north. These diatom blooms initially support hordes of copepods and krill that are important food for fish, and then the diatoms sediment en masse to fuel the worms and clams that nourish crabs, bottom-feeding fish, gray whales, and walrus. The concern is that as climate change intensifies, the winter extent of the ice pack will lessen, and sea ice will melt sooner in the spring, disrupting the intensity and timing of the spring diatom blooms on which Bering Sea fisheries depend.

Even though we were supposed to evaluate the effects of diminished sea ice, during the March–April cruises of 2008 and 2009 there was unusually extensive sea ice, with only early-stage diatom blooms. When the spring 2010 cruise was scheduled later in the season (May–June), we were able to sample intense open water diatom blooms. As for the findings from our SBI project in the western Arctic Ocean, the BEST data underscored the importance of both diatoms and microzooplankton to Bering Sea food webs. A clear result from both these projects was that large, non-pigmented dinoflagellates were important consumers of bloom-forming diatoms, both as single cells and chains (Sherr et al. 2009, 2013). Most investigations of microzooplankton as food for zooplankton had considered ciliates (e.g. Fig. 1b, c) as the major group of herbivorous protists. We found that in the western Arctic Ocean, in the Bering Sea, and also in upwelling diatom blooms in Oregon coastal waters, herbivorous dinoflagellates (e.g. Fig. 1d, e) often composed 50 % or more of microzooplankton biomass, and were major grazers of diatoms in the sea (Sherr and Sherr 2007; Sherr et al. 2013).

Looking to the Future

Research on the quantitative impact of phagotrophic protists on recycling and trophic transfers in aquatic ecosystems has expanded to include assessment of the genetic diversity of heterotrophic protists (Caron et al. 2012), investigation of chemical cues that affect prey ingestion (Roberts et al. 2011), and study of the biochemistry of prey detection in phagotrophic protists (Hartz et al. 2008, 2011). Results of molecular genetic studies suggest that the taxonomic diversity of both autotrophic and heterotrophic protists in marine systems is extraordinary. One interesting finding is the ubiquitous presence in the sea of protists related to a group of dinoflagellates, the Syndiniales, that parasitize other protists (Guillou et al. 2008; Siano et al. 2011). Protist parasitism is an alternate mode of predator–prey interaction that is ripe for further study; we noted parasitoid flagellates infecting diatoms in the Bering Sea (Sherr et al. 2013; Fig. 1a).

Investigation of environmental cues that affect prey detection or avoidance is another new and exciting research field (Roberts et al. 2011). The initial discovery that the metabolite dimethyl sulfide (DMS) could be released by haptophyte algae as a defense against protist predation was made by Gordon Wolfe (Wolfe et al. 1994, 1997), a post-doctoral colleague working in our laboratory at OSU. Our graduate student, Aaron Hartz (Hartz et al. 2008), subsequently demonstrated that inhibition of intracellular signal transduction pathways decreased chemosensory response to prey in a marine dinoflagellate and ciliate. He also found that the non-pigmented marine dinoflagellate, *Oxyrrhis marina*, expressed rhodopsin and could orient to various wavelengths of light. We speculated that both chemical and light signals could be used by protists to find prey in situ (Hartz et al. 2011).

Evelyn B. Sherr and Barry F. Sherr

We bonded over sewage sludge. In 1975–1977, Ev, a new research scientist at the University of Georgia Marine Institute on Sapelo Island, and Barry, a graduate student at the University of Georgia, collaborated on a project to evaluate whether salt marsh microbes could break down organic wastes and remove nitrogen via denitrification from secondarily treated sewage sludge applied to a cordgrass marsh. We asked Larry Pomeroy, who was directing a major project in the salt marshes around Sapelo, if we could borrow his research van to deliver sludge from the Athens, GA, waste treatment plant to the island. We may not have been entirely clear about what we were going to transport. After loading up the sludge in trash barrels, we drove to the coast in time for the afternoon ferry to the island. It was a warm day, and with windows open we didn't notice the winged creatures emerging from the barrels. But when we got to the coast and opened the van, a cloud of sludge flies exploded from the interior. We tried to shoo out as many flies as we could, without great success. Pomeroy later had the van fumigated. We were married in 1979, after Barry finished his Ph.D., and went to Israel for a year and a half research "honeymoon." We returned to work at the UGMI until 1990 when we relocated to Oregon State University with our two young sons. Projects in the Arctic and in the upwelling system off the Oregon coast occupied us until our retirement in 2012.

References

Azam F, Fenchel T, Field JG, Gray JS, Meyer-Reil LA, Thingstad F (1983) The ecological role of water-column microbes in the sea. Mar Ecol Prog Ser 10:257–263

Andersen P, Fenchel T (1985) Bacterivory by microheterotrophlc flagellates in seawater samples. Limnol Oceanogr 30:198–202

Calbet A, Landry MR (2004) Phytoplankton growth, microzooplankton grazing, and carbon cycling in marine systems. Limnol Oceanogr 40:51–57

Campbell RG, Sherr EB, Ashjian CJ, Plourde S, Sherr BF, Hill V, Stockwell DA (2009) Mesozooplankton prey preference and grazing impact in the Western Arctic Ocean. Deep Sea Res II 56:1274–1289

Caron DA, Goldman JC (1988) Dynamics of protistan carbon and nutrient cycling. J Protozool 35:247–249
Caron DA, Countway PD, Jones AC, Kim DY, Schnetzer A (2012) Marine protistan diversity. Ann Rev Mar Sci 4:467–493
Ducklow HW (1983) Production and fate of bacteria in the oceans. Bioscience 33:494–501
Fenchel T (1982) Ecology of heterotrophic flagellates. IV. Quantitative occurrence and importance as bacterial consumers. Mar Ecol Prog Ser 9:35–42
Flynn K, Stoecker D, Mitra A, Raven J, Glibert P, Hansen P, Graneli E, Burkholder J (2012) Misuse of the phytoplankton-zooplankton dichotomy: the need to assign organisms as mixotrophs within plankton functional types. J Plankton Res. doi:10.1093/plankt/fbs062
Glibert PM, Lipschultz F, McCarthy JJ, Altabet MA (1982) Isotope dilution models of uptake and remineralization of ammonium by marine plankton. Limnol Oceanogr 27:639–650
Glibert PM, Miller CA, Garside C, Roman MRR, McManus GB (1992) 14NH4+ regeneration and grazing: interdependent processes in size-fractionated 15NH4+ experiments. Mar Ecol Prog Ser 35:462–476
Gonzalez JM, Sherr EB, Sherr BF (1993) Differential feeding by marine flagellates on growing vs starving bacteria, and on motile vs non-motile bacteria. Mar Ecol Prog Ser 102:257–267
Guillou L, Viprey M, Chambouvet A, Welsh RM, Kirkham AR, Massana R, Scanlan DJ, Worden AZ (2008) Widespread occurrence and genetic diversity of marine parasitoids belonging to Syndiniales (Alveolata). Environ Microbiol 10:3349–3365
Harrison WG (1978) Experimental measurements of nitrogen remineralization in coastal waters. Limnol Oceanogr 23:684–694
Hartz AJ, Sherr BF, Sherr EB (2008) Using inhibitors to investigate the involvement of cell signaling in predation by marine phagotrophic protists. J Euk Microbiol 55:18–21
Hartz AJ, Sherr BF, Sherr EB (2011) Photoresponse in the heterotrophic marine dinoflagellate *Oxyrrhis marina*. J Eukaryot Microbiol 58:171–177
Landry MR, Hassett RP (1982) Estimating the grazing impact of marine microzooplankton. Mar Biol 67:283–288
Pomeroy LR (1974) The ocean's food web: a changing paradigm. Bioscience 24:499–504
Roberts EC, Legrand C, Steinke M, Wootton EC (2011) Mechanisms underlying chemical interactions between predatory planktonic protists and their prey. J Plankton Res 33:833–841
Sherr EB, Sherr BF (1987) High rates of consumption of bacteria by pelagic ciliates. Nature 325:710–711
Sherr EB, Sherr BF (1988) Role of microbes in pelagic food webs: a revised concept. Limnol Oceanogr 33:1225–1227
Sherr BF, Sherr EB (1991) Proportional distribution of total numbers, biovolume, and bacterivory among size classes of 2-20 mm nonpigmented marine flagellates. Mar Microb Food Webs 5:227–237
Sherr EB, Sherr BF (1992) Trophic roles of pelagic protists: phagotrophic flagellates as herbivores. Arch Hydrobiol Beih 37:165–172
Sherr EB, Sherr BF (1993) Protistan grazing rates via uptake of fluorescently labeled prey. In: Kemp P, Sherr B, Sherr E, Cole J (eds) Current methods in aquatic microbial ecology. Lewis Publ, NY, pp 695–702
Sherr EB, Sherr BF (1994) Bacterivory and herbivory: key roles of phagotrophic protists in pelagic food webs. Microb Ecol 28:223–235
Sherr EB, Sherr BF (2002) Significant of predation by protists in aquatic microbial food webs. Anton Leeuw Int J G Mol Microbiol 81:293–308
Sherr BF, Sherr EB (2003) Community respiration/production and bacterial activity in the upper water column of the central Arctic Ocean. Deep Sea Res I 50:529–542
Sherr EB, Sherr BF (2007) Heterotrophic dinoflagellates: a significant component of microzooplankton biomass and major grazers of diatoms in the sea. Mar Ecol Prog Ser 352:187–197
Sherr EB, Sherr BF (2008) Understanding roles of microbes in marine pelagic food webs: a brief history (Chapter 2). In: Kirchman D (ed) Microbial ecology of the oceans, 2nd edn. Wiley, New York, NY, pp 27–44

Sherr EB, Sherr BF (2009) Role of herbivorous protists in controlling initiation and development of marine phytoplankton blooms. Aquat Microb Ecol 57:253–262
Sherr BF, Sherr EB, Berman T (1982) Decomposition of organic detritus: a selective role for microflagellate protozoa. Limnol Oceanogr 27:765–769
Sherr BF, Sherr EB, Berman T (1983) Grazing, growth, and ammonia excretion rates of a heterotrophic microflagellate fed four species of bacteria. Appl Environ Microbiol 45:1196–1201
Sherr EB, Sherr BF, Fallon RD, Newell SY (1986a) Small aloricate ciliates as a major component of the marine heterotrophic nanoplankton. Limnol Oceanogr 31:177–183
Sherr EB, Sherr BF, Paffenhofer G (1986b) Phagotrophic protozoa as food for metazoans: a "missing" link in marine pelagic food webs? Mar Microb Food Webs 1:61–80
Sherr BF, Sherr EB, Fallon RD (1987) Use of monodispersed, fluorescently labeled bacteria to estimate in situ protozoan bacterivory. Appl Environ Microbiol 53:958–965
Sherr BF, Rassoulzadegan F, Sherr EB (1989) Bacterivory by pelagic choreotrichous ciliates in coastal waters of the NW Mediterranean Sea. Mar Ecol Prog Ser 55:235–240
Sherr BF, Sherr EB, McDaniel J (1992) Effect of protistan grazing on the frequency of dividing cells (FDC) in bacterioplankton assemblages. Appl Environ Microbiol 58:2381–2385
Sherr EB, Caron DA, Sherr BF (1993) Staining of heterotrophic protists for visualization via epifluorescence microscopy. In: Kemp P, Sherr B, Sherr E, Cole J (eds) Current methods in aquatic microbial ecology. Lewis Publ, New York, NY, pp 213–228
Sherr EB, Sherr BF, Fessenden L (1997) Heterotrophic protists in the central Arctic Ocean. Deep Sea Res II 44:1665–1682
Sherr EB, Sherr BF, Wheeler PA, Thompson K (2003) Temporal and spatial variation in stocks of autotrophic and heterotrophic microbes in the upper water column of the central Arctic Ocean. Deep Sea Res I 50:557–571
Sherr EB, Sherr BF, Hartz AJ (2009) Microzooplankton grazing impact in the Western Arctic Ocean. Deep Sea Res II 56:1264–1273
Sherr EB, Sherr BF, Ross C (2013) Microzooplankton grazing impact in the Bering sea during spring sea ice conditions. Deep Sea Res II 94:57–67
Siano R, Alves-de-Souza C, Foulon E, Bendif EM, Simon N, Guillou L, Not F (2011) Distribution and host diversity of Amoebophryidae parasites across oligotrophic waters of the Mediterranean Sea. Biogeosciences 8:267–278
Sieburth JNM (1981) Protozoan bacterivory in pelagic marine waters. In: Hobbie PJB (ed) Heterotrophic activity in the sea. Plenum, New York, NY, pp 405–444
Sieburth JMN, Smetacek V, Lenz J (1978) Pelagic ecosystem structure: heterotrophic compartments of the plankton and their relationship to plankton size fractions. Limnol Oceanogr 23:1256–1263
Stoecker DK, Johnson MD, de Vargas C, Not F (2009) Acquired phototrophy in aquatic protists. Aquat Microb Ecol 57(3):279–310
Strom SL (2000) Bacterivory: interactions between bacteria and their grazers. In: Kirchman DL (ed) Microbial ecology of the oceans. Wiley-Liss, New York, pp 351–386
Suzuki MT, Sherr EB, Sherr BF (1996) Estimation of ammonium regeneration efficiencies associated with bacterivory in pelagic food webs via a ^{15}N tracer method. J Plankton Res 18: 411–428
Tillmann U, Hesse K-J, Tillmann A (1999) Large-scale parasitic infection of diatoms in the Northfrisian Wadden Sea. J Sea Res 42:255–261
Wolfe GV, Sherr EB, Sherr BF (1994) Release and consumption of DMSP from *Emiliania huxleyi* during grazing by *Oxyrrhis marina*. Mar Ecol Prog Ser 111:111–119
Wolfe GV, Steinke M, Kirst GO (1997) Grazing-activated chemical defence in a unicellular marine alga. Nature 387:894–897
Zubkov MV, Tarran GA (2008) High bacterivory by the smallest phytoplankton in the North Atlantic Ocean. Nature 455:224–226

Drivers That Structure Biodiversity in the Plankton

Tatiana A. Rynearson and Susanne Menden-Deuer

Plankton Biodiversity

Plankton (Fig. 1) made Earth habitable by initially oxygenating the atmosphere some 2–3 billion years ago (Canfield 2005; Crowe et al. 2013). Currently, plankton generate as much oxygen and organic matter as all terrestrial plants combined (Field et al. 1998), fueling fisheries production, driving gas exchange between the atmosphere and oceans, and serving as key agents in global climate cycles (Falkowski et al. 2008). Plankton are the base of the marine food web and thus support virtually all protein derived from the ocean. These microbes represent sentinels of climate change and may harbor solutions to providing renewable energy sources.

A quantitative understanding of the abundance, distribution, and diversity of plankton is essential for estimating their biogeochemical impact, particularly under a changing environment. Yet none of these basic metrics are well known. Even the number of extant planktonic species is unknown, although current estimates yield staggering numbers. A recent global survey of eukaryotic plankton species estimated 150,000 extant species (de Vargas et al. 2015) and it has been estimated that the number of prokaryotic species reaches into the millions (Ward 2002). The "paradox of the plankton," coined by Hutchinson (1961), describes the conundrum that the observed high biodiversity of planktonic organisms is not reconcilable with the competitive exclusion principle (i.e. winner takes all) (Hardin 1960). The basic argument has been that the surface ocean is a well-mixed, homogeneous environment that only permits the strongest competitor for a specific resource to persist, ultimately leading to only few coexisting species and a low level of biodiversity in the plankton. Thus, the paradox lies in the fact that many planktonic species coexist where few should remain.

T.A. Rynearson (✉) • S. Menden-Deuer
Graduate School of Oceanography, University of Rhode Island,
South Ferry Rd., Narragansett, RI 02882, USA
e-mail: rynearson@uri.edu; smenden@uri.edu

P.M. Glibert, T.M. Kana (eds.), *Aquatic Microbial Ecology and Biogeochemistry: A Dual Perspective*, DOI 10.1007/978-3-319-30259-1_2

Fig. 1 Plankton make life on Earth habitable. They are genetically, morphologically, physiologically, and behaviorally diverse and stunningly beautiful. Illustration by Dean Jacobson, by permission of the artist

Alternate Hypotheses That Explain the Paradox of the Plankton

Naturally, the paradox has found numerous explanations over time, including the role of resource competition (Tilman 1994), emergence of chaotic oscillations in species abundances (Huisman and Weissing 1999), chaotic fluid motion (Károlyi et al. 2000), spatial structure and localized competition (Levin 1992; Kerr et al. 2002), competition and predation as structuring factors (Record et al. 2013; Cropp and Norbury 2012) as well as large-scale spatial factors such as latitudinal and seasonal gradients (Barton et al. 2010). Gause (1934) made early observations of the competitive exclusion principle, suggesting that as long as environmental conditions were variable, then biodiversity could be enhanced, e.g. by offering a spatial refuge in his predator–prey experiments through the addition of a sediment base in his test tube cultures. Most likely, all these factors contribute to some degree to the overall high biodiversity in the plankton.

Most studies that have aimed to explain the paradox of the plankton rely on some external, environmental factor that provides a disturbance to the assumed homogeneous environment. Consequently, the disturbance (e.g. fluid mixing or selective predation) results in enhancement of biodiversity by providing an opportunity for hanger-on species to emerge. And indeed spatial heterogeneity and temporal disturbances are easy to invoke as key drivers of biodiversity. Unlike the assumption of homogeneity, the ocean is a complex and heterogeneous environment, characterized by steep gradients in physical, chemical, and biological properties over a continuum from microscopic to global scales. An organism's location relative to these gradients is critical for its survival. For example, because light decays exponentially with depth, light-dependent photosynthetic organisms can be exposed to vastly different solar irradiances, depending on their depth in the water column, from high irradiance that can be harmful to insufficient amounts that prohibit survival. In fact, environmental heterogeneity is thought to be an important factor structuring planktonic communities (Levin 1976, 1992). However, are external drivers the sole factors that can be invoked to maintain planktonic biodiversity? Are there inherent characteristics of the organisms themselves that lead to the maintenance of multiple species in the water column?

One explanation of the paradox of the plankton that is based on inherent, species-specific characteristics rather than external, environmental factors draws on the resource competition theory developed by Tilman (1994). This theory was applied by Huisman and Weissing (1999) who showed that in model simulations, non-equilibrium conditions permitted coexistence of more species than the number of limiting resources in those conditions. The validity of these model predictions has subsequently been demonstrated through long-term observations of laboratory cultures of mixed plankton species that included a multi-trophic food web and was void of external disturbances (Benincà et al 2008). The work by Huisman and Weissing (1999) and Benincà et al (2008) clearly demonstrate that coexistence of multiple plankton species is possible in the absence of external disturbances. One limitation of both model and empirical evidence is that the number of species supported is still vastly smaller than the high levels of species diversity observed in the ocean (Ward 2002; de Vargas et al. 2015).

An Organismal Perspective on the Paradox of the Plankton: A Biodiversity Explosion from Within?

A crucial element that is missing from explanations of the paradox of the plankton is a focus on individual organisms. This element takes into account the fact that organisms compete and are selected upon as individuals, not as species. In addition to the immense biodiversity of plankton, the vast time and space scales of ocean ecology require an integrative view, and necessitate assessing the outcomes of ecological interactions—resource uptake, predation, and reproduction—at the population level. However, virtually all processes that affect the abundance, distribution, or production rates of a species occur at the individual level, including feeding, motility, resource uptake, reproduction, and mortality.

Shifting the level of investigation from that of a species to that of an individual brings opportunities and challenges. Putting the individual at the center shifts the time and space scales over which ecological interactions affect the survival of the species and refocuses them on the level of cellular interactions with the abiotic and biotic environment (Fig. 2). In this individual-centric perspective, cellular interactions that occur on time scales of seconds to hours and over spatial scales of microns to centimeters become the drivers of the ecosystem functions ultimately of interest, such as abundance, distribution, and production. Predicting biological function

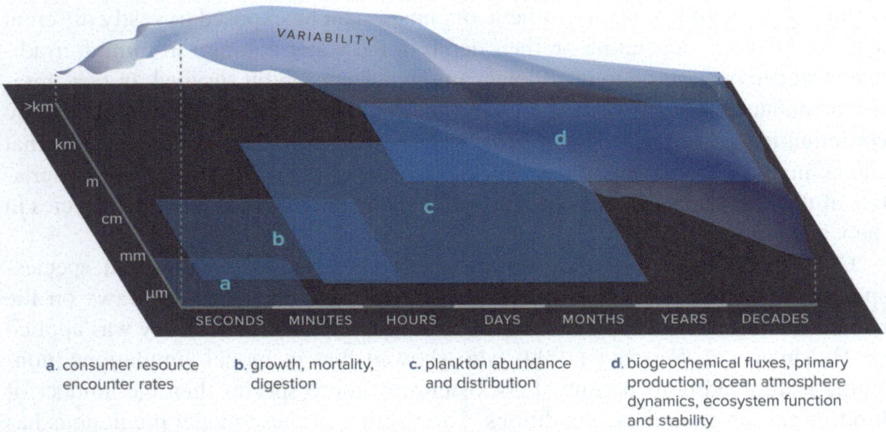

Fig. 2 Modified version of a Stommel diagram, showing the linkages of biological processes in the ocean over ecologically relevant time and space scales. Overlain is the degree of associated variability. It is noteworthy that large-scale processes emerge from underlying, often poorly characterized smaller-scale processes. Integrating small-scale processes to characterize biogeochemically important, large-scale events (e.g. the annual cycle of phytoplankton) constitutes a key challenge for oceanographers. Illustration by Josh Wood

within a system that changes on scales from micrometers and seconds to kilometers and years defies available approaches. While it is intellectually and analytically challenging to link a myriad of micro-scale processes to their large-scale ramifications, taking such a scale-integrative approach would also provide the opportunity to identify the underlying mechanisms driving processes (e.g. Kiørboe 2008) and ultimately provide a predictive understanding across spatio-temporal scales.

Linking Individual Level Behaviors with Plankton Ecology

To incorporate the importance of cell–cell interactions into both plankton ecology and investigations of the mechanisms underlying the high biodiversity of planktonic species, we recently developed a mathematical theory that suggests a unifying explanation and the causal mechanism for the staggering species diversity of plankton (Menden-Deuer and Rowlett 2014). This approach was motivated by laboratory and field observations of high, inherent, intra-specific variability in the genetics, physiology, and behaviors of plankton. To examine the importance of these observed characteristics, we created a new theoretical model and tested this model with high-resolution simulations across a robust parameter space. The results showed both theoretically, in the form of a mathematical theorem, and numerically, in the form of competition simulations across a robust parameter space, that (1) incorporation of individual variability supports arbitrarily high biodiversity among plankton, (2) elimination of behavioral or physiological heterogeneity produces results consistent with the competitive exclusion principle, and that (3) spatial structure can delay extinction of species with invariant physiological or behavioral characteristics but extinction is inevitable. Using the modeling approach described in Menden-Deuer and Rowlett (2014) we explored the importance of intra-specific variation, including in competition experiments with superior and inferior competitors. In agreement with the findings in Menden-Deuer and Rowlett (2014), our results consistently show that intra-specific variability is sufficient to explain high species diversity in planktonic organisms, independent of specific external factors, providing a mechanistic underpinning for previously advanced explanations of the paradox of the plankton (Fig. 3).

Pervasive Intra-specific Variability in the Genetic Diversity, Physiological Capacity, and Behavioral Repertoire of Plankton

If our theoretical considerations are correct, intra-specific variability is highly adaptive for plankton, and indeed variation is commonly observed in a range of physiological, demographic, and morphological traits among and within phylogenetically distinct plankton species. Anywhere researchers have looked, they have found intra-specific variability, including in global, inter- and intra-specific patterns of temperature regulation (Thomas et al. 2012; Boyd et al 2013), responses to elevated pCO_2 concentrations (Schaum et al. 2013), tolerance of environmental conditions (Brand 1984), elemental composition (Moal et al. 1987), and growth rates (Rynearson and Armbrust 2004). A comprehensive study of intra-specific variability in one phytoplankton species showed distinguishing characteristics among strains in terms of cell size, maximum growth and photosynthesis rates, tolerance of low salinities, resource use, and toxicity (Fredrickson et al. 2011). Linked empirical and theoretical analyses have shown that intra-specific variability in motility can enhance species dispersal, with downstream ramification for organism distributions and ecological function (Menden-Deuer 2010). Molecular analyses have shown that physiological variability is associated with genetically distinct strains (Rynearson and Armbrust 2000, 2004). Blooms of the otherwise slow growing dinoflagellate species *Akashiwo sanguinea* have been tracked by satellites over vast spatial (100 s of kms along the US West Coast) and temporal (months to years) scales (Du et al. 2011; White et al. 2014). Laboratory examination of multiple *A. sanguinea* strains revealed that a high degree of intra-specific variability in the temperature tolerance, movement behaviors, and growth rates of *A. sanguinea* appears to be an important factor in broadening the species' niche, suggesting that intra-specific variability supports persistent blooms of this dinoflagellate species (Menden-Deuer and Montalbano 2015).

The ecological function of intra-specific variability had been determined only in very few cases, partly because intra-specific variability in the physiology and behavior of planktonic organisms has, for the most part, only recently been recognized and quantified. Thus we suggest that pervasive intra-specific genetic, behavioral, and physiological variability provides a fundamental organizing principle in the ecology of unicellular, self-replicating organisms.

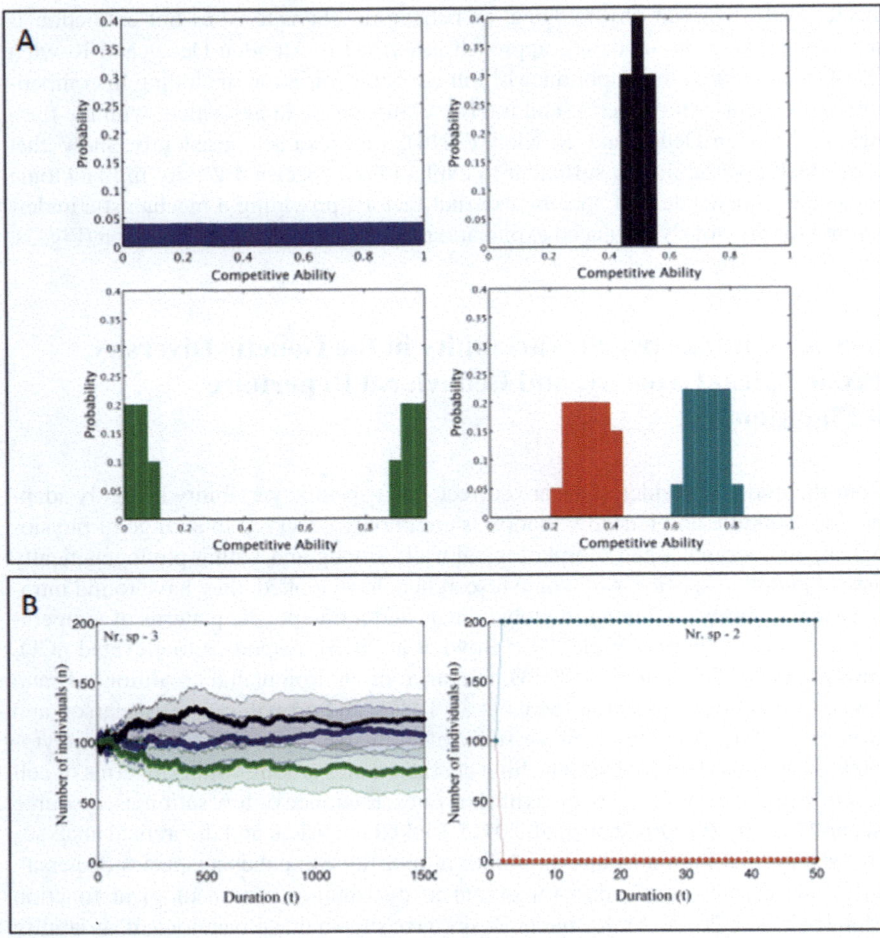

Fig. 3 Modeling approaches provide insights into the role of intra-specific variation in maintaining inter-specific diversity in the plankton. Panel (**a**) shows three different probability distributions of competitive ability, each with an average competitive ability of 0.5 and, bottom right, two probability distributions representing unequal competitive abilities (average competitive abilities of 0.3 (red) and 0.7 (green)). Panel (**b**) shows abundance over time of species representing the outcomes of a species competition model simulation (after Menden-Deuer and Rowlett 2014) including three competing species with on average identical but differently shaped probability distributions (*left figure*) and two competing species (*right figure*) with grossly different competitive abilities. Color of species abundance in panel (**b**) matches underlying probability distribution of competitive abilities from panel (**a**). Incorporation of intra-specific variability permits persistent species coexistence (panel (**b**), *left figure*), whereas competitive exclusion is accurately reproduced only when species are grossly different in their competitive ability (panel (**b**), *right figure*). Note rapid extinction indicated by difference in simulation durations in panel (**b**)

The mechanisms that maintain intra-specific variability are not well understood but may be related to the complexity of cellular morphology (Gray et al. 2010; Huisman et al. 2001). Moreover, specific traits and their variability are not easily eliminated within a microbial population (e.g. "the plankton cloud" sensu Smetacek 2012) due to the vast number of independent clones. Some species even retain possibly maladaptive traits, reflecting ancient, rather than current environmental conditions (Hutchins et al. 2013). Plasticity, or the ability for a single strain to vary its physiology, has been identified as a key characteristic for adapting to changing or novel conditions, can be observed in the phytoplankton (e.g. Schaum et al. 2013), and is associated with elevated success in range expansion of invasive species (Lee et al. 2003).

Evolution: Generating and Structuring Diversity over the Long Term

Thus far, we have discussed the structuring function of intra-specific variability in *maintaining* planktonic species diversity. Here, we focus on the processes that *generate* diversity and how those factors play into the structure and function of planktonic ecosystems. De novo mutation is the primary generator of diversity and has been observed in cultured phytoplankton (Collins and Bell 2004; Lakeman and Cattolico 2007; Collins 2011). There is growing evidence for the role of horizontal gene flow in transferring chunks of exogenous DNA into planktonic cells, even in eukaryotic plankton, where bacterial DNA has been found embedded in diatom genomes (Bowler et al. 2008).

Novel mutations and newly incorporated exogenous DNA are then acted on by natural selection. Depending on the type of selection pressure, evolution may maintain, reduce, or even eliminate that diversity. In fact, most mutation is deleterious (e.g. Sanjuan et al. 2004). Planktonic microbes may be able to avoid the negative impacts of deleterious mutation through enormous population and census sizes. For example, a spring bloom of the diatom *Ditylum brightwellii* likely contained some 2400 genetically different clonal lineages (Rynearson and Armbrust 2005). That same bloom had cell numbers of >10,000 cells L^{-1}, highlighting that the blooming population also had a very large census size. In addition to maintaining diversity (as discussed above), the large number of clonal lineages provides enormous diversity for selection to act upon.

Organism life cycles also influence diversity. For example, sexual reproduction events release new diversity into populations. Other components of plankton life cycles may be equally important although their import has, for the most part, not been quantified. For example, resting spores are essentially an archive of extant diversity for future selection to act upon (Härnström et al. 2011; Rynearson et al. 2013). Of course, asexual reproduction, the most common means of replication in the plankton, provides successful cells with the opportunity to generate many copies of themselves (except see references above showing that asexual reproduction has less than perfect fidelity). Some phytoplankton, such as coccolithophorids alter their ploidy, changing from diploid to haploid with life stage (Green et al. 1996). By having a range of reproductive modes on hand, it is likely that planktonic species can influence how selection acts on their gene pool.

Opportunities for Progress

Intra-specific Variability and Its Ramifications for Plankton Ecology Need to Be Quantified

Just as an experiment lacking within-treatment replication is difficult to interpret in terms of among-treatment differences, attempts to characterize among-species differences are limited if no information is available regarding within-species variation. The observation of high, intra-specific variability in physiology, behavior, and genetics of planktonic species suggests that approaches that try to distinguish species, or even genera or classes based on single characteristics (e.g. dinoflagellates are slow growers) need to first quantify the degree of within-species variability and determine if variation among strains of one species is sufficiently small to permit distinction among species and higher taxonomic levels. Otherwise, multiple traits will be necessary to place species in multi-niche space.

One aspect of within-species variation that has only recently been identified is the existence of genetically distinct populations. There is robust evidence showing that the gene pools of dinoflagellates (Richlen et al. 2012), diatoms (Rynearson and Armbrust 2004; Rynearson et al. 2006; Casteleyn et al. 2010), and coccolithophores (Iglesias-Rodriguez et al. 2006; Gäbler-Schwarz et al. 2015) can be subdivided into genetically distinct populations, despite the enormous potential for both continuous dispersal of individuals and mixing between populations. There is some evidence that these populations are physiologically distinct, allowing populations to diverge and perhaps adapt to local conditions (Rynearson and Armbrust 2004). This is supported by data from the diatom *Skeletonema marinoi* showing that a genetically unique population persisted in one fjord for at least 100 years (Härnström et al. 2011). These observations suggest that the potential to adapt to changing conditions is present in marine phytoplankton, though the relative importance of adaptation by local populations versus replacement by immigrant types has yet to be established and is an important avenue of research.

At the same time, further development of mathematical theory and exploration using model simulations will provide opportunities to formulate testable hypotheses on the role of intra-specific variability in the maintenance of biodiversity. Our own simulations (Menden-Deuer and Rowlett 2014), for example, can be enhanced from current assumptions of general "competitive abilities" on which strains and species are compared by incorporating biological realism and complexity.

Plankton Ecology, Now and in the Future

An important application of enhanced understanding of intra-specific variability is the prediction of plankton responses to environmental conditions. Such predictive capacity is urgently needed in order to better understand how planktonic communities will respond to increases in temperature, pCO_2 and other variables and conditions related to climate change (e.g. Hoegh-Guldberg and Bruno 2010). Currently, our ability to conduct appropriate measurements of plankton ecology in the context of climate change is stifled by the fact that climate change is forcing environmental change that is gradual but sustained (e.g. rates of ocean acidification and temperature change). One approach toward understanding the response of plankton to these changes is to use elegant but time-consuming, long-term studies using experimental evolution (reviewed in Collins et al. 2014) which have shown that at least for some climate change variables, such as changing pCO_2, phytoplankton may evolve by adapting to decreasing pH.

Experimental evolution experiments are typically done with single clones, pointing to a second limitation plankton ecologists currently face: our inability to conduct incubation measurements with *mixed* plankton assemblages that have been acclimated to target conditions (e.g. temperature). Acclimation of mixed plankton communities to target treatments poses a particular challenge because the issues of examining long-term responses on a single species (most often a single clone) are multiplied by the varied growth requirements of multiple species and interactions among species. Identification of appropriate acclimation procedures is thus far an unsolved challenge. It is unclear how a mixed assemblage of diverse species can be acclimated to target conditions. Each species likely requires a different acclimation rate and type (e.g. gradient vs. step functions), which may vary depending on the process (e.g. enzymatic activity vs. growth). Acclimation is challenging and in itself may induce biases. For example, for mono-specific phytoplankton laboratory cultures, Brand et al. (1981) found that in order to achieve constancy of a single metric (growth rate), the required acclimation period was 1–3 weeks, depending on species. Thus, acclimation of diverse plankton communities to a target condition would require a prolonged incubation, while at the same time maintaining initial biotic and abiotic conditions, including species composition and nutrient concentrations to ensure applicability of results. Nevertheless, such challenges need to be solved, so that physiological and community responses to changing temperatures and other climate variables can be examined experimentally, leaving behind the current practice of keeping acclimation undefined or inadequate in most studies.

Ultimately, the inclusion of an organismal focus that incorporates fundamental individual-level variation should provide deep insights into the factors driving marine biodiversity, strengthen the theoretical underpinnings of ecology, and enhance our understanding of the population dynamics of microbes. Identification of structuring mechanisms is not only scientifically fascinating but also has significant implications for how we understand the function of planktonic ecosystems, and our ability to predict how these ecosystems may respond to changing climate conditions.

Tatiana A. Rynearson and Susanne Menden-Deuer

Since the New York Times featured our wedding announcement, there is little information that is publicly inaccessible about our relationship. Thankfully, the intricate details of our intersecting science interests were of minor interest to the NYT fact checker, so a few things are left to be told. We both had our first immersion in oceanography on a long cruise to the southern ocean polar frontal zone as part of the Southern Ocean JGOFS effort, led by mentor Professor V. Smetacek. During the cruise, a look through the microscope at the phytoplankton community composition gave a more accurate account of location and water mass than most other metrics measured on the cruise. This left us both with a lasting impression of the importance of an organismal perspective and a deep desire to understand the factors driving these distributions. As any good couple, we subsequently took vastly different approaches to pursue our scientific interests. One of us uses molecular tools to investigate the evolution, speciation, and biogeochemical function of phytoplankton and feeding ecology of zooplankton, while the other has focused on linking microscopic predator-prey behaviors with their population-level ramifications of plankton production, food web structure, and patchiness. With the help of supportive mentors, we have been fortunate to spend long stretches of our career in the same place, first at the Alfred Wegener Institute as technician and M.Sc. student, then as graduate students at the University of Washington, and finally as faculty members at the Graduate School of Oceanography, University of Rhode Island. Although we do not try to work together formally, sometimes it is unavoidable and we have coauthored a few papers. Nonetheless, our most productive and joyful collaboration matured in December 2013 when our daughter was born.

References

Barton AD, Dutkiewicz S, Flierl G, Bragg J, Follows MJ (2010) Patterns of diversity in marine phytoplankton. Science 327:1509–1511

Benincà E, Huisman J, Heerkloss R, Johnk KD, Branco P, Van Nes EH, Scheffer M, Ellner SP (2008) Chaos in a long-term experiment with a plankton community. Nature 451:822–825

Bowler C, Allen AE, Badger JH, Grimwood J, Jabbari K, Kuo A, Maheswari U, Martens C, Maumus F, Otillar RP, Rayko E, Salamov A, Vandepoele K, Beszteri B, Gruber A, Heijde M, Katinka M, Mock T, Valentin K, Verret F, Berges JA, Brownlee C, Cadoret J-P, Chiovitti A, Choi CJ, Coesel S, De Martino A, Detter JC, Durkin C, Falciatore A, Fournet J, Haruta M, Huysman MJJ, Jenkins BD, Jiroutova K, Jorgensen RE, Joubert Y, Kaplan A, Kroger N, Kroth PG, La Roche J, Lindquist E, Lommer M, Martin-Jezequel V, Lopez PJ, Lucas S, Mangogna M, McGinnis K, Medlin LK, Montsant A, Secq M-PO-L, Napoli C, Obornik M, Parker MS, Petit J-L, Porcel BM, Poulsen N, Robison M, Rychlewski L, Rynearson TA, Schmutz J, Shapiro H, Siaut M, Stanley M, Sussman MR, Taylor AR, Vardi A, von Dassow P, Vyverman W, Willis A, Wyrwicz LS, Rokhsar DS, Weissenbach J, Armbrust EV, Green BR, Van de Peer Y, Grigoriev IV (2008) The Phaeodactylum genome reveals the evolutionary history of diatom genomes. Nature 456:239–244

Boyd PW, Rynearson TA, Armstrong EA, Fu F, Hayashi K, Hu Z, Hutchins DA, Kudela RM, Litchman E, Mulholland MR, Passow U, Strzepek RF, Whittaker KA, Yu E, Thomas MK (2013) Marine phytoplankton temperature versus growth responses from polar to tropical waters – outcome of a scientific community-wide study. PLoS One 8, e63091

Brand LE (1984) The salinity tolerance of 46 marine-phytoplankton isolates. Estuar Coast Shelf Sci 18:543–556
Brand LE, Guillard RRL, Murphy LS (1981) A method for the rapid and precise determination of acclimated phytoplankton reproduction rates. J Plankton Res 3:193–201
Canfield DE (2005) The early history of atmospheric oxygen: Homage to Robert M. Garrels. Annu Rev Earth Planet Sci 33(1):1–36
Casteleyn G, Leliaert F, Backeljau T, Debeer A-E, Kotaki Y, Rhodes L, Lundholm N, Sabbe K, Vyverman W (2010) Limits to gene flow in a cosmopolitan marine planktonic diatom. Proc Natl Acad Sci 107:12952–12957
Collins S (2011) Competition limits adaptation and productivity in a photosynthetic alga at elevated CO_2. Proc R Soc B Biol Sci 278:247–255
Collins S, Bell G (2004) Phenotypic consequences of 1,000 generations of selection at elevated CO_2 in a green alga. Nature 431:566–569
Collins S, Rost B, Rynearson TA (2014) Evolutionary potential of marine phytoplankton under ocean acidification. Evol Appl 7:140–155
Cropp RA, Norbury J (2012) The mechanisms of coexistence and competitive exclusion in complex plankton ecosystem models. Ecosystems 15:200–212
Crowe SA, Dossing LN, Beukes NJ, Bau M, Kruger SJ, Frei R, Canfield DE (2013) Atmospheric oxygenation three billion years ago. Nature 501:535–538
de Vargas C, Audic S, Henry N, Decelle J, Mahé F, Logares R, Lara E, Berney C, Le Bescot N, Probert I, Carmichael M, Poulain J, Romac S, Colin S, Aury J-M, Bittner L, Chaffron S, Dunthorn M, Engelen S, Flegontova O, Guidi L, Horák A, Jaillon O, Lima-Mendez G, Lukeš J, Malviya S, Morard R, Mulot M, Scalco E, Siano R, Vincent F, Zingone A, Dimier C, Picheral M, Searson S, Kandels-Lewis S, Coordinators TO, Acinas SG, Bork P, Bowler C, Gorsky G, Grimsley N, Hingamp P, Iudicone D, Not F, Ogata H, Pesant S, Raes J, Sieracki ME, Speich S, Stemmann L, Sunagawa S, Weissenbach J, Wincker P, Karsenti E (2015) Eukaryotic plankton diversity in the sunlit ocean. Science 348(6237): DOI:10.1126/science.1261605
Du X, Peterson W, McCulloch A, Liu G (2011) An unusual bloom of the dinoflagellate *Akashiwo sanguinea* off the central Oregon, USA, coast in autumn 2009. Harmful Algae 10(6):784–793
Falkowski PG, Fenchel T, DeLong EF (2008) The microbial engines that drive Earth's biogeochemical cycles. Science 320(5879):1034–1039
Field CB, Behrenfeld MJ, Randerson JT, Falkowski P (1998) Primary production of the biosphere: integrating terrestrial and oceanic components. Science 281:237–240
Fredrickson KA, Strom SL, Crim R, Coyne KJ (2011) Interstrain variability in physiology and genetics of *Heterosigma akashiwo* (Raphidophycae) from the west coast of North America. J Phycol 47(1):25–35
Gäbler-Schwarz S, Medlin LK, Leese F (2015) A puzzle with many pieces: the genetic structure and diversity of *Phaeocystis antarctica* Karsten (Prymnesiophyta). Eur J Phycol 50:112–124
Gause GF (1934) The struggle for existence. Hafner Press, New York, NY
Gray MW, Lukes J, Archibald JM, Keeling PJ, Doolittle FW (2010) Irremediable complexity? Science 330(5005):920–921
Green JC, Course PA, Tarran GA (1996) The life-cycle of *Emiliania huxleyi*: a brief review and a study of relative ploidy levels analysed by flow cytometry. J Mar Syst 9:33–44
Hardin G (1960) The competitive exclusion principle. Science 131:1292–1297
Härnström K, Ellegaard M, Andersen TJ, Godhe A (2011) Hundred years of genetic structure in a sediment revived diatom population. Proc Natl Acad Sci 108:4252–4257
Hoegh-Guldberg O, Bruno JF (2010) The impact of climate change on the world's marine ecosystems. Science 328(5985):1523–1528
Huisman J, Weissing FJ (1999) Biodiversity of plankton by species oscillations and chaos. Nature 402:407–410
Huisman J, Johansson AM, Folmer EO, Weissing FJ (2001) Towards a solution of the plankton paradox: the importance of physiology and life history. Ecol Lett 4:408–411
Hutchins DA, Fu F-X, Webb EA, Walworth N, Tagliabue A (2013) Taxon-specific response of marine nitrogen fixers to elevated carbon dioxide concentrations. Nat Geosci 6:790–795
Hutchinson GE (1961) The paradox of the plankton. Am Nat 95(882):137–145

Iglesias-Rodriguez D, Schofield OM, Batley J, Medlin LK, Hayes PK (2006) Intraspecific genetic diversity in the marine cocolithophore *Emiliania huxleyi* (Prymnesiophyceae): the use of microsatellite analysis in marine phytoplankton population studies. J Phycol 42:526–536
Károlyi G, Péntek A, Scheuring I, Tél T, Toroczkai Z (2000) Chaotic flow: the physics of species coexistence. Proc Natl Acad Sci 97(25):13661–13665. doi:10.1073/pnas.240242797
Kerr B, Riley MA, Feldman MW, Bohannan BJM (2002) Local dispersal promotes biodiversity in a real-life game of rock-paper-scissors. Nature 418:171–174
Kiørboe T (2008) A mechanistic approach to plankton ecology. Princeton University Press, Princeton, NJ, p 224
Lakeman M, Cattolico RA (2007) Cryptic diversity in phytoplankton cultures is revealed using a simple plating technique. J Phycol 43:663–674
Lee CE, Remfert JL, Gelembiuk GW (2003) Evolution of physiological tolerance and performance during freshwater invasions. Integr Comp Biol 43(3):439–449
Levin S (1976) Population dynamic models in heterogeneous environments. Annu Rev Ecol Syst 7:287–310
Levin S (1992) The problem of pattern and scale in ecology: the Robert H. MacArthur award lecture. Ecology 73(6):1943–1967
Menden-Deuer S (2010) High-correlation of individual motility enhances population dispersal rates in a heterotrophic protist. PLoS Comput Biol 6(10):e1000943
Menden-Deuer S, Montalbano A (2015) Bloom formation potential in the toxic dinoflagellate *Akashiwo sanguinea*: clues from movement behaviors and growth characteristics. Harmful Algae 47:75–85. doi:10.1016/j.hal.2015.06.001
Menden-Deuer S, Rowlett J (2014) Many ways to stay in the game: individual variability maintains high biodiversity in planktonic microorganisms. J R Soc Interface 11:20140031
Moal J, Martin-Jezequel V, Harris RP, Samain JF, Poulet SA (1987) Inter- specific and intraspecific variability of the chemical-composition of marine-phytoplankton. Oceanol Acta 10(3):339–346
Record NR, Pershing AJ, Maps F (2013) The paradox of "the paradox of the plankton". ICES J Mar Sci 70:1–5
Richlen ML, Erdner DL, McCauley LAR, Libera K, Anderson DM (2012) Extensive genetic diversity and rapid population differentiation during blooms of *Alexandrium fundyense* (Dinophyceae) in an isolated salt pond on Cape Cod, MA, USA. Ecol Evol 2:2583–2594
Rynearson TA, Armbrust EV (2000) DNA fingerprinting reveals extensive genetic diversity in a field population of the centric diatom *Ditylum brightwellii*. Limnol Oceanogr 45:1329–1340
Rynearson TA, Armbrust EV (2004) Genetic differentiation among populations of the planktonic marine diatom *Ditylum brightwellii* (Bacillariophyceae). J Phycol 40:34–43
Rynearson TA, Armbrust EV (2005) Maintenance of clonal diversity during a spring bloom of the centric diatom *Ditylum brightwellii*. Mol Ecol 14:1631–1640
Rynearson TA, Newton JA, Armbrust EV (2006) Spring bloom development, genetic variation and population succession in the planktonic diatom *Ditylum brightwellii*. Limnol Oceanogr 51:1249–1261
Rynearson TA, Richardson K, Lampitt RS, Sieracki ME, Poulton AJ, Lyngsgaard MM, Perry MJ (2013) Major contribution of diatom resting spores to vertical flux in the sub-polar North Atlantic. Deep Sea Res I Oceanogr Res Papers 82:60–71
Sanjuan R, Moya A, Elena S (2004) The distribution of fitness effects caused by single-nucleotide substitutions in an RNA virus. Proc Natl Acad Sci U S A 101:8396–8401
Schaum E, Rost B, Millar AJ, Collins C (2013) Variation in plastic responses of a globally distributed picoplankton species to ocean acidification. Nat Clim Change 3:298–302
Smetacek V (2012) Making sense of ocean biota: how evolution and biodiversity of land organisms differ from that of the plankton. J Biosci 37:589–607
Thomas MK, Kremer CT, Klausmeier CA, Litchman E (2012) A global pattern of thermal adaptation in marine phytoplankton. Science 338(6110):1085–1088
Tilman D (1994) Competition and biodiversity in spatially structured habitats. Ecology 75:2–16
Ward B (2002) How many species of prokaryotes are there? Proc Natl Acad Sci U S A 99(16): 10234–10236
White AE, Watkins-Brandt KE, McKibben SM, Wood AM, Hunter M, Forster Z, Du X, Peterson WT (2014) Large-scale bloom of *Akashiwo sanguinea* in the Northern California current system in 2009. Harmful Algae 37:38–46

The Elongated, the Squat and the Spherical: Selective Pressures for Phytoplankton Shape

Lee Karp-Boss and Emmanuel Boss

Introduction

The size and shape of a phytoplankter are important morphological traits that impact resource acquisition (light, nutrients), vertical motion (sinking), and interactions with grazers. Consequently, selection for size and form is reflected in phytoplankton community structure (Kruk et al. 2010; Fraisse et al. 2013). The ecological significance of cell size has been the subject of numerous studies (e.g., Chisholm 1992; Karp-Boss et al. 1996; Litchman and Klausmeier 2008; Litchman et al. 2009; Finkel 2007; Maranon 2015 and many others) and will not be discussed further here. As for cell shape, it can justly be said that it has been studied a great deal but understood very little. Shape has been used to classify phytoplankton since the nineteenth century, and taxonomists have provided detailed drawings and descriptions of phytoplankton forms. What is not clear is how the shape of a phytoplankter might affect its performance and contribute to natural selection. Several hypotheses have been put forth as to the potential selective value(s) of cell shape (Smayda 1970; Lewis 1976; Sournia 1982), but relatively few studies have approached this question mechanistically, and empirical evidence from fitness experiments is difficult to obtain.

Phytoplankton exhibit striking morphological diversity (e.g., in shape, solitary vs. colonial growth, presence/absence of spines and other projections, types of cell coverings and ultrastructural features), and classification schemes of "morphological types" have been produced since the early days of oceanography (reviewed in Sournia 1982). The general shape of cells, though, appears to be restricted to a finite number of geometric forms (Lewis 1976; Hillebrand et al. 1999). As a first approximation, phytoplankton shape can be represented by a series of spheroids and cylinders with varying aspect ratios. For example, Nguyen et al. (2011) showed that the motion of a cell with spines or projections (e.g., *Thalassiosira* sp.) could be well

L. Karp-Boss (✉) • E. Boss
School of Marine Science, University of Maine, Orono, ME 04469, USA
e-mail: lee.karp-boss@maine.edu

P.M. Glibert, T.M. Kana (eds.), *Aquatic Microbial Ecology and Biogeochemistry: A Dual Perspective*, DOI 10.1007/978-3-319-30259-1_3

predicted from theory of the motion of spheroids in a simple shear flow, providing that the cell is described by the smallest inscribing spheroid that encompasses both the cell and its spines. The aspect ratio of a spheroid (E) is defined as the ratio between the principal axis of revolution and the maximum diameter perpendicular to this axis; thus two dimensions are sufficient to describe these shapes. An analysis of the distribution of aspect ratios of more than 8000 coastal microphytoplankton cells, representing different seasons and environmental conditions, indicates that attenuated shapes such as prolate spheroids and elongated cylinders ($E>1$) are much more prevalent than spheres ($E=0$), oblate spheroids or disks ($E<1$; Fig. 1). These observations beg the question: Why is a prolate or rod-like shape much more common among microphytoplankton compared to oblate and discoid shapes? Assuming that shape selection is driven by environmental pressures, what are the selective advantages associated with elongated shapes compared to other shapes with similar volume and surface area?

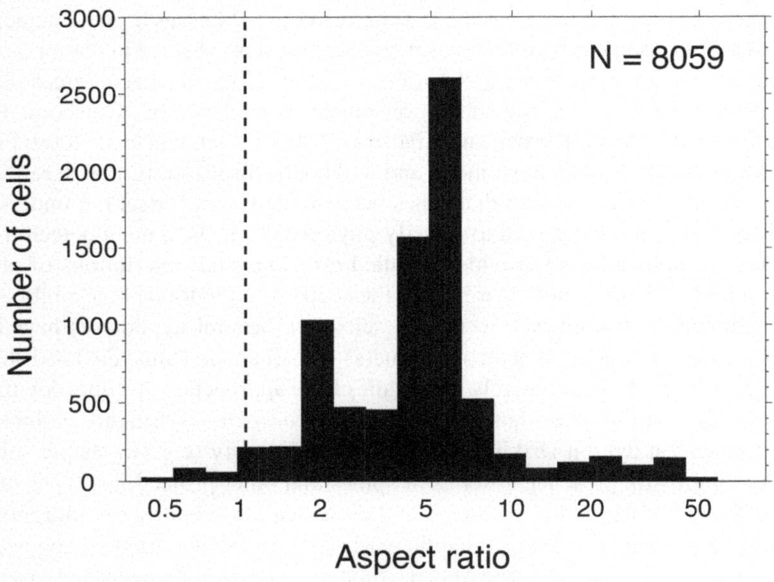

Fig. 1 Frequency distribution of aspect ratios of phytoplankton. Figure modified from Clavano et al. (2007). *Data source*: California State Department of Water Resources and the U.S. Bureau of Reclamation. A subset of the data (2002–2003) was randomly selected for the analysis, representing a variety of aquatic habitats: from fresh water in the Sacramento-San Joaquin delta to estuarine environments in the Suisun and San Pablo Bays (California, USA). Data include phytoplankton from five different classes, including Bacillariophyceae (diatoms), Chlorophyceae, Cryptophyceae, Dynophyceae, and Cyanophyceae ($N=8059$ cells). The aspect ratio is calculated as the ratio in length along the rotational and equatorial axes of a cell, based on the three-dimensional shape associated with each species as provided in Hillebrand et al. (1999). *Vertical, dashed line* highlights the position of spheres along the X-axis (Aspect ratio = 1)

Effects of Shape on Diffusion

To address these questions we first look at diffusion, a fundamental physical factor in the life of a phytoplankter and a primary selective pressure that drives natural selection in organisms that live on solutes. Diffusion governs the exchange of solutes between a cell and its surrounding environment, including nutrients, dissolved gasses, info-chemicals, toxins, and other metabolites. While relative flow may enhance nutrient delivery to large phytoplankton cells (Karp-Boss et al. 1996), in the vicinity of the cell diffusion is always the dominant transport mechanism. Dependence on diffusion sets a constraint on cell size and, as we argue here, has also influenced cell shape. Of all convex shapes, a sphere has the lowest surface area-to-volume ratio for a given unit of volume (from here on we denote surface area, S, and volume, V, and their ratio S/V). Hence, deviation from sphericity will confer competitive advantage, especially in the case of larger cells. Since S/V decreases with size for any given shape, one would expect transport limitation to increase with size. Observations of 57 species of unicellular algae show that size-dependent variation in surface area and volume deviates from the one expected for any series of objects that differ in size but share the same geometry ($S \alpha V^{2/3}$), indicating that cell shape changes with body size (Niklas 2000). Increasing surface area, however, is costly since the cell has to invest additional material and energy in constructing and maintaining a larger envelope per unit of volume (i.e., membranes and organic or mineralized cell walls). Competitive advantage will therefore be conferred only when the enhancement of diffusive flux will exceed the costs associated with added surface area. Different combinations of shapes and sizes can converge to the same V and S/V. If maximizing S/V was the only guiding principle that influences shape, then elongated, rod-like and flat, disk-like shapes would be equally likely to be found in nature. Yet, disk-like shapes are significantly less common among phytoplankton (Fig. 1), or other osmotrophs such as bacteria (Dusenbery 1998; Young 2006). Dusenbery (1998), addressing the same question from the perspective of a bacterium, suggested that a disk-shaped cell is the optimal form with respect to increasing surface area because a disk has the largest surface area for a given minimum radius of curvature. Because bacteria are rarely shaped as disks, he concluded that increasing surface area is not a major component of fitness for bacteria (Dusenbery 1998).

The philosopher Marcus Aurelius (121–180 AD) said, "Look beneath the surface; let not the several quality of a thing nor its worth escape thee." For phytoplankton (including photosynthetic bacteria), elaboration of surface area is important for both nutrient and light acquisition, but S/V per se is not the best predictor of whether a specific shape contributes or not to the fitness of phytoplankton. To demonstrate this point, we compare fluxes of solutes to a series of spheroids and cylinders of identical surface areas and volumes. The rate of mass transfer to a particle of a given shape, F, is given by

$$F = \phi\left(C_{\infty} - C_0\right)D, \quad (1)$$

where C_{∞} and C_0 are the ambient concentration and the concentration at the surface of the cell, respectively, D is the diffusion coefficient of the molecule of interest (dimensions: $L^2 T^{-1}$) and ϕ is a shape factor that has dimensions of length, L (also called external conductance because the Laplace equation applies to both electrical conductivity and molecular diffusion of mass; Clift et al. 1978; Murray and Jumars 2002). Analytical solutions have been derived for a number of geometric forms including spheroids, cylinders, rectangular plates, and cubes (Clift et al. 1978). For the purpose of this study, we applied solutions for prolate and oblate spheroids and cylinders (Table 1). If the diffusion coefficient, ambient concentration and concentration at the cell surface are the same for all shapes, the flux ratio is equal to the ratio of their respective shape (conductance) factors.

We first examine the flux to spheroids and cylinders of varying aspect ratios relative to the flux to a sphere with an equal volume (a proxy for mass). We refer to the ratio between these fluxes as the "relative flux." The range of aspect ratios in these calculations was determined from observations (i.e., Fig. 1). As expected, a spherical cell experiences the lowest flux compared to oblate or prolate spheroids, disks, and elongated cylinders, because this shape has the lowest S/V for a unit mass (Fig. 2). Next, we compare "relative fluxes" between prolate and oblate spheroids of the equal surface area and volume and elongated cylinders and disks of equal surface area and volume. Depending on the aspect ratio, the relative flux to a prolate spheroid and cylinders is up to 30 % higher than that to oblate spheroids and disks of the same S/V (Fig. 3). Differences in fluxes between shapes of equal S/V arise from the fact that diffusion is not uniform over the surface area of an object of an arbitrary shape, but is affected by edges, corners, and the curvature of the object (Clift et al. 1978); prolate spheroids have, on average, higher curvature than flat disks of the same volume.

This result strongly suggests selective pressure for specific shapes rather than a convergence of different shapes that maximize S/V. For the same unit volume and investment in surface area, an elongated cell is expected to have a competitive advantage over a squat or flat one. Among elongated shapes, within the range of aspect ratios that is representative of phytoplankton and again for equal volume, cigar-like cells are predicted to experience slightly higher fluxes than cylindrical cells (Fig. 3). This theoretical prediction is not only consistent with the observation of the prevalence of rod-shaped cells compared to disk-shaped cells among phytoplankton and bacteria, but it is further supported by experimental results. Sommer (1998) conducted competition exclusion experiments with natural phytoplankton assemblages, grown under different silicate:nitrate (Si:N) ratios. Non-diatom species in these natural populations tended to be spherical, whereas diatoms were morphologically diverse. At (quasi) equilibrium, four of the six most successful diatom competitors were species with thin, elongated shapes; none of the diatom species that were excluded was very elongated (Sommer 1998).

The diffusion model presented here does not take into account cell physiology that by itself is under selective pressure. Using a physiological model, Grover (1989) examined the influence of cell shape on competition for phosphate. His model combines the Droop model for algal growth (growth as a function of internal stores of nutrients; Droop 1973) and empirical relationships between a physiological process (uptake rate) and cell volume and surface area that were derived from the freshwater literature. These allometric relationships show that while maximal uptake is proportional to surface area (Smith and Kalff 1983), the minimal nutrient

Table 1 (A) Geometric relations for spheroids and cylinders (Beyer, 1987) and (B) diffusion shape factor (conductance), after Table 4.2 in Clift et al. (1978)

A

Shape	*Surface area*	*Volume*
Oblate spheroid ($b/a<1$)	$\frac{2\pi}{\sqrt{a^2-b^2}}\left[a^2\sqrt{a^2-b^2}+ab^2\ln\left(\frac{a+\sqrt{a^2-b^2}}{b}\right)\right]$	$\frac{4\neq}{3}a^2b$
Sphere ($a=b$)	$4\pi a^2$	$\frac{4\neq}{3}a^3$
Prolate spheroid ($b/a>1$)	$2\pi a^2+\frac{2\pi ab^2}{\sqrt{b^2-a^2}}\sin^{-1}\left(\frac{\sqrt{b^2-a^2}}{b}\right)$	$\frac{4\neq}{3}a^2b$
Cylinder	$2\pi a(a+h)$	$4\pi a^2h$

B

Shape	*Shape factor (conductance)*
Oblate spheroid ($b/a<1$)	$\frac{4\pi a\sqrt{1-E^2}}{cos^{-1}(E)}, E\equiv\frac{b}{a}<1$
Sphere ($a=b$)	$4\pi a$
Prolate spheroid ($b/a>1$)	$\frac{4\pi a\sqrt{E^2-1}}{\ln\left(E+\sqrt{E^2-1}\right)}, E\equiv\frac{b}{a}>1$
Cylinder	$a\left[8+6.95\left(\frac{h}{2a}\right)^{0.76}\right]$

E is the aspect ratio (b/a) for spheroids where "b" is the axis of revolution and "a" is the equatorial axis. For cylinders, "h" is the height and "a" is the radius (see illustration below)

Prolate spheroid Oblate spheroid Cylinder

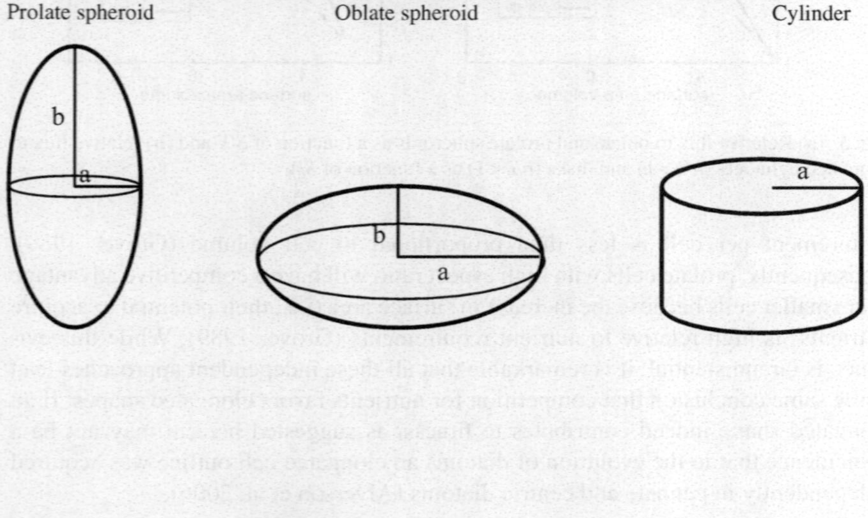

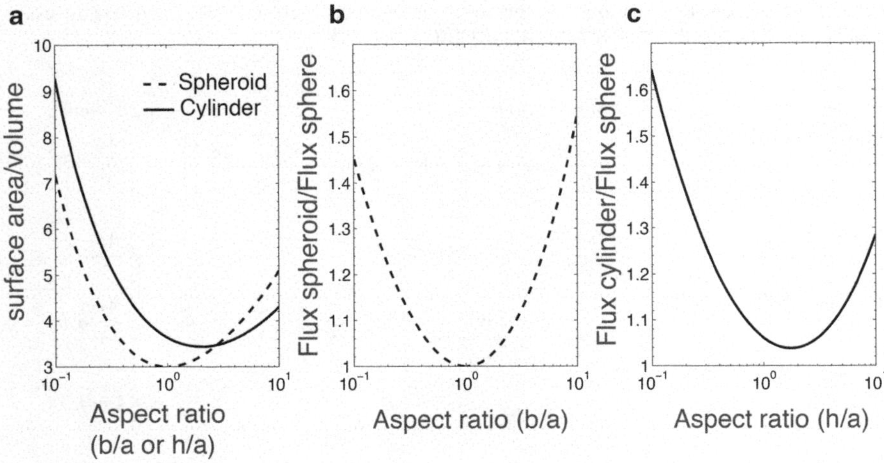

Fig. 2 (**a**) A comparison of *S*/*V* for spheroids of different aspect ratios. (**b**) "Relative flux" for spheroids as a function of aspect ratio (the flux to a spheroid relative to that of a sphere of equal volume). (**c**) "Relative flux" for cylinders as a function of aspect ratio (the flux to a cylinder relative to that of an equal volume sphere)

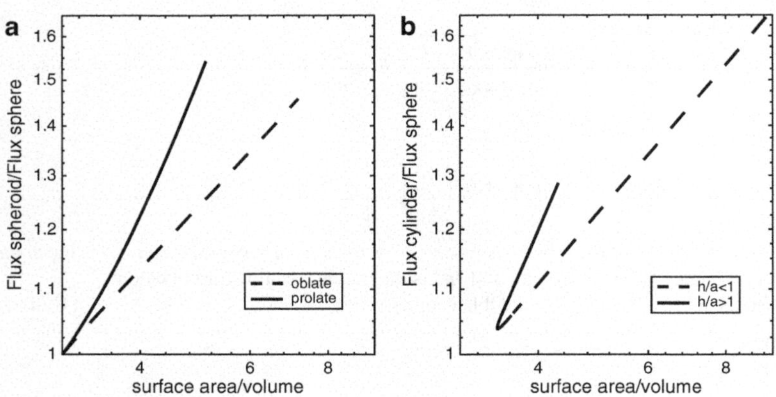

Fig. 3 (**a**) Relative flux to oblate and prolate spheroids as a function of *S*/*V* and (**b**) relative flux to elongated cylinders ($h/a > 1$) and disks ($h/a < 1$) as a function of *S*/*V*

requirement per cell is less than proportional to cell volume (Grover 1989). Consequently, prolate cells with high aspect ratio will have a competitive advantage over smaller cells because the increase in surface area (i.e., their potential to acquire nutrients) is high relative to nutrient requirements (Grover 1989). While this evidence is circumstantial, it is remarkable that all these independent approaches lead to the same conclusion that competition for nutrients favors elongated shapes. If an elongated shape indeed contributes to fitness, as suggested here, it may not be a coincidence that in the evolution of diatoms an elongated cell outline was acquired independently in pennate and centric diatoms (Alverson et al. 2006).

Other Selective Pressures

Nutrient acquisition reflects only one of the selective pressures experienced by phytoplankton. Other selective pressures may favor other shapes or be neutral with respect to shape, resulting in the coexistence of different shapes in nature. Light absorption, for example, depends on the cross-sectional area of a particle perpendicular to the path of light. Of all convex shapes a sphere has the smallest cross-sectional area-to-volume ratio, therefore deviation from a spherical shape may again enhance fitness. Clavano et al. (2007) used a suite of numerical methods to study light absorption by randomly oriented spheroids of varying aspect ratios and compared them to spheres of equal volumes. Refractive index was chosen to be representative of phytoplankton and the derived approximate solutions cover optical regimes that span the range of cell sizes exhibited by phytoplankton. These numerical results suggest that for small particle diameters (<1 μm), absorption is similar for a spheroid and an equal volume sphere while for large phytoplankton-like particles (diameter >50 μm) absorption by a spheroid is always larger than that of a sphere of the same volume, but is similar for oblate and prolate spheroids with the same S and V (Fig. 8c of Clavano et al. 2007). For intermediate size particles (1–50 μm), the bias in absorption between a spheroid and an equal volume sphere increases monotonically with size, without an apparent advantage to one form (prolate or oblate) over the other (Fig. 8c of Clavano et al. 2007). Thus, for picophytoplankton absorption might be neutral with respect to shape selection whereas for microphytoplankton, deviation from sphericity enhance fitness relative to an equal volume sphere, but prolate and disk-like cells, of the same S and V, will have similar fitness.

Another possibly important evolutionary pressure operating on phytoplankton shape is grazing. Experiments showing morphological changes in response to grazers strongly suggest that morphology is a significant, selectable trait with respect to grazing, but thus far these studies have focused on colony or spine formation (Long et al. 2007; Selander et al. 2011; Bergkvist et al. 2012; Van Donk 1997). For solitary cells, the adaptive value of shape remains speculative with no theoretical or empirical arguments for the potential fitness of one shape over another. Encounter rates and the handling efficiency of a predator depend on the orientation of the prey item in the flow field and when intercepting the capture area of a predator. Spherical particles allow the greatest curvature of the foregut and can be swallowed without the need to re-orient the particles. Copepods can successfully re-orient elongated cells by means of their feeding currents but the efficiency of re-orientation is predicted to decrease in the presence of turbulence (Visser and Jonsson 2000). Turbulence may therefore select for rod-like shapes by reducing grazing pressures on these forms. More observations, theoretical and empirical models are needed to elucidate how grazing pressures may affect selection for shape.

Environmental forces act simultaneously, and their interactions result in a complex combination of responses, making predictions of the selective value of shape difficult. Using a quantitative, mechanistic analysis, we can begin to highlight a few general trends about cell shape. Most notably, competition for nutrients selects for elongated shapes, while competition for light appears to be neutral with respect to shape for small cells and equally selective for prolate and oblate spheroids for larger cells. Here we touched on only three selective factors, but left out other important ones such as interaction with ambient flows, namely turbulence, gravity (the need to remain suspended), motility, and physiological benefits associated with polar differentiation of cells. More studies, including those that address evolutionary lineages of morphology, are needed to evaluate the relative contributions of these diverse selective forces to shape selection in phytoplankton.

Acknowledgements We thank Pete Jumars for helpful comments that greatly improved the manuscript.

Lee Karp-Boss and Emmanuel Boss

We met as undergrads at the Hebrew University (Jerusalem, Israel), in Dr. Boaz Luz's "Introduction to Oceanography" class. At the end of the semester, we went diving in the Red Sea, and the rest is history. After completing our M.Sc. degrees in 1991 (Lee studied flow effects on corals while Emmanuel worked on flow instabilities), we left Israel and moved to pursue our Ph.D.s at the University of Washington. Lee became interested in biophysical interactions in the plankton and Emmanuel continued to work on problems in geophysical fluid dynamics. Serendipitously, our mentors were dual-career couples in Oceanography. Emmanuel was a student of Luanne Thompson (married to Greg Johnson) and Lee was a student of Pete Jumars (married to Mary Jane Perry who became a mentor for both of us). The Ocean Optics summer class at Friday Harbor was a turning point for Emmanuel who saw a future full of light, changed course and went to postdoc with Mary Jane Perry in the field of bio-optics. Our collaborations with Perry/Jumars continue to this day. We moved to Oregon in 1998 for postdoc positions at OSU and in 2002 we took faculty positions at the University of Maine. Over the years, we have collaborated on several papers that address phytoplankton-turbulence interactions, science education, optical properties of phytoplankton, and on field campaigns (Tara Oceans and currently NAAMES). Our most fulfilling collaboration is raising our three sons, Yuval, Tom, and Itai. The first two were born while in graduate school, both during Ocean Optics summer classes (Yuval when Emmanuel was a student in the class and Tom when Emmanuel was a TA), and took their first ferry rides to the Friday Harbor lab at age 2 days. Our kids enjoyed spending countless hours with our office mates and colleagues, "participated" in many Ocean Optics summer classes, endured Ocean Sciences meetings, and got their sea legs during a research cruise in the Red Sea and a sleepover on the R/V Tara in the middle of Paris. Not surprisingly, none is considering a career in science.

References

Alverson AJ, Cannon JJ, Gutell RR, Theriot EC (2006) The evolution of elongate shape in diatoms. J Phycol 42:655–668
Bergkvist J, Thor P, Jakobsen HH, Wängberg SÅ, Selander E (2012) Grazer-induced chain length plasticity reduces grazing risk in a marine diatom. Limnol Oceanogr 57:318–324
Beyer WA, Fawcett LR, Hartem LP, Swartz BK (1987) The volume common to two congruent circular cylinders, Journal of Symbolic Computation 4(2):381–390.
Chisholm SW (1992) Phytoplankton size. In: Falkowski P (ed) Primary productivity and biogeochemical cycles in the sea. Springer, New York, NY, pp 213–237
Clavano WR, Boss E, Karp-Boss L (2007) Inherent optical properties of non-spherical marine-like particles – from theory to observation. Oceanogr Mar Biol Annu Rev 45:1–38
Clift R, Grace JR, Weber ME (1978) Bubbles, drops and particles. Academic, New York, NY, p 380
Droop MR (1973) Some thoughts on nutrient limitation in algae. J Phycol 9:264–272
Dusenbery DB (1998) Fitness landscapes for effects of shape on chemotaxis and other behaviors of bacteria. J Bacteriol 180:978–983
Finkel ZV (2007) Does phytoplankton cell size matter? The evolution of modern marine food webs. In: Falkowski PG, Knoll AH (eds) Evolution of Aquatic Photoautotrophs. Academic, San Diego, pp 333–350
Fraisse S, Bormans M, Lagadeuc Y (2013) Morphofunctional traits reflect differences in phytoplankton community between rivers of contrasting flow regime. Aquat Ecol 47:315–327
Grover JP (1989) Influence of cell shape and size on algal competitive ability. J Phycol 25:402–405
Hillebrand H, Dürselen C-D, Kirschtel D, Pollingher U, Zohary T (1999) Biovolume calculation for pelagic and benthic microalgae. J Phycol 35:403–424
Karp-Boss L, Boss E, Jumars PA (1996) Nutrient fluxes to planktonic osmotrophs in the presence of fluid motion. Oceanogr Mar Biol Annu Rev 34:71–107
Kruk C, Huszar VLM, Peeters ETHM, Bonilla S, Costa L, Lürling M, Reynolds CS, Scheffer M (2010) A morphological classification capturing functional variation in phytoplankton. Freshwater Biol 55:614–627
Lewis WM (1976) Surface/volume ratio: implications for phytoplankton morphology. Science 192:885–887
Litchman E, Klausmeier CA (2008) Trait-based community ecology of phytoplankton. Annu Rev Ecol Evol Syst 39:615–639
Litchman E, Klausmeier CA, Yoshiyama K (2009) Contrasting size evolution in marine and freshwater diatoms. Proc Natl Acad Sci U S A 106:2665–2670
Long JD, Smalley GW, Barsby T, Anderson JT, Hay ME (2007) Chemical cues induce consumer-specific defenses in a bloom-forming marine phytoplankton. Proc Natl Acad Sci U S A 104:10512–10517
Maranon E (2015) Cell size as a key determinant of phytoplankton metabolism and community structure. Ann Rev Mar Sci 7:241–264
Murray JLS, Jumars PA (2002) Clonal fitness of bacteria predicted by analog modeling. Biosciences 52:343–355
Nguyen HV, Karp-Boss L, Jumars PA, Fauci L (2011) Hydrodynamic effects of spines: a different spin. Limnol Oceanogr Fluid Environ 1:110–119
Niklas KJ (2000) The evolution of plant body plans – a biomechanical perspective. Ann Bot 85:411–438
Selander E, Jakobsen HH, Lombard F, Kiörboe T (2011) Grazer cues induce stealth behavior in marine dinoflagellates. Proc Natl Acad Sci 108:4030–4034
Smayda T (1970) The suspension and sinking of phytoplankton in the sea. Oceanogr Mar Biol Ann Rev 8:53–414

Smith REH, Kalff J (1983) Competition for phosphorous among co-occurring freshwater phytoplankton. Limnol Oceanogr 28:440–444

Sommer U (1998) Silicate and the functional geometry of phytoplankton. J Plankton Res 20:1853–1859

Sournia A (1982) Form and function in marine phytoplankton. Bio Rev 57:347–394

Van Donk E (1997) Defenses in phytoplankton against grazing induced by nutrient limitation, UV-B stress and infochemicals. Aquat Ecol 31:53–58

Visser AW, Jonsson PR (2000) On the reorientation of non-spherical particles in a feeding current. J Plankton Res 22:761–777

Young KD (2006) The selective value of bacterial shape. Microbiol Mol Biol Rev 70:660–703

Crossing the Freshwater/Saline Barrier: A Phylogenetic Analysis of Bacteria Inhabiting Both Freshwater and Marine Ecosystems

Mina Bižić-Ionescu and Danny Ionescu

Introduction

Saline and freshwater body systems are usually separate entities, but transition of microorganisms from one to another constantly occurs through various mechanisms. Rivers unidirectionally transport biomass from freshwater to the marine system. Bidirectional transport mechanisms include aerosols, plants, euryhaline (anadromous) fish, birds, and on an evolutionary more recent timescale there is also an increasingly anthropogenic contribution.

The separation between limnic and marine bacteria can be driven by several factors. Salinity of natural waters is one of the main environmental factors with a major contribution to structure and functional characteristics of aquatic microbiota. Hence, it can be an important factor for the separation of bacterial communities (Bouvier and del Giorgio 2002; Selje and Simon 2003; Telesh et al. 2013). To cope with external salinity, bacteria use two different mechanisms for osmoregulation: the "salt in" mechanism, utilized mostly but not exclusively by extreme halophiles and "compatible solutes" used by moderate halophiles and halotolerant bacteria (Wood 2011). Freshwater bacteria have been shown to survive in marine systems since the 1950s when the fate of sewage outflow into the ocean was investigated (Carlucci and Pramer 1959; Székely et al. 2013). Recent metagenomic studies have shown that while freshwater bacteria probably use the "salt in" osmoregulatory pathways, they have a significantly lower number of Na pumps in comparison to marine bacteria (Oh et al. 2011). This would impair their ability to deal with the Na influx in saline waters or increase its energetic cost. Results showing that competition with resident communities have the strongest influence (Székely et al. 2013)

M. Bižić-Ionescu • D. Ionescu (✉)
Department of Experimental Limnology, Leibniz Institute for Freshwater Ecology and Inland Fisheries, IGB, Alte Fischerhütte 2 OT, Neuglobsow 16775, Stechlin, Germany
e-mail: mbizic@igb-berlin.de

P.M. Glibert, T.M. Kana (eds.), *Aquatic Microbial Ecology and Biogeochemistry: A Dual Perspective*, DOI 10.1007/978-3-319-30259-1_4

would favor for the latter option. An additional factor that may play a role in the transition of bacteria from freshwater to a marine environment is pH. While the pH in freshwater may vary dramatically between systems, ranging between <2 and >12 (Joint et al. 2011), the oceanic pH is rather stable at slightly alkaline values of about 8.1–8.3.

Biogeographically, this continuous transport leads over time to a ubiquitous distribution of aquatic bacteria thus fulfilling the "Everything is everywhere" clause from the Baas-Becking (1934) theorem. The establishment and lasting of different microbial communities in freshwater and saline aquatic systems proves that indeed "Environment selects" and fulfills the second clause of the theorem. Nevertheless, most biogeographical studies neglect the time of transfer as a possible significant factor in the absence of an organism from an ecosystem. In this study, we searched for organisms with identical 16S rRNA gene sequence inhabiting both marine and freshwater systems and addressed the time issue by compiling sequence data from a database containing several decades of data.

Recent Data on Shared Taxa

Comparing samples collected over a prolonged period of time from two different limnic and two different marine systems, we tried to evaluate the frequency of taxa common to both types of environments (Bižić-Ionescu et al. 2014). Interestingly, only 68 Operational Taxonomic Units (OTU) (clustered at 99 % similarity) out of 35,554 were shared between the limnic and marine systems. Out of these, 41 OTUs belonged to particle-associated (PA) bacteria, 19 of which were known as potential pathogens. These include bacteria such as *Staphylococcus aureus*, *S. epidermidis*, *S. pasteuri*, *Brevundimonas diminuta*, *Bosea massiliensis*, and *Vibrio* sp. Although we cannot rule out that these sequences are anthropogenic contaminations that occurred during sampling or sample preparation, particles are known to serve as a refuge for pathogenic bacteria (Grossart et al. 2010; Tang et al. 2011). It has been described before that important pathogens like *Vibrio cholera* survive and spread on particles (Colwell et al. 2003; Danovaro et al. 2009). Nevertheless, this is the first time that 16S rRNA gene sequences identical to that of pathogenic bacteria are described simultaneously, on particles from both limnic and marine systems. Our findings suggest that particles could serve as transfer vehicle between different ecosystems. It has been observed that in estuaries, environments with transitional properties between limnic and marine systems, >50 % of the bacteria are found to be associated with particles (Crump et al. 1998; Simon et al. 2002). Therefore, it is possible that the microniche of the particle protects the bacteria against sudden changes in the environmental conditions. A similar observation was made for bacteria "hiding" from UV irradiation in wastewater treatment plants (Tang et al. 2011). While a particle can shade the UV light thus offering the bacteria a protective environment, it remains to be determined if it can also provide a safe passage through a strong salinity gradient.

Synthesis of Published Sequence Data

To establish the validity of these results on a global scale, we conducted a small-scale comparison using the SILVA PARC small rRNA subunit database (release 115). We extracted 367,000 marine and freshwater sequences (out of 3,800,000), using text identifiers applied to the sequence isolation location. A similar comparison was done by Barberán and Casamayor (2010) using >7000 surface water sequences which clustered at 97 % similarity. When clustering the sequences obtained from the SILVA database (Quast et al. 2013) at different similarity levels, we observed the omnipresence of a certain set of sequences between marine and freshwater systems (Fig. 1). Interestingly, the number of obtained OTUs (with 90–99 % sequence similarity) shared between marine and freshwater systems appear to be rather conserved, ranging between 4850 and 5000 OTUs, and only dropping to 2279 at 100 % similarity. Nevertheless, these OTUs show a decrease in their proportional significance from the entire dataset decreasing from 25 to 0.81 % out of the total OTUs when the similarity criterion was increased from 90 to 100 %, respectively. In our study (Bižić-Ionescu et al. 2014), we were able to provide a PA/FL (free-living) distribution of the taxa shared between limnic and marine habitats; however, this was not possible for the SILVA database as the information is not available.

The number of globally shared OTUs at 99 % similarity clustering is much higher than obtained in our localized study (4850 vs. 68 OTUs making up 4.3 and 0.2 % of all OTUs, respectively). Part of the large number of shared taxa could be the result

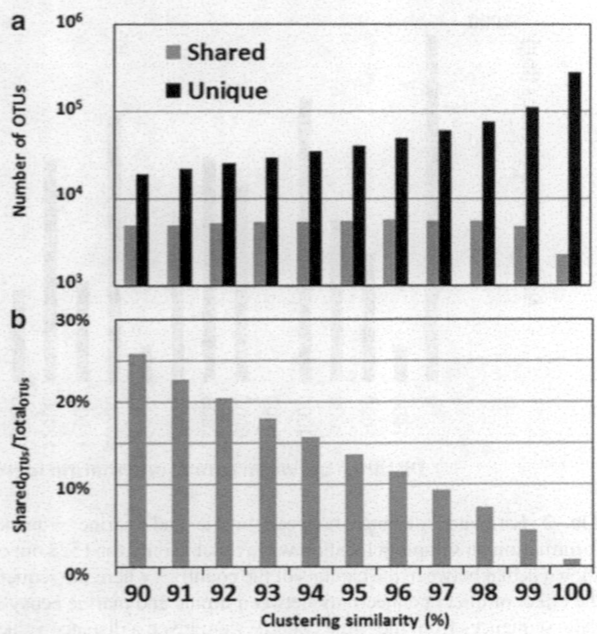

Fig. 1 Total number of operational taxonomic units (OTUs) obtained by clustering sequences from marine and freshwater environments available in the SILVA SSU PARC 115 database alongside the number of shared OTUs between the two environments (**a**). Percentage of shared OTUs out of the total for each sequence similarity criterion (**b**). The shared OTUs span over the three domains of life: *Archaea*, *Bacteria*, and *Eukaryota*

of wrong data mining coupled with incomplete data submission for the deposited sequence; however, manual verification of part of the results showed that this does not account for the entire phenomenon. Interestingly, the grouping of these taxonomic entities is conserved up to 99 % sequence similarity, suggesting that these sequences belong to conserved clusters.

A second possible explanation for the relatively high number of shared sequences between limnic and marine systems can be that the sequences originate from directly connected systems (i.e. lakes, rivers, and their deltas). To evaluate the nature of connectivity between identical sequences, we calculated the distance between the capitals of the countries from which the sequences were obtained. To minimize the effect of erroneous or minimalistic meta-data submission, distance between sequences from the same country was set by definition to 0. As this includes large countries like the USA and China, the results presented in Fig. 2 are probably an underestimation of the actual number of geographically distanced sequences. Nevertheless, nearly half (738/1528) of the sequences for which geographical information was available were identical to a marine or limnic counterpart sampled over 1000 km away.

The distribution of distances between identical marine and limnic sequences is of biogeographical significance supporting the Baas Becking hypothesis of "Everything is everywhere but the environment selects" (Baas-Becking 1934). Due to the limited duration of standard projects, biogeography studies omit the time factor involved in transport of organisms across distant ecosystems. Therefore, these studies provide temporally concentrated snapshots that may result in an apparent pattern (e.g. Ghiglione et al. 2012; Yan et al. 2009; Brandsma et al. 2012) whose presence may or

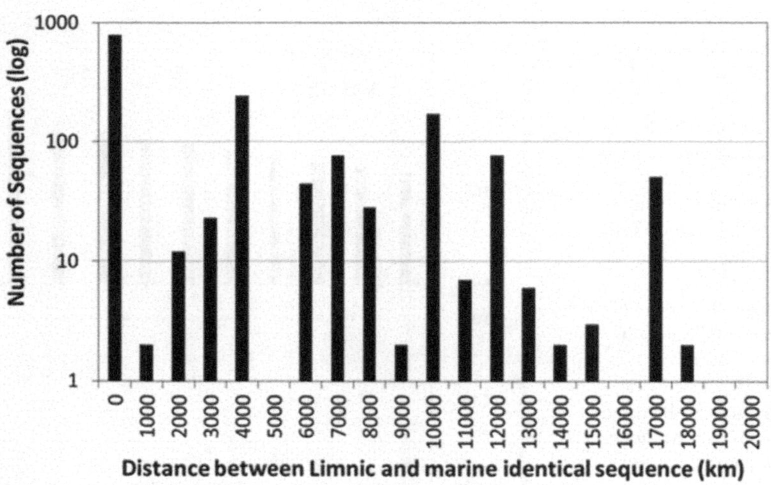

Fig. 2 Estimated distance between limnic and marine sequences within individual clusters. Information on sampling location was available only for 1528 out of 8192 sequences. The distance shown is that between the capitals of the countries where the sequences were isolated. To minimize the effect of direct connectivity between limnic and marine ecosystems in the absence of accurate data, sequences from the same country were given a distance value of 0

may not withstand the test of time. Nevertheless, some studies do acknowledge the problem of short sampling durations (e.g. Ionescu et al. 2010; Ghiglione et al. 2012). Thus, such studies obtain information relatively simultaneously from several locations. In contrast, a long-term, continuously expanding database such as the SILVA SSU database contains sequences submitted during at least the last two decades. Therefore, finding marine- and limnic-associated sequences in distant opposing environments (Fig. 3) suggests transport and thus supports that indeed "Everything (has the chance to be) everywhere." The low number of shared sequences out of the overall data is strong evidence that certainly "Nature selects" for those organisms most adapted to the environment. This is in line with what was postulated by Ionescu et al. (2010) stating that two distanced environments sharing both identical and non-identical sequences result from transfer of organisms followed by local environmental (biotic as well as abiotic factors) selection. Accordingly, sampling at a different time point may reveal a different environmental snapshot due to (a) shift in dominant communities; (b) accumulation of random mutations in once-identical, dominant species (as opposed to evolutionary stagnant cryptic species).

Based on the above, we hypothesize that the identified transitions between limnic and marine systems (or vice versa) represent recent evolutionary events. If indeed identical 16S rRNA sequences originate in this case from identical organisms, the shared clusters represent either recent or unsuccessful transitions that did not yet accumulate any mutations. Given that short sequences generated by high-throughput methods are not included in the SILVA database, it is reasonable to believe that the analyzed data do not belong to the rare biosphere. This leaves the option of recent transition as the most plausible.

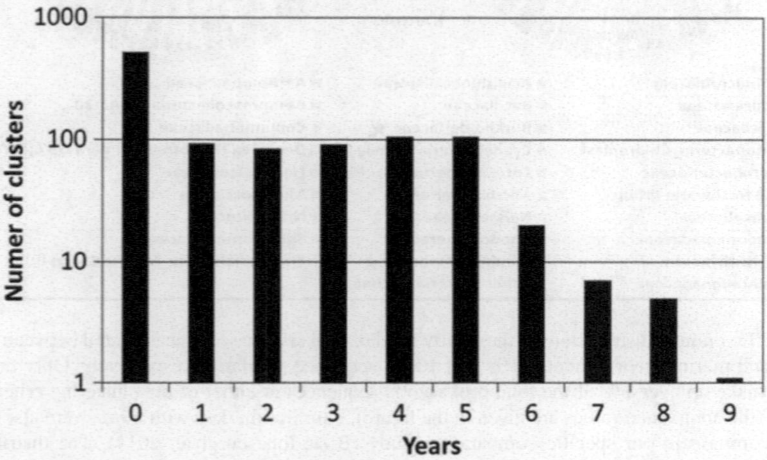

Fig. 3 Maximum time difference (in years) between submission dates of limnic and marine sequences within individual clusters. Data were available for 2373 sequences forming 1027 clusters

The shared OTUs are spread across numerous families with those making over 1 % of the prokaryotic sequences shown in Fig. 4. Families with the largest number of shared sequences include *Acidimicrobiaceae* (*Actinobacteria*) *Acidithio-bacillaceae* (*Gamma-proteobacteria*), *Burkhold-eriaceae* and *Comamonadaceae* (*Betaproteobacteria*), and the gram-positive *Bacillaceae*. The *Betaproteobacteria* have been suggested before to have made several transitions between freshwater and marine environments (Rappé et al. 2000), therefore it is expected that they make the largest fraction of the shared prokaryotic sequences. Nevertheless, Walsh et al. (2013) point out only the *Methylophilalles* (OM43 clade) as shared taxa. Interestingly, other groups that were suggested to be shared between freshwater and marine environments such as the *Planctomycetes* (Glöckner et al. 1999; Pizzetti et al. 2011), *Verrucomicrobia* (Lee et al. 2009; Arnds et al. 2010) make up only a very small fraction (<1 %) of the shared taxa. An interesting result comes from the SAR11 clade. This clade harbors the ubiquitous marine SAR11 (Giovannoni et al. 2005) and its

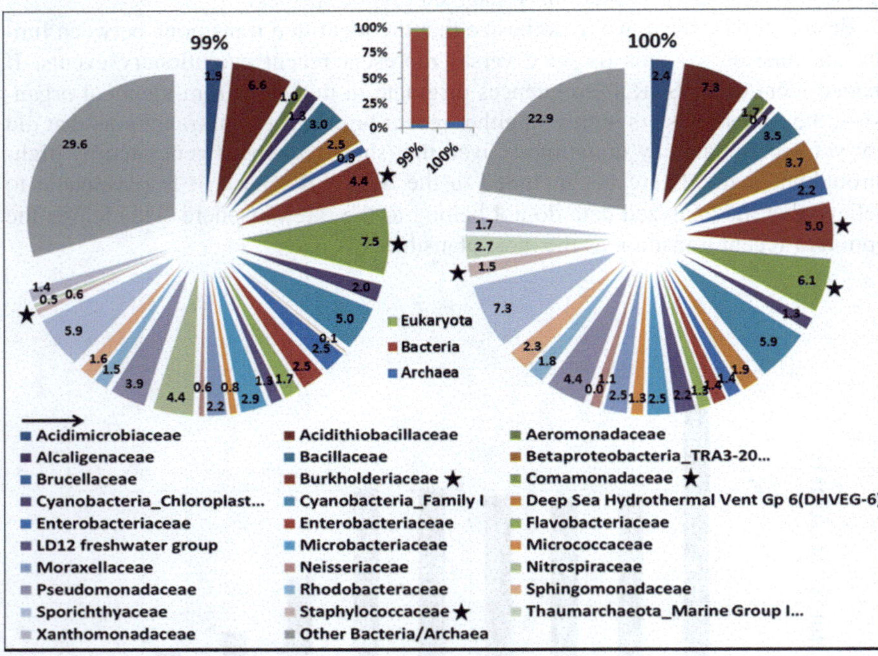

Fig. 4 Taxonomic distributions at the family level of prokaryotic sequences shared between freshwater and marine environments at 99 and 100 % sequence similarity, respectively. Only families which make up over 1 % of the total prokaryotic sequences in either of the clustering criteria are shown (the abundance values are given in the figure). Families marked with a *star* were also found to be common in our specific comparative study (Bižić-Ionescu et al. 2014). The distribution among the three domains of life of the sequences shared between marine and freshwater environments is shown in the *central panel* for each clustering criteria. Groups for which taxonomic data are not available at the family level are represented by the last two defined taxonomic levels followed by *three dots*

freshwater counterpart LD12 (Newton et al. 2011). At 97 % similarity this bacterial group spans over all aquatic environments (Logares et al. 2010) yet at higher resolution the marine and freshwater species are believed to be separated (Logares et al. 2010). Interestingly, while we do not find marine designated SAR11 sequences to be shared with freshwater environments, OTUs recognized as LD12 make up 1.3 and 2.2 % of the shared sequences at 99 and 100 % similarity.

If this result holds true, when sequences are chosen from better defined locations, it suggests that some members of the freshwater adapted SAR11 bacteria maintained their ability to survive in saline waters. Interestingly, Zaremba-Niedzwiedzka et al. (2013) suggest following the analysis of 57 LD12 clade genomes that these organisms are highly adapted to their freshwater environment, unlike their marine counterparts which show a large genetic variability. This is in contrast to our hypothesis where some LD12 bacteria maintained their ability to return to the sea.

When expanding the borders of the study to the global scale, two things become evident: (1) Indeed, transitions between freshwater and marine environments are either not frequent or have not left a strong evolutionary footprint on the microbial community; (2) Despite their low contribution to the overall aquatic communities, omnipresent bacterial groups do exist, and they span across a large part of the phylogenetic tree.

Walsh et al. (2013) suggest that a border-crossing bacteria may on the one hand benefit from the lack of competition from his source community, as known from invasion events of higher organisms (e.g. fish, mussels or copepods; Lee 1999; Lee and Bell 1999), while on the other hand it may perish, as it is not well adapted to the new environment. The latter however, may place the invader among the less dominant, yet detectable, members of the biosphere accounting for the relatively low abundance of shared freshwater-marine taxa.

Future Perspectives

The 16S rRNA sequence comparison based on the SILVA database serves as a proof of principle and a calls for an extended study. Metagenomic data from both limnic and marine ecosystems are freely available and coupled to analytical tools through platforms such as MG-RAST (Meyer et al. 2008) and IMG-MER (Markowitz et al. 2012). The comparative analysis of these data will show whether the phenomenon of shared limnic and marine taxa holds beyond the 16S rRNA level. Expanding the similarity to longer sequences up to the genomic level will allow for a more accurate determination of separation time between the twin limnic and marine organisms. Additionally, the same data can provide answers regarding survival strategies of these organisms in the new limnic or marine environment (e.g. increased number of Na pumps).

Bižić-Ionescu et al. (2014) have shown that the majority of the OTUs they identified as shared between limnic and marine systems were found associated with particulate organic matter. These particles served as nutrient and carbon oasis in an often oligotrophic aquatic environment and as such are rapidly colonized by bacteria. Such particles were suggested to be hotspots for lateral gene transfer via several mechanisms (Riemann and Grossart 2008; Walsh et al. 2013; Bižić-Ionescu et al. 2014) including phages and uptake of naked DNA from lysed cells. It is possible therefore that particles hold a significant role in the adaptation of limnic and marine organisms to counter-environments by facilitating rapid acquisition of necessary genes. This hypothesis remains to be evaluated in future studies.

Acknowledgments We thank Hans-Peter Grossart, Rudolf Amann, and Pelin Yilmaz for their support and help on this project. M.B.I. thanks the DAAD for sponsoring her Ph.D., in which this work has been commenced.

Mina Bižić-Ionescu and Danny Ionescu

After months of politely refusing suggestions to meet from a mutual friend, we eventually had a coffee together in January 2009. Mina, who finished her M.Sc. studies in Belgrade, Serbia, in 2005 and has since finished a second M.Sc. on modern science, and the kosher laws in Judaism in Stockholm, Sweden, was at the time a research assistant at the Kinneret Limnological Laboratory in Israel, working with Werner Eckert. Danny was in his last year of his Ph.D. with Aharon Oren at the Hebrew University of Jerusalem. After a short hello, we both confessed that we are heading for Germany for our next career step, Mina for a Ph.D. and Danny for a postdoc. This was already a sign that this coffee is only the first of many. Our following dates were a series of 24 h experiments that made us realize how well we function together as a research team. Our passion for the sea took us to the next level, as in September 2009 at a depth of 36 m in the Red Sea, Danny proposed; a step that culminated a year later with a sunset wedding on the Mediterranean coast. Mina's Ph.D. was a collaborative research between the Leibniz Institute for Freshwater Ecology and Inland Fisheries and the Max Planck Institute for Marine Microbiology in Bremen, where Danny did his postdoc. This allowed us to collaborate at work and build a home and a family in Bremen. We have been officially and unofficially involved in each other's projects resulting in several common publications. However, the part of our collaboration we are most proud of is Duan El'azar born in 2012, making Mina's Ph.D. "slightly" more exciting, and Eva-Mai Leah born in 2015. Today we are both postdocs at the Leibniz Institute for Freshwater Ecology and Inland Fisheries working on several projects, among them methane production in oxygenated water column and particulate organic matter degradation and succession of associated microbial communities.

References

Arnds J, Knittel K, Buck U, Winkel M, Amann R (2010) Development of a 16S rRNA-targeted probe set for Verrucomicrobia and its application for fluorescence in situ hybridization in a humic lake. Syst Appl Microbiol 33:139–148

Baas-Becking LGM (1934) Geobiologie of inleiding tot de milieukunde. W.P. Van Stockum & Zoon, The Hague

Barberán A, Casamayor E (2010) Global phylogenetic community structure and β-diversity patterns in surface bacterioplankton metacommunities. Aquat Microb Ecol 59:1–10

Bižić-Ionescu M, Zeder M, Ionescu D, Orlić S, Fuchs BM, Grossart H-P, Amann R (2014) Comparison of bacterial communities on limnic versus coastal marine particles reveals profound differences in colonization. Environ Microbiol. doi:10.1111/1462-2920.12466

Bouvier T, del Giorgio P (2002) Compositional changes in free-living bacterial communities along a salinity gradient in two temperate estuaries. Limnol Oceanogr 47:453–470

Brandsma J, Martínez JM, Slagter H, Evans C, Brussaard CPD (2012) Microbial biogeography of the North Sea during summer. Biogeochemistry 113:119–136

Carlucci A, Pramer D (1959) Factors affecting the survival of bacteria in sea water. Appl Microbiol 7:388–392

Colwell RR, Huq A, Islam MS, Aziz KMA, Yunus M, Khan NH et al (2003) Reduction of cholera in Bangladeshi villages by simple filtration. Proc Natl Acad Sci U S A 100:1051–1055

Crump BC, Baross J, Simenstad C (1998) Dominance of particle-attached bacteria in the Columbia River estuary, USA. Aquat Microb Ecol 14:7–18

Danovaro R, Fonda Umani S, Pusceddu A (2009) Climate change and the potential spreading of marine mucilage and microbial pathogens in the Mediterranean Sea. PLoS One 4, e7006

Ghiglione J-F, Galand PE, Pommier T, Pedrós-Alió C, Maas EW, Bakker K et al (2012) Pole-to-pole biogeography of surface and deep marine bacterial communities. Proc Natl Acad Sci U S A 109:17633–17638

Giovannoni SJ, Tripp HJ, Givan S, Podar M, Vergin KL, Baptista D et al (2005) Genome streamlining in a cosmopolitan oceanic bacterium. Science 309:1242–1245

Glöckner F, Fuchs B, Amann R (1999) Bacterioplankton compositions of lakes and oceans: a first comparison based on fluorescence in situ hybridization. Appl Environ Microbiol 65:3721–3726

Grossart H-P, Dziallas C, Leunert F, Tang KW (2010) Bacteria dispersal by hitchhiking on zooplankton. Proc Natl Acad Sci U S A 107:11959–11964

Ionescu D, Hindiyeh M, Malkawi H, Oren A (2010) Biogeography of thermophilic cyanobacteria: insights from the Zerka Ma'in hot springs (Jordan). FEMS Microbiol Ecol 72:103–113

Joint I, Doney SC, Karl DM (2011) Will ocean acidification affect marine microbes? ISME J 5:1–7

Lee CE (1999) Rapid and repeated invasions of fresh water by copepod. Evolution (N Y) 53:1423–1434

Lee CE, Bell MA (1999) Causes and consequences of recent freshwater invasions by saltwater animals. Trends Ecol Evol 14:284–288

Lee K-C, Webb RI, Janssen PH, Sangwan P, Romeo T, Staley JT, Fuerst J (2009) Phylum Verrucomicrobia representatives share a compartmentalized cell plan with members of bacterial phylum Planctomycetes. BMC Microbiol 9:5

Logares R, Bråte J, Heinrich F, Shalchian-Tabrizi K, Bertilsson S (2010) Infrequent transitions between saline and fresh waters in one of the most abundant microbial lineages (SAR11). Mol Biol Evol 27:347–357

Markowitz VM, Chen IMA, Palaniappan K, Chu K, Szeto E, Grechkin Y et al (2012) IMG: the integrated microbial genomes database and comparative analysis system. Nucleic Acids Res 40:115–122

Meyer F, Paarmann D, D'Souza M (2008) The metagenomics RAST server–a public resource for the automatic phylogenetic and functional analysis of metagenomes. BMC Bioinformatics 9:386

Newton RJ, Jones SE, Eiler A, McMahon KD, Bertilsson S (2011) A guide to the natural history of freshwater lake bacteria. Microbiol Mol Biol Rev 75:14–49

Oh S, Caro-Quintero A, Tsementzi D, DeLeon-Rodriguez N, Luo C, Poretsky R, Konstantinidis KT (2011) Metagenomic insights into the evolution, function, and complexity of the planktonic microbial community of Lake Lanier, a temperate freshwater ecosystem. Appl Environ Microbiol 77:6000–6011

Pizzetti I, Fuchs BM, Gerdts G, Wichels A, Wiltshire KH, Amann R (2011) Temporal variability of coastal Planctomycetes clades at Kabeltonne station, North Sea. Appl Environ Microbiol 77:5009–5017

Quast C, Pruesse E, Yilmaz P, Gerken J, Schweer T, Yarza P et al (2013) The SILVA ribosomal RNA gene database project: improved data processing and web-based tools. Nucleic Acids Res 41:D590–D596

Rappé M, Vergin K, Giovannoni S (2000) Phylogenetic comparisons of a coastal bacterioplankton community with its counterparts in open ocean and freshwater systems. FEMS Microbiol Ecol 33:219–232

Riemann L, Grossart H-P (2008) Elevated lytic phage production as a consequence of particle colonization by a marine Flavobacterium (*Cellulophaga* sp.). Microb Ecol 56:505–512

Selje N, Simon M (2003) Composition and dynamics of particle-associated and free-living bacterial communities in the Weser estuary, Germany. Aquat Microb Ecol 30:221–237

Simon M, Grossart HP, Schweitzer B, Ploug H (2002) Microbial ecology of organic aggregates in aquatic ecosystems. Aquat Microb Ecol 28:175–211

Székely AJ, Berga M, Langenheder S (2013) Mechanisms determining the fate of dispersed bacterial communities in new environments. ISME J 7:61–71

Tang KW, Dziallas C, Grossart H-P (2011) Zooplankton and aggregates as refuge for aquatic bacteria: protection from UV, heat and ozone stresses used for water treatment. Environ Microbiol 13:378–390

Telesh I, Schubert H, Skarlato S (2013) Life in the salinity gradient: discovering mechanisms behind a new biodiversity pattern. Estuar Coast Shelf Sci 135:317–327

Walsh DA, Lafontaine J, Grossart H-P (2013) On the eco-evolutionary relationships of fresh and salt water bacteria. In: Gophna U (ed) Lateral gene transfer in evolution. Springer, New York, NY, p 443

Wood JM (2011) Bacterial osmoregulation: a paradigm for the study of cellular homeostasis. Annu Rev Microbiol 65:215–238

Yan S, Fuchs BM, Lenk S, Harder J, Wulf J, Jiao N-Z, Amann R (2009) Biogeography and phylogeny of the NOR5/OM60 clade of Gammaproteobacteria. Syst Appl Microbiol 32:124–139

Zaremba-Niedzwiedzka K, Viklund J, Zhao W, Ast J, Sczyrba A, Woyke T et al (2013) Single-cell genomics reveal low recombination frequencies in freshwater bacteria of the SAR11 clade. Genome Biol 14:R130

Approaches and Challenges for Linking Marine Biogeochemical Models with the "Omics" Revolution

Victoria J. Coles and Raleigh R. Hood

Introduction

Since 2002, exponential growth in the number of marine science journal articles that used the term "omics revolution" (Fig. 1) attests to a general recognition of how genomics is reshaping marine science. The revolution is fueled by rapid advances in genomic, transcriptomic, and proteomic methods that are providing increasingly rapid, detailed, and reliable information about the genetic composition and variability of marine organisms, and the degree to which this potential is expressed. Moreover, this information is being generated at such high rates due to advances in sequencing technologies (Fig. 1) that marine microbial ecologists have struggled to learn the bioinformatics tools associated with analysis, interpretation, and visualization to keep up with the data onslaught. New organisms, metabolic processes and pathways that were previously unknown have been discovered, and subtle metabolic variations within and among species and across environmental gradients have also been revealed (Chisholm et al. 1988; Proctor and Fuhrman 1989; Zehr et al. 1998; Beja et al. 2000; Morris et al. 2002; Rynearson and Armbrust 2004). There is every reason to believe that this astonishing rate of discovery will continue for years to come. The obvious challenge is how to make sense out of the torrent of new information to bring insights to marine ecology and biogeochemistry. What role do all these new genes, transcripts, and proteins play in driving marine ecosystem dynamics and biogeochemical cycles?

Marine ecosystem and biogeochemical modeling techniques provide one approach to address this question. Traditionally, these are either steady state or prognostic numerical models comprising a set of state variables representing various ecosystem and biogeochemical constituents with exchanges between them

V.J. Coles (✉) • R.R. Hood
Horn Point Laboratory, University of Maryland Center for Environmental Science,
2020 Horns Point Road, Cambridge, MD 21613, USA
e-mail: vcoles@umces.edu

P.M. Glibert, T.M. Kana (eds.), *Aquatic Microbial Ecology and Biogeochemistry: A Dual Perspective*, DOI 10.1007/978-3-319-30259-1_5

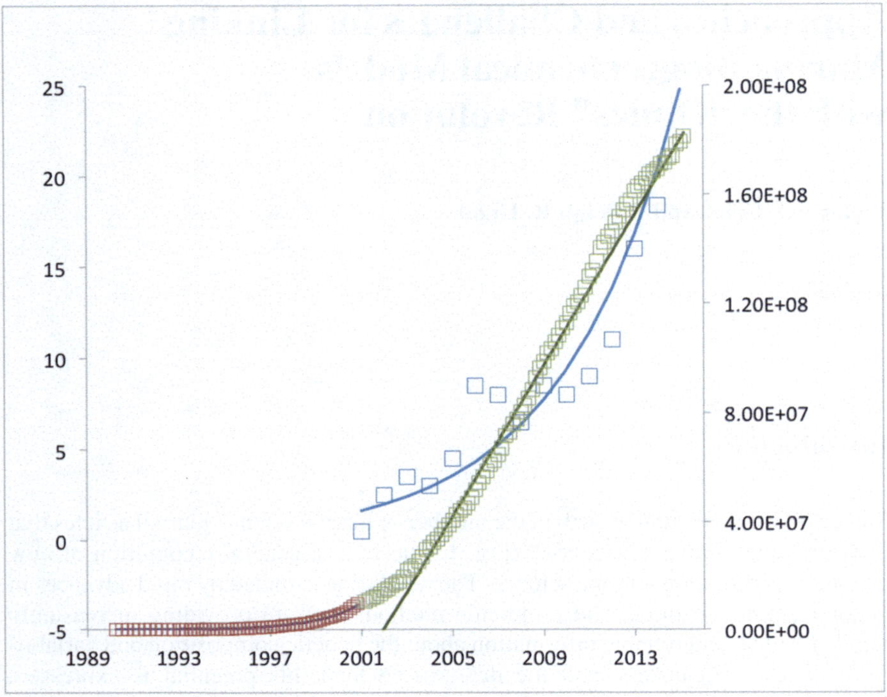

Fig. 1 Number of Google scholar citations including the term "omics revolution" in relation to marine science (*blue*). Growth in GenBank sequence number prior to 2001 (*red*) and after 2001 (*green*). Note that the term "omics revolution" didn't come into use until the rate of sequence addition began to level off to a linearincrease

specified by a set of equations (Hood et al. 2006; Hood and Christian 2008). They synthesize information and fold it into a quantitative mathematical framework that can be coupled with a physical representation of the marine environment and used for prediction. Models attempt to encapsulate, in a mathematical and mechanistic framework, the state of our knowledge of the marine ecosystem and microbial dynamics that control marine biogeochemical cycles. However, at present, little progress has been made toward incorporating "omics" data into either the development or validation of these models. Rather, we continue to depend upon traditional measurements of bulk biogeochemical properties and rates, even though these data are usually not sufficiently abundant to adequately constrain our models (Hood et al. 2006; Friedrichs et al. 2007). The key question addressed in this chapter is: How can we use genomic, transcriptomic, and proteomic data to help us develop and validate marine biogeochemical models?

The challenge is daunting. Our conceptual understanding of microbial diversity and metabolic function is evolving much faster than marine ecosystem biogeochemical model development. This is dramatically illustrated by the fact that even though hundreds of new bacterial and archaeal phylotypes have been discovered in recent years, many large-scale prognostic biogeochemical model formulations still do not include explicit representations of bacteria. Indeed, only one of seven ecosystem models run in the recent IPCC CMIP5 ensemble and analyzed by Bopp et al. (2013) included explicitly a heterotrophic bacterial functional group, and only two of the models included a diazotrophic cyanobacterial group. Moreover, a scientific cultural divide has emerged between the marine biogeochemical modeling and omics research communities that we suggest has been, to a large degree, created and reinforced by differences in scientific methods, terminology, and reporting.

We argue that to bring these two separated fields of marine research together we have to: (1) bridge the scientific cultural and structural divide that has emerged between the marine biogeochemical modeling and omics research communities that prevents meaningful scientific exchange and collaboration; (2) find novel methods for relating omics data to our current biogeochemical models that both allow useful comparisons between models and data and that provide insights into how one might alter a more traditional model structure to make it more realistic; and (3) move toward development of models that are more representative of the diverse microbial community composition and the associated diversity of metabolic functions that exists in marine waters so that this mapping can be more fully realized. Thus, the solutions we propose are not revolutionary. Rather, they are evolutionary. We argue that we simply need to work on bridging the divide through mapping genes to model structures and functions and vice versa in increasingly sophisticated ecosystem and biogeochemical models, and through fostering better communication that will facilitate comparisons and understanding. We hope that this chapter represents a substantial first step.

Bridging the Cultural and Structural Divide

The scientific divide that has emerged between the marine biogeochemical modeling and omics research communities is created and reinforced by differences in scientific methods and terminology. Indeed, a marine microbiologist likely communicates more effectively with a medical microbiologist than a marine biogeochemical modeler. Breaking down barriers to communication requires educating both modelers and microbiologists. It also requires structural changes in biogeochemical models to incorporate the diversity of prokaryotic biogeochemical function in addition to the eukaryotes generally simulated in ecosystem models.

Omics Measurements for the Modeler

Omics research can be classified into investigation of DNA (genomics), RNA (transcriptomics) and protein synthesis (Proteomics), as well as a host of further divides such as metabolomics, lipidomics, and more. Here, we will generally use the term "genomics" broadly to refer both to the genetic potential encoded in an organism's DNA as well as to the study of how that DNA is transcribed and encoded into proteins that influence metabolic processes. However, each branch of omics research typically targets a different function. Studies of the genome target the gene sequences of organisms. In marine microbial research this includes efforts to determine the DNA sequences of microorganisms and the proteins and metabolic functions for which these sequences encode. Thus genes provide information about the physiological and metabolic potential of marine microbes. Metagenomics provides this information about communities as a whole. In addition, differences in the genomes of microbes are used to distinguish clades or genotypes (intraspecies variability), phylotypes (species), and taxonomic relationships. Much of the effort in environmental genomics has focused on taxonomic relationships, as well as measures of the similarities of lineages. These measurements allow for measuring and assessing the diversity in a community as a function of space and/or time (Fuhrman et al. 2008; Ladau et al. 2013; Fortunato et al. 2012). In addition to assessing community composition, new technologies and expanding databases allow for searching metagenomes for genes that code for specific metabolic and biochemical processes. Thus, the genetic potential of a community to evolve and adapt in a changing environment can be assessed, though whether this genetic potential will be expressed is not determined.

Transcriptomics investigates the transient mRNA that is actively transcribed within the cells of marine microbes. This provides information about the degree to which the genomic potential of a microbe is being expressed at a given time. Metatranscriptomics measures transcription over the whole microbial community (typically for eukaryotes and prokaryotes separately). In contrast to genomic assays of an individual or static community, transcription varies with environmental and cellular cues. Thus, the genome provides potential, whereas transcription of the gene indicates that an organism is building or regulating the protein structures required for a biochemical transformation. However, quantification of how transcription relates to a biochemical rate is a challenge. Some cells may transcribe mRNA only when actively completing a biochemical transformation while others may transcribe mRNA continuously whether the substrates or need for a biochemical transformation occurs or not. The direct link is also weakened because the half lives of proteins (~20 h) are much longer than mRNA (~2–5 min) (Moran et al. 2013 and references within), and because cells, particularly eukaryotes, may transcribe mRNA but then not complete protein synthesis. However, averaged over many cells in a community, transcription rates should broadly relate to protein expression rates (Moran et al. 2013).

Proteomics is the study of the expression, structure, and function of proteins, which provides information about the degree to which mRNA is translated into proteins that carry out cellular and metabolic functions. This is important because, as described above, transcription and actual construction and activation of a protein may not be tightly linked. Therefore, proteomics provides a characterization of the degree to which the genomic potential of an organism is actually realized. As proteomics methods evolve, this field may provide a more direct link to rates of biochemical transformation, though the large number of observed proteins whose biochemical function remains uncharacterized and further complexity associated with protein transformation after generation as a function of the state of the cell, or the environmental variability will have to be better understood.

Because of the challenge of linking these types of genomics data to biochemical transformations and rates, it is critical for integration into models that the collection of genomics data be coincident with measures of the biogeochemical environment and that they be as quantitative as possible. From the perspective of identifying measurements of utility for ocean modelers, the importance of time-series programs is difficult to overstate. These programs provide long-term consistent nutrient, biomass, and rate measurements which are often directly comparable to processes and quantities derived by biogeochemical models (Lomas et al. 2013). Discoveries associated with time-series genomics measurements have resulted in new methods, new discoveries, and will continue to inform new studies (Giovannoni et al. 2014a, b; Karl and Church 2014). Ensuring that genomics (in the broad sense of the term) measurements continue to be collected at these time series is critical to better understanding and characterizing how gene expression, genetic potential, and protein synthesis relate to biogeochemical process and environmental variability and therefore also for linking genomics measurements to biogeochemical processes and environmental variability simulated by models.

Biogeochemical Models for Microbial Ecologists

Before we can relate genomic analyses to models, it is helpful to understand how most coupled physical and biogeochemical models are constructed. For many applications, ecological models are embedded within a circulation model that specifies the environment, e.g., temperature, salinity, advection, and diffusion. In this framework, biogeochemical "state variables" are specified such as nutrients, phytoplankton, zooplankton, and detritus (NPZD). Each variable is acted upon at each location (typically 50,000–50,000,000 cells) at each time (typically each 1–0.04 h). An ecological model determines the transformations and fluxes between state variables. The computational expense of adding state variables has led to a trend of specifying "functional groups" of organisms rather than to specifying multiple species of diatoms for example (Hood et al. 2006). Biogeochemical models with fewer than 10–20 types of inorganic, particulate and dissolved organic substrates, and living groups generally strain computation resources to their limits. Carrying significantly more than 10–20 state variables in a three-dimensional coupled model is often unfeasible.

Each functional group has sources and sinks, such as growth, respiration, mortality, and grazing. Thus,

$$\frac{\partial M}{\partial t} = \mu H(I,S)T(N)K(M,I,S)M - \left(R(Z)+e\right)$$

represents the evolution over time of the concentration of an organism, M. The growth term is a function of some maximum growth rate, μ, a function $H(I,S)$

which limits growth based on the availability of an energy resource such as light, I, or substrate, S, which could be prey or dissolved substrate availability, a function $T(N)$ which limits growth based on the availability of biochemical building blocks, N, a function K which is unique to the organism, and represents its growth efficiency for a given energy and substrate, and the organism's present biomass. The loss term is a function of predation, $R(Z)$, by heterotrophs (Z), as well as a growth independent respiration, e. A number of increasingly complex functions can be applied to these equations to more subtly represent luxury consumption of nutrients, or grazing preferences, however most biogeochemical models embedded within physical circulation models have this basic form.

In a functional group model, each group's environmental response is carefully tuned to exploit a specific niche within an environmental gradient such as the tropical Atlantic and Amazon River plume (Fig. 2). The model (Stukel et al. 2014), based on prior modeling efforts (Hood and Coles 2004; Yoshikawa et al. 2013), explicitly represents functional groups, particularly nitrogen fixers, thought to be important in biochemical transformations of inorganic and organic nutrients delivered from the terrestrial to the marine environment. It is important to emphasize that this type of biogeochemical model is limited to representing the consequences of aggregated genomic processes. Most cellular processes are not explicitly represented in biogeochemical models, much less genes (genomics) and their transcription (transcriptomics) and translation (proteomics). From the equations above, it is clear that there are relatively few terms for representing the cellular metabolism described in genomics level studies. Of course, there are

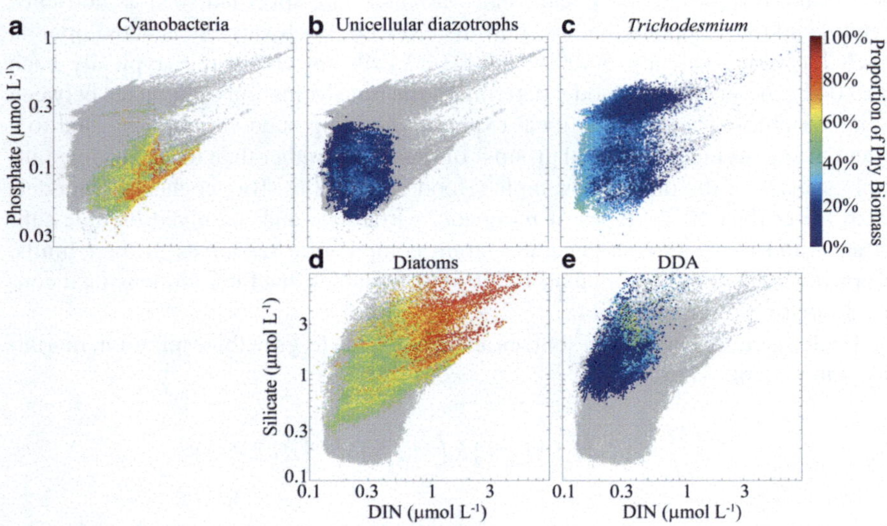

Fig. 2 Fraction of biomass associated with each phytoplankton functional group as a function of the limiting nutrient for each organism. (**a**) cyanobacteria, (**b**) unicellular diazotrophs, (**c**) *Trichodesmium* spp., (**d**) diatoms, (**e**) diatom-diazotroph assemblages. From Stukel et al. (2014)

biogeochemical models that are explicitly designed to represent more complex cellular processes (e.g. Flynn 2008; Flynn and Mitra 2009); however, these are not applied in large-scale biogeochemical modeling studies because they are so computationally demanding and the data sets for constraining these models are limited. The potential for incorporating more complex cellular processes in large-scale biogeochemical models so as to make them more amenable to comparison with genomics data is discussed further in section Near-Term Innovation.

These models are generally used to predict primary production, carbon import and export from the water column, air–sea climate interactions, ocean deoxygenation, and a host of processes that are influenced by processes at the cellular level, but indirectly. Models may also be used to test hypotheses relating to community dynamics, but typically at the functional type level, rather than a specific organismal or phenotype level. Stukel et al. (2014), for example, use their model (Fig. 2) to understand whether community structure along the Amazon plume is mediated primarily by the evolution of the biochemistry of the plume, or whether grazing control drives the community evolution. Thus, there is a broad divide between the detailed process information that emerges from genomics research and what the biogeochemical modeling community can currently incorporate into existing models.

The Structural Divide

Another important challenge to integrating the information emerging from genomics research with our current biogeochemical ocean models is that genomics analyses have been more focused on prokaryotes than eukaryotes, whereas the reverse is true in biogeochemical models. While efforts such as the Marine Microbial Eukaryote Transcriptome Project are aimed at increasing the number of eukaryote sequences, information on eukaryote gene structure and function in marine systems is much scarcer than for prokaryotes (e.g. Fig. 3). In contrast, most large-scale three-dimensional coupled physical–biogeochemical models lack any dynamic representation of heterotrophic bacteria, though some unique functions associated with cyanobacteria such as nitrogen fixation have become more common in models (Fig. 3; Moore et al. 2004; Stock et al. 2014; Fennel et al. 2008; Follows et al. 2007; Friedrichs et al. 2007). This mismatch between modeling effort, which has targeted mostly eukaryotic phytoplankton and zooplankton, and genomics effort, which has been more focused on prokaryotic organisms, means that there is surprisingly little overlap between the organisms for which we have some understanding of genes and the biochemical processes for which they encode, and the organisms whose functions are represented in marine biogeochemical models. More investigations where prokaryote and eukaryote genomics are measured in tandem with conventional rate measurements should be motivated, and model development efforts should place more emphasis on prokaryotic organisms which are the primary agents of biogeochemical transformation in marine systems.

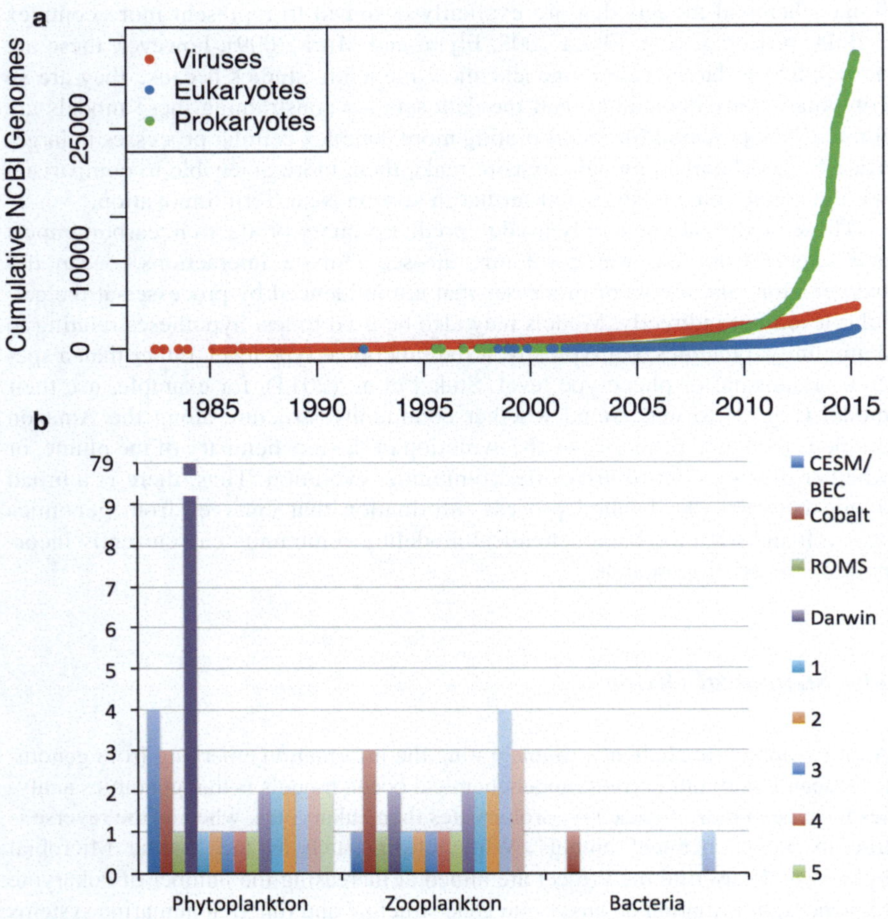

Fig. 3 (**a**) number of partial genomes available at the National Center for Biotechnology Information (NCBI) for eukaryotes, prokaryotes, and viruses. (**b**) histogram of the number ofmodel state variables dedicated to phytoplankton, zooplankton, and bacterial processes for a selection of biogeochemical models. Models referenced in the text

Relating Existing Omics to Current Biogeochemical Models

Taxonomy and Diversity

Initially, environmental metagenomics focused on assessing and characterizing the vast array of microbial diversity, and how organisms or operational taxonomic units (OTUs) were related taxonomically (e.g. Muyzer et al. 1993; Crump et al. 1999). However, this is not how organisms are generally represented in marine biogeochemical models. In a model, the "taxonomic" groupings tend to be structured around biogeochemical processes (Hood et al. 2006). Both conceptual and numerical biogeochemical models tend to collapse diversity to the minimum number of organisms able to exploit the relatively low temporal, spatial, and structural variability in model physics and biochemistry. For typical biogeochemical models, "diversity" exists only where a process of interest, such as calcification or nitrogen fixation is being expressed in addition to the other "background" autotrophic processes (e.g. Stukel et al. 2014; Fig. 2). New efforts to develop models better poised to explore the role of diversity are reviewed in Follows and Dutkiewicz (2011) who cite examples from their Darwin model (Follows et al. 2007) with 78 autotroph species that explore the spatial characteristics that drive large-scale diversity in the ocean. The Darwin approach represents a very important step toward developing models that better represent the taxonomic diversity that is observed in the marine systems, which facilitates comparisons between genomics data and model output (discussed below).

Statistical techniques for reducing complex data sets (e.g. multi-dimensional scaling, clustering) that are widely used in ecology are now commonly applied to samples where operational taxonomic units have been identified using genomic methods (e.g., 16S Ribosomal RNA). Communities are then arranged along non-dimensional axes, based on their similarity. Environmental factors such as salinity or nutrients can then be projected onto the axes to suggest factors that correlate with, and thus might structure similar or dissimilar communities. While this typically requires a broad array of species, it is not outside the range of possibility for models to include such information. Furthermore, it is not necessarily required that a model has a large number of species to quantify the similarity of different communities, thus the technique can be used in the analysis of current generation biogeochemical models that have multiple functional groups. Applying the same kinds of statistical analyses, that are now being widely applied to analyze genomic data, to the output of biogeochemical models enables meaningful comparisons between genomic and modeling studies and it can help bridge the gap between the genomics and modeling research communities. The following is an example of how this can be done using data and model results from a recent study of changes in microbial community structure and biogeochemistry in the Amazon River plume.

Microscope counts of phytoplankton abundance in the western tropical North Atlantic (Goes et al. 2013) reveal patterns of similarity that appear to be related first to salinity, and secondly to nutrient concentrations (Fig. 4a). Environmental covariates have been linearly projected onto the non-dimensional axes to suggest physical gradients that could structure the community variability. In this case, salinity and nutrients are on the same line, reflecting their tight correlation along the river plume salinity gradient. The opposing axis is not well defined by the environmental factors shown. Similar patterns emerge from an analysis that uses 16s ribosomal RNA to delineate the community structure within bacteria (Fig 4b, B. Crump, unpublished data). This is important because, it suggests that the bacterial community structure changes congruently with the phytoplankton community structure. Interestingly, the bacterial community structure is more tightly defined by salinity class than is the eukaryotic phytoplankton structure. Modeled community structure (Stukel et al. 2014; Fig. 2) is much more tightly constrained with less variability than both bacterial and phytoplankton observations, but shows also a community structured by

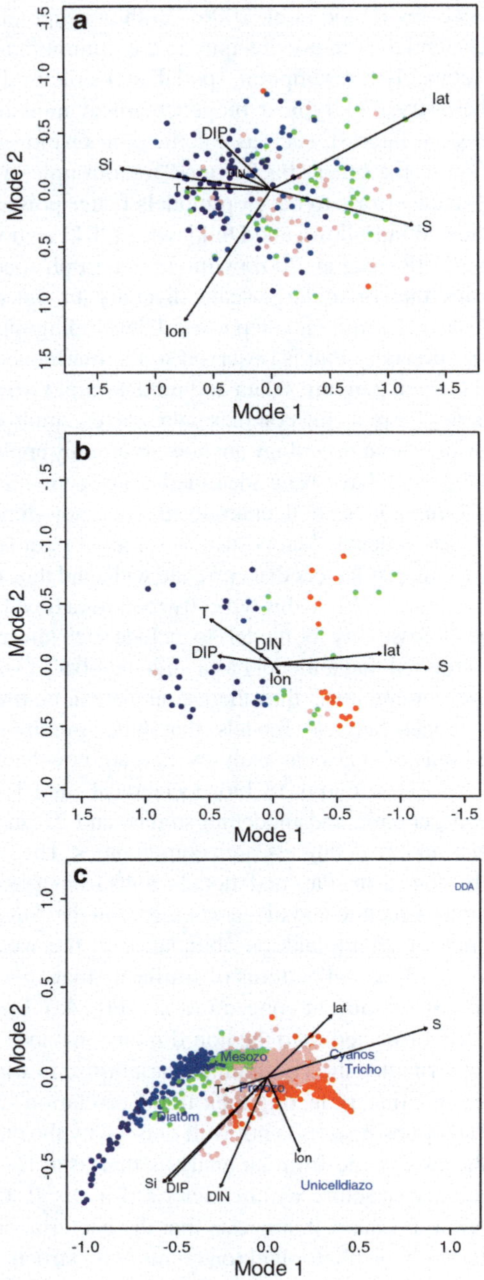

Fig. 4 All panels show data from June 2010 or model runs from June in the western tropical North Atlantic. Points show the site distribution, and they are colored according to their salinity class. (**a**) multidimensional scaling plot based on the microscopically identified plankton community. (**b**) multidimensional scaling plot based on pyrosequencing of the 16s ribosomal DNA. (**c**) multidimensional scaling plot based on a three dimensional coupled biological and physical model with 5 autotrophs and 2 zooplankton (Stukel et al. 2013)

salinity (Fig. 4c). There is no direct feedback between model salinity and the model ecology, thus we know that salinity is a proxy for other environmental factors in the model. Though the model does not explicitly resolve the bacterial community, it is interesting that the three analyses reveal similar gradients in community structure as the environment varies suggesting a very tight coupling between bacterial and phytoplankton community structure.

We can also learn from the mismatch between models and data. In this case, the model representation of diatom–diazotroph symbiosis appears to be too tightly linked with phosphorus concentration in contrast to observations. The model also appears to span the salinity nutrient axis broadly, but has a limited projection onto the second axis suggesting that the model fails to capture enough physical or biogeochemical variability. Taken together, these patterns strongly suggest that the bacterial community is evolving in response to similar environmental cues as the eukaryotes and larger organisms (e.g. Pommier et al. 2007), and this further suggests that expression of genes is likely to be tightly coupled with environmental variability. These pointers indicate that models may be able to capture patterns of gene expression, because they are generally able to simulate biogeochemical gradients.

Targeting Genes and/or Pathways

In many instances, expression of a single gene to code for synthesis of a protein may be manifesting one link in a long and complex biochemical chain with many concurrently expressed gene groups coding for individual transformations. For example, *nifH* codes for a subunit of the nitrogenase enzyme, which is just one component of the pathway for conversion of N_2 gas to ammonium. It is helpful in connecting genomics data to biogeochemical function expressed in a model to consolidate or map the genomic information in these pathways in order to reduce the vast amount of information to a more manageable level that relates more directly to biogeochemical function. Identifying genes in these pathways for further study was one goal of the Microbiological Targets for Ocean Observing Laboratories workshops (Participants 2010). Care must be taken however to consider potential bottlenecks in the biochemical process or dependence on external environmental conditions, such as the high iron requirement of the nitrogenase enzyme. The costs and benefits associated with the marker gene for models must include the full pathway's environmental sensitivity, or rate limitation.

One functional group that many biogeochemical models include are diazotrophs, or organisms that break the triple bond of dinitrogen gas, to render the nitrogen bioavailable (Moore and Doney 2007; Coles and Hood 2007; Monteiro et al. 2010; Yoshikawa et al. 2013) (see also Fig. 2). The expression of characteristic genes like *nifH* can illustrate the ubiquity of a process in the ocean, and contribute to model representation of that ubiquity. When these genes are observed in transcriptomics sampling, they can be directly related to the presence or absence of nitrogen fixation in models. Where a number of observations of expression of the nitrogenase genes have been conducted over space and/or time, we can begin to estimate the relationship between observed expression and modeled nitrogen fixation rate and the spatial or temporal scales in the model. These targeted uses of gene expression already exist (e.g. Luo et al. 2012) and have been compared with current generation models (Monteiro et al. 2010; Yoshikawa et al. 2013; Stukel et al. 2014).

More generally, genetically controlled biochemical transformations can be linked to models through understanding how gene expression influences cellular growth and mortality—the two terms in the equations described above. The housekeeping metabolic processes that sustain a cell independently of the cellular growth or response to external environmental conditions are included through prescription of an organism's respiration or mortality rate. Categorizing cellular processes regulated by gene expression in terms of their benefits and costs, while understanding the dependence of each on internal (i.e., maximum growth rate or basal respiration) versus external (nutrient concentration, light availability, predation) factors is the first step to representing them in current generation biogeochemical models. The benefits of environmentally dependent functions such as expression of genes for nutrient transporters or repair of light harvesting proteins will be represented in the growth part of the model as a function of the external stimulus, whereas the benefits of expression of genes regulating motility leading to particle attachment or expression of genes for chitin synthase leading to regulation of cell buoyancy may be growth independent, and more appropriately represented through respiration or mortality terms. Costs of these functions must be considered independently, as they may have a different regulation.

Quantifying the costs and benefits to an organism of expression of a gene, leading to protein synthesis and a biochemical reaction, is a major challenge for marine biogeochemical modeling efforts which seek to specify the competitive impacts of carrying few versus many metabolic capabilities for exploiting the environment. Apparently, there are costs associated with carrying the genes and/or maintaining the ability to express a very diverse array of metabolic capabilities, otherwise this strategy would be ubiquitous in marine microbes. Indeed, this is an active area of research (see review by Giovannoni et al. 2014a). One approach might be to determine the free energies of major biochemical transformation pathways, and to relate them to housekeeping functions to normalize them to cell growth rate for example. This approach was used in a pioneering model study of chemolithoautotrophs and the cryptic sulfur cycle in a low oxygen environment that explicitly modeled gene copy number as a function of biomass (Reed et al. 2014). Apparently, this approach of targeting the specific energy associated with a chemolithoautotroph-mediated reaction is not yet viable for a heterotroph or photoautotroph (Reed et al. 2014). Nonetheless, it provides tremendous insight and potential for relating modeled gene number to observed gene number and perhaps ultimately to gene transcription and protein expression. Even knowing the costs of biochemical pathways relative to each other would be a step forward to estimating the costs and benefits of carrying the genes and cellular machinery to complete a biochemical transformation.

Other methods for estimating the cost–benefit tradeoffs for gene expression might include culture studies. One approach is to eliminate specific genes, and then identify the influence of this elimination on growth rate. Other approaches could involve culture manipulations that influence growth rate and for which transcription responses can be measured (e.g., You et al. 2013). While these approaches are limited to single, culturable organisms, they may provide direct evidence of the gene impact for a given environmental regime that is directly relatable to a model parameter—growth rate. Alternative approaches might consider mesocosms—simplified physical environments in which mixed community response to external stimulus may be determined in the context of gene expression. Data assimilation techniques could then be used to estimate quantitatively the optimal model parameters for matching the observations (e.g., Vallino 2000). Such an approach could be used to estimate cost–benefit tradeoffs for numerical models that include gene expression explicitly.

A pragmatic approach to estimating cost–benefit tradeoffs is simply to estimate cost and benefit parameters, then run the model and compare the resulting spatial and temporal patterns with observations. The cost–benefit tradeoffs are then altered and the process is iterated forward. As genomics measurements become more widely distributed in space and time, this approach could be formalized in an inverse approach to the problem of cost–benefit estimation that can then be tested in culture or mesocosm studies.

Near-Term Innovation

More effectively bridging the gap between genomics and models may require new modeling approaches (Hood et al. 2007). These include increasing the complexity and flexibility of existing model systems to represent an increasing number of possible biochemical transformations. For example, Flynn et al. (1997) developed a biogeochemical model that explicitly simulates intracellular biochemical reactions related to nutrient transport and assimilation. This type of model therefore provides a much more detailed representation of the intracellular pathways and dynamics within phytoplankton cells that can be directly related to genomic and, particularly, transcriptomic data (e.g., explicit induction or repression of the synthesis of enzymes). These types of interactions are computationally expensive to specify, however, so this approach may be more appropriate in an individual-based framework. These complex models have not yet been tested in three-dimensional systems.

A logical extension of this idea is to attempt to simulate all of the biochemical reactions and the attendant up- and down-regulation of DNA transcription and translation that controls the synthesis of proteins within the cells of microbes, i.e., construct fully functional cellular genetic models. Indeed, a huge amount of research and model development has already been carried out along these lines (e.g., Karr et al. 2012). In theory, scaling this approach up using three-dimensional individual-based modeling approaches would allow direct, quantitative comparisons between model simulations and genomic, transcriptomic, and proteomic data because these measured quantities would all be explicitly simulated in (presumably) billions of virtual cells. In practice, however, it is likely that computational limitations will make large-scale simulations along these lines unfeasible for the foreseeable future. Moreover, our current understanding of the biochemical reactions and their genetic regulation in marine microbes is limited. Indeed, it is not clear that we have enough information to attempt such a simulation for even one marine microbial species, much less the vast diversity of microorganisms in the sea.

An empirical approach based on environmental envelope modeling has been applied by Larsen et al. (2012). Using neural networks, they statistically relate microbial communities based on observations of the metagenome to environmental factors that covary with the communities. Assuming that they have measured all the relevant parameters, and that the relationships do not vary in time, they can predict community metagenome based on predictions of the environment. This approach allows direct comparison with genomic data and has the benefit of being computationally feasible. Careful inspection of the environmental relationships that emerge from the neural network can lead to insights about the factors regulating the community structure, and could feed into more mechanistic model efforts.

A different approach pioneered by Follows et al. (2007) involves initializing circulation models with a large number of autotrophic organisms with semi-random model parameter assignments based on predetermined bounds and using known cost–benefit tradeoffs to relate the various parameter choices. This model and follow on versions (e.g., Goebel et al. 2010; Ward et al. 2013) have succeeded in generating an emergent phytoplankton community structure and biogeography consistent with observed global phytoplankton distributions. Such models may potentially be used to generate significantly greater microbial species diversity in simulated marine environments compared to more traditional model formulations that set species diversity a priori.

Recently, we have developed a model inspired by this self-organizing approach (Coles et al. 2016) that explicitly encodes metabolic functionality through the random assignment of functional genes to a large ensemble of organisms. The model generates: realistic communities of organisms with size structure similar to observations, gene abundance based on a single gene per cell, and spatial gradients in transcription throughout the subtropical and tropical Atlantic. The fundamental horizontal and vertical nutrient gradients in the Atlantic are realized regardless of the specific emergent community suggesting that metabolic diversity rather than taxonomic diversity structures the primary biogeochemical gradients in the ocean. Comparison of modeled gene abundances and transcription with observations remains challenging because of the difficulty in synthesizing and curating information on genes that encode for a step in a specificv process such as nitrate assimilation, and because transcription rates in the model are more directly related to biogeochemical transformations than may be observed in the marine environment. However, spatial patterns in gene abundance and transcription are similar between the model and observations suggesting that this approach has some predictive power.

Alternatively, more traditional modeling approaches can also be used to increase microbial diversity. For example, the plankTOM5 model incorporates 32 state variables, three phytoplankton functional types, two zooplankton function types, and a wide range of biogeochemical specialization including production of climatically active sulfur compounds, calcification, and ballasted export (Vogt et al. 2010). These approaches do not, however, provide a means to make direct comparisons between models and genomics data. In order to do this, one must explicitly incorporate metabolic, physiological, or behavioral capabilities into models that are linked to the presence or absence of specific genes. However, models with increased microbial diversity can facilitate indirect comparison with omics data if they provide predictions of species or functional groups that can be validated with genomics data.

An important caveat here is that these approaches lead directly to the challenge of dealing with increasingly complex models (Hood et al. 2006, 2007; Hood and Christian 2008). The trend toward building increasingly complex three-dimensional coupled biogeochemical–physical models presents computational challenges because each new state variable has to be numerically advected and diffused in three dimensions. Moore's law (or computer science) will provide a solution to this problem only if the rate of increase in computing power is greater than the rate of increase of model complexity and computational demand. Moreover, more complex models must be properly constrained with data, i.e., if they have too many degrees of freedom, then they can be tuned to fit noise in the data, which will result in reduced predictive skill (Hood et al. 2007). The implication is that continuing to add new organisms and metabolic processes that are discovered through microbiological and genomic studies to models may not be useful unless validation data relevant to these processes (e.g., time-series or spatial data) can also be obtained. Perhaps the avalanche of genomic information that is driving the need for increased model complexity will ultimately provide the solution to this dilemma through enabling new methods to effectively assimilate information into biogeochemical models.

Conclusions

In this chapter, we argue that a scientific divide has emerged between the marine biogeochemical modeling and genomics research communities that has impeded the utilization of genomics data in the development and validation of large-scale, three-dimensional, coupled physical–biogeochemical models. In order to close this divide we first need to bridge the communication gap that has emerged between the marine biogeochemical modeling and genomics research communities that prevents meaningful scientific exchange and collaboration. However, penetrating the jargon is only the first hurdle. There is also a mismatch between modeling efforts, which have targeted mostly eukaryotic phytoplankton and zooplankton, and genomics efforts, which have been more focused on prokaryotic organisms. More investigations where prokaryote and eukaryote genomics are measured in tandem with conventional rate measurements should be motivated, and model development efforts should place more emphasis on prokaryotic organisms.

Differences in the formats for presenting data and results also create obstacles to effective communication between the biogeochemical modeling and genomics research communities. Techniques as simple as using the same plotting methods that allow us to directly compare models with genomic data can help bridge this gap and reveal new insights into errors in model structure as well as misinterpretations of genomic data. New modeling approaches that allow representation of increased microbial diversity and/or cellular physiology and metabolism have also emerged in recent years that are facilitating indirect comparisons between models and genomics. These new approaches do not yet, however, provide a means to make direct comparisons between models and genomics data (with the exception of Reed et al. 2014). In order to do this, one must explicitly incorporate metabolic, physiological, or behavioral capabilities into models that are linked to the presence or absence of specific genes.

An important caveat here is that these approaches lead directly to the challenge of dealing with increasingly complex models, which creates both computational and validation challenges. Hopefully, Moore's law (or computer science) will solve the computational problem. We are also optimistic that the avalanche of genomics data that is motivating the development of increasingly complex models will, ultimately, provide enough data on species and metabolic diversity and variability to constrain the complex models that they spawned.

Acknowledgments This chapter emerged from ongoing interactions between biogeochemical modelers (the authors) and microbial ecologists that were fostered by projects funded by NSF and the Gordon and Betty Moore Foundation (Amazon iNfluence on the Atlantic: CarbOn export from Nitrogen fixation by DiAtom Symbioses or ANACONDAS and River Ocean Continuum of the Amazon or ROCA). ANACONDAS was funded as part of the NSF Emerging Topics in Biogeochemical Cycles (ETBC) program (Directorate for Geosciences, NSF 07-049) that was explicitly intended to support emerging areas of interdisciplinary research. In particular, we would like to thank M.A. Moran, B. Satinsky, J. Paul, B. Zielinkski, P. Yager, B. Crump, C. Fortunato, and M. Doherty for patiently explaining many of the genomics concepts and challenges that are discussed in this chapter. We would also like to acknowledge the contributions of M. Stukel and M. Brooks who, though many lab group discussions, helped to formulate the ideas presented in this chapter. V. Coles and R. Hood were supported by research grants from the National Science Foundation (OCE-0933975) and the Gordon and Betty Moore Foundation. This is UMCES contribution number 5193.

Victoria J. Coles and Raleigh R. Hood

We met at the UCSD SeaDeucers scuba club over a shared interest in diving and Baja expeditions more than 5 years before we became a couple. Acting as cheap labor on Raleigh's graduate cruises in the California Current, Victoria became interested in oceanography—but not if it required filtering. As Raleigh finished up his Ph.D. work, and headed to a postdoc at OSU, we became engaged and decided to look for a joint graduate and postdoc location. RSMAS at the University of Miami became a joint option, so we married and moved there. At the time, we had no intention of collaborating scientifically, and indeed, we didn't. Victoria's graduate work was on observations and models of climate signals in Antarctic Bottom Water, and Raleigh headed off for cruises in the South Atlantic and Arctic, while learning biological modeling. Ultimately a faculty position at UMCES opened up for Raleigh, and Victoria took a postdoc opportunity at NASA GSFC. An opportunity developed after a few years to work collaboratively at Horn Point Laboratory on a coupled physical and biogeochemical model of nitrogen fixation in the tropical Atlantic. Over the years since, we've generally maintained some common research projects while working on other projects solo.

References

Beja O, Aravind L, Koonin EV, Suzuki MT, Hadd A, Nguyen LP, Jovanovich SB, Gates CM, Feldman RA, Spudich JL, Spudich EN, DeLong EF (2000) Bacterial rhodopsin: evidence for a new type of phototrophy in the sea. Science 289:1902–1906. doi:10.1126/science.289.5486.1902

Bopp L, Resplandy L, Orr JC, Doney SC, Dunne JP, Gehlen M, Halloran P, Heinze C, Ilyina T, Séférian R, Tjiputra J, Vichi M (2013) Multiple stressors of ocean ecosystems in the 21st century: projections with CMIP5 models. Biogeosciences 10:6225–6245. doi:10.5194/bg-10-6225-2013

Chisholm SW, Olson RJ, Zettler ER, Goericke R, Waterbury NA, Welschmeyer JB (1988) A novel free-living prochlorophyte abundant in the oceanic euphotic zone. Nature 334:340–343. doi:10.1038/334340a0

Coles VJ, Hood RR (2007) Modeling the impact of iron and phosphorus limitations on nitrogen fixation in the Atlantic Ocean. Biogeosciences 4:455–479. doi:10.5194/bg-4-455-2007

Coles VJ, Hood RR, Stukel MR, Brooks, Burd A, Moran MA, Paul JH, Satinsky BM, Zielinski BL, Yager PL (2016) Emergent microbial metagenomes and metatranscriptomes reproduce genetic streamlining and biogeochemical gradients in a model ocean. Science (in prep)

Crump BC, Armbrust EV, Baross JA (1999) Phylogenetic analysis of particle-attached and free-living bacterial communities in the Columbia river, its estuary, and the adjacent coastal ocean. Appl Environ Microbiol 65:3192–3204

Fennel K, Wilkin J, Previdi M, Najjar R (2008) Denitrification effects on air-sea CO_2 flux in the coastal ocean: simulations for the northwest North Atlantic. Geophys Res Lett 35, L24608. doi:10.1029/2008GL036147

Flynn KJ (2008) The importance of the form of the quota curve and control of non-limiting nutrient transport in phytoplankton models. J Plankton Res 30:423–438. doi:10.1093/plankt/fbn007

Flynn KJ, Mitra A (2009) Building the "perfect beast": modelling mixotrophic plankton. J Plankton Res 31:965–992. doi:10.1093/plankt/fbp044

Flynn KJ, Fasham MJR, Hipkin CR (1997) Modelling the interactions between ammonium and nitrate uptake in marine phytoplankton. Philos Trans R Soc B Biol Sci 352:1625–1645. doi:10.1098/rstb.1997.0145

Follows MJ, Dutkiewicz S (2011) Modeling diverse communities of marine microbes. Ann Rev Mar Sci 3:427–451. doi:10.1146/annurev-marine-120709-142848

Follows MJ, Dutkiewicz S, Grant S, Chisholm SW (2007) Emergent biogeography of microbial communities in a model ocean. Science 315:1843–1846. doi:10.1126/science.1138544

Fortunato CS, Herfort L, Zuber P, Baptista AM, Crump BC (2012) Spatial variability overwhelms seasonal patterns in bacterioplankton communities across a river to ocean gradient. ISME J 6:554–563. doi:10.1038/ismej.2011.135

Friedrichs MAM, Dusenberry JA, Anderson LA, Armstrong RA, Chai F, Christian JR, Doney SC, Dunne J, Fujii M, Hood R, McGillicuddy DJ, Moore JK, Schartau M, Spitz YH, Wiggert JD (2007) Assessment of skill and portability in regional marine biogeochemical models: role of multiple planktonic groups. J Geophys Res 112, C08001. doi:10.1029/2006JC003852

Fuhrman JA, Steele JA, Hewson I, Schwalbach MS, Brown MV, Green JL, Brown JH (2008) A latitudinal diversity gradient in planktonic marine bacteria. Proc Natl Acad Sci U S A 105:7774–7778. doi:10.1073/pnas.0803070105

Giovannoni SJ, Thrash JC, Temperton B (2014a) Implications of streamlining theory for microbial ecology. ISME J. doi:10.1038/ismej.2014.60

Giovannoni SJ, Vergin KL, Carlson CA (2014b) Twenty-five years of omics at BATS. Ocean Carbon Biogeochem News 7:1–6

Goebel NL, Edwards CA, Zehr JP, Follows MJ (2010) An emergent community ecosystem model applied to the California current system. J Mar Syst 83:221–241. doi:10.1016/j.jmarsys.2010.05.002

Goes JI, Gomes HDR, Chekalyuk AM, Carpenter EJ, Montoya JP, Coles VJ, Yager PL, Berelson WM, Capone DG, Foster RA, Steinberg DK, Subramaniam A, Hafez MA, do Rosario Gomes H (2013) Influence of the Amazon River discharge on the biogeography of phytoplankton communities in the western tropical North Atlantic. Prog Oceanogr 120:29–40. doi:10.1016/j.pocean.2013.07.010

Hood RR, Christian JR (2008) Ocean nitrogen cycle modeling. In: Capone DG, Bronk DA, Mulholland MR, Carpenter EJ (eds) Nitrogen in the marine environment. Elsevier, Amsterdam, pp 1445–1495

Hood RR, Coles VJ (2004) Modeling the distribution of *Trichodesmium* and nitrogen fixation in the Atlantic Ocean. J Geophys Res 109, C06006. doi:10.1029/2002JC001753

Hood RR, Laws EA, Armstrong RA, Bates NR, Brown CW, Carlson CA, Chai F, Doney SC, Falkowski PG, Feely RA, Friedrichs MAM, Landry MR, Keith Moore J, Nelson DM, Richardson TL, Salihoglu B, Schartau M, Toole DA, Wiggert JD (2006) Pelagic functional group modeling: progress, challenges and prospects. Deep Sea Res Part II Top Stud Oceanogr 53:459–512. doi:10.1016/j.dsr2.2006.01.025

Hood RR, Laws EA, Follows MJ, Siegel DA (2007) Modeling and prediction of marine microbial populations in the genomic era. Oceanography 20:155–165

Karl DM, Church MJ (2014) Microbial oceanography and the Hawaii Ocean Time-series programme. Nat Rev Microbiol 12:699–713

Karr JR, Sanghvi JC, Macklin DN, Gutschow MV, Jacobs JM, Bolival B, Assad-Garcia N, Glass JI, Covert MW (2012) A whole-cell computational model predicts phenotype from genotype. Cell 150:389–401. doi:10.1016/j.cell.2012.05.044

Ladau J, Sharpton TJ, Finucane MM, Jospin G, Kembel SW, O'Dwyer J, Koeppel AF, Green JL, Pollard KS (2013) Global marine bacterial diversity peaks at high latitudes in winter. ISME J 7:1669–1677. doi:10.1038/ismej.2013.37

Larsen PE, Field D, Gilbert JA (2012) Predicting bacterial community assemblages using an artificial neural network approach. Methods Mol Biol 1260:33–43. doi:10.1038/NMETH.1975

Lomas MW, Bates NR, Johnson RJ, Knap AH, Steinberg DK, Carlson CA (2013) Two decades and counting: 24-years of sustained open ocean biogeochemical measurements in the Sargasso Sea. Deep Sea Res Part II Top Stud Oceanogr 93:16–32. doi:10.1016/j.dsr2.2013.01.008

Luo Y-W, Doney SC, Anderson LA, Benavides M, Berman-Frank I, Bode A, Bonnet S, Boström KH, Böttjer D, Capone DG, Carpenter EJ, Chen YL, Church MJ, Dore JE, Falcón LI, Fernández A, Foster RA, Furuya K, Gómez F, Gundersen K, Hynes AM, Karl DM, Kitajima S, Langlois RJ, LaRoche J, Letelier RM, Marañón E, McGillicuddy DJ, Moisander PH, Moore CM, Mouriño-Carballido B, Mulholland MR, Needoba JA, Orcutt KM, Poulton AJ, Rahav E, Raimbault P, Rees AP, Riemann L, Shiozaki T, Subramaniam A, Tyrrell T, Turk-Kubo KA, Varela M, Villareal TA, Webb EA, White AE, Wu J, Zehr JP (2012) Database of diazotrophs in global ocean: abundance, biomass and nitrogen fixation rates. Earth Syst Sci Data 4:47–73. doi:10.5194/essd-4-47-2012

Monteiro FM, Follows MJ, Dutkiewicz S (2010) Distribution of diverse nitrogen fixers in the global ocean. Global Biogeochem Cycles 24:no. 3. doi:10.1029/2009GB003731

Moore JK, Doney SC (2007) Iron availability limits the ocean nitrogen inventory stabilizing feedbacks between marine denitrification and nitrogen fixation. Global Biogeochem Cycles 21:GB2030. doi:10.1029/2006GB002762

Moore JK, Doney SC, Lindsay K (2004) Upper ocean ecosystem dynamics and iron cycling in a global three-dimensional model. Global Biogeochem Cycles 18:GB4028. doi:10.1029/2004GB002220

Moran MA, Satinsky B, Gifford SM, Luo H, Rivers A, Chan L-K, Meng J, Durham BP, Shen C, Varaljay VA, Smith CB, Yager PL, Hopkinson BM (2013) Sizing up metatranscriptomics. ISME J 7:237–243. doi:10.1038/ismej.2012.94

Morris RM, Rappe MS, Connon SA, Vergin KL, Siebold WA, Carlson CA, Giovannoni SJ (2002) SAR11 clade dominates ocean surface bacterioplankton communities. Nature 420:806–810. doi:10.1038/nature01281.1

Muyzer G, de Waal EC, Uitterlinden AG (1993) Profiling of complex microbial populations by denaturing gradient gel electrophoresis analysis of polymerase chain reaction-amplified genes coding for 16S rRNA. Appl Environ Microbiol 59:695–700

Participants W (2010) Microbiological targets for ocean observing laboratories workshop I

Pommier T, Canbäck B, Riemann L, Boström KH, Simu K, Lundberg P, Tunlid A, Hagström A (2007) Global patterns of diversity and community structure in marine bacterioplankton. Mol Ecol 16:867–880

Proctor LM, Fuhrman JA (1989) Viral mortality of marine bacteria and cyanobacteria. Nature 343:60–62

Reed DC, Algar CK, Huber JA, Dick GJ (2014) Gene-centric approach to integrating environmental genomics and biogeochemical models. Proc Natl Acad Sci U S A 111:1879–1884. doi:10.1073/pnas.1313713111

Rynearson TA, Armbrust EV (2004) Genetic differentiation among populations of the planktonic marine diatom *Ditylum brightwellii* (Bacillariophyceae)1. J Phycol 40:34–43. doi:10.1046/j.1529-8817.2004.03089.x

Stock CA, Dunne JP, John JG (2014) Global-scale carbon and energy flows through the marine planktonic food web: an analysis with a coupled physical-biological model. Prog Oceanogr 120:1–28. doi:10.1016/j.pocean.2013.07.001

Stukel MR, Coles VJ, Brooks MT, Hood RR (2014) Top-down, bottom-up and physical controls on diatom-diazotroph assemblage growth in the Amazon River plume. Biogeosciences 11:3259–3278. doi:10.5194/bg-11-3259-2014

Vallino JJ (2000) Improving marine ecosystem models: use of data assimilation and mesocosm experiments. J Mar Res 58:117–164. doi:10.1357/002224000321511223

Vogt M, Vallina SM, Buitenhuis ET, Bopp L, Le Quéré C (2010) Simulating dimethylsulphide seasonality with the dynamic Green Ocean Model PlankTOM5. J Geophys Res 115, C06021. doi:10.1029/2009JC005529

Ward BA, Dutkiewicz S, Follows MJ (2013) Modelling spatial and temporal patterns in size-structured marine plankton communities: top-down and bottom-up controls. J Plankton Res 36:31–47. doi:10.1093/plankt/fbt097

Yoshikawa C, Coles VJ, Hood RR, Capone DG, Yoshida N (2013) Modeling how surface nitrogen fixation influences subsurface nutrient patterns in the North Atlantic. J Geophys Res 118:2520–2534. doi:10.1002/jgrc.20165

You C, Okano H, Hui S, Zhang Z, Kim M, Gunderson CW, Wang Y-P, Lenz P, Yan D, Hwa T (2013) Coordination of bacterial proteome with metabolism by cAMP signalling. Nature 500:301–306. doi:10.1038/nature12446

Zehr JP, Mellon MT, Zani S (1998) New nitrogen-fixing microorganisms detected in oligotrophic oceans by amplification of nitrogenase (nifH) genes. Appl Environ Microbiol 64:5067

Part II
Viewing Growth and Trophodynamics Through a Stoichiometric Lens

Out of Africa and into Stoichiometry

Susan S. Kilham and Peter Kilham†

The development of ideas that led to what is now known as the concept of ecological stoichiometry (Sterner and Elser 2002) began for us in Africa with Peter Kilham's dissertation on water chemistry of lakes and rivers related to the geology of East and Central Africa (Kilham 1971a). His ultimate intention was to relate water chemistry of lakes to changes in diatom assemblages in long sediment cores from lakes to better understand aspects of climate change in the Pleistocene of Africa, but very little was then known about species-specific diatom requirements for chemical constituents. He noticed that core sections that had sponge spicules had quite different diatom assemblages than did sections without spicules. He reasoned that this was likely the result of differences in silica concentrations. A search of the literature for data relating silica concentrations to diatoms resulted in the hypothesis paper about silica and planktonic diatoms (Kilham 1971b). This was the beginning of the melding of microbial ecology to biogeochemistry.

To test the silica hypothesis it was necessary to establish a culture collection of freshwater diatoms and this is where our separate talents merged in a scientifically productive way. A freshwater algal culture collection was established and a set of inexpensive but effective methods for assessing the resource physiology of algal species was developed (Kilham 1978). Tilman joined our lab as a graduate student. Our aim was to study algal nutrient physiology and explore competitive interactions among species to sort out their species-specific resource requirements. Tilman concentrated on phosphorus-limited diatoms and I on silica-limited diatoms (Kilham 1975; Tilman and Kilham 1976). This set us up for one of those rare "eureka moments." Tilman and I noticed that *Asterionella formosa* had different colony morphologies under Si (more than 8 cells/colony) and P limitation (fewer than 8

S.S. Kilham (✉)
Department of Biodiversity, Earth and Environmental Science, Drexel University, Philadelphia, PA 19104, USA
e-mail: kilhams@drexel.edu

P. Kilham† (deceased)

P.M. Glibert, T.M. Kana (eds.), *Aquatic Microbial Ecology and Biogeochemistry: A Dual Perspective*, DOI 10.1007/978-3-319-30259-1_6

cells/colony; Titman et al. 1976). We discussed with Peter the conditions under which these changes happened. Peter's training as a biogeochemist led him to naturally think in ratios and we all suddenly realized that it was the ratio of Si:P that was important and not the concentrations per se. *Asterionella* had fewer cells/colony above a Si:P ratio of ca. 80:1 and more cells/colony below that ratio. This species had an "optimum" ratio at which it used Si and P for growth. We also observed that *Cyclotella meneghiniana* had a very different optimal Si:P ratio of ca. 6:1. This was the first clue that different diatom species had unique resource ratio signatures. Tilman redesigned his competition experiments and developed equations to explain the resource ratio theory (Titman 1976; Tilman 1977). Thus, stoichiometry became a byword in our laboratory. This eventually led to a review paper (Tilman et al. 1982) and a book Tilman 1982) and 20 years later to a book on ecological stoichiometry co-authored by one of Tilman's students, Robert Sterner (Sterner and Elser 2002).

The nutrient requirements of other phytoplankton species were characterized, especially diatoms (Mechling and Kilham 1982; van Donk and Kilham 1990; Kilham et al. 1997a, b; Lynn et al. 2000). It was remarkable how distinct species actually were in their nutrient requirements, sometimes by orders of magnitude. Natural phytoplankton communities were used to test resource ratio theory which demonstrated that species assorted themselves in predictable ways along ratio gradients based on their individual physiological requirements (Kilham and Kilham 1978; Sommer and Kilham 1985; Kilham 1986; Tilman et al. 1986). These ideas were extended in a more general ecological framework in a review paper on the evolutionary ecology of phytoplankton (Kilham and Kilham 1980). Work also continued on relating diatom communities to water chemistry in African lakes (Hecky and Kilham 1973; Gasse et al. 1983; Kilham et al. 1986; Kilham 1990a). A pair of papers reviewed differences in marine and freshwater phytoplankton ecology and nutrient requirements (Kilham and Hecky 1988; Hecky and Kilham 1988).

Kilham summarized the geochemical mechanisms controlling the chemistry of lakes and rivers in Africa (Kilham 1990b). He demonstrated that the water chemistry of African lakes and rivers was driven by the processes of atmospheric precipitation (Fig. 1, lower right), rock weathering (central area), and evaporation/mineral precipitation (upper right). All of the outlier points were unusual springs in unique geochemical formations.

Together, we explored ideas about how limnological processes are fundamentally different in tropical and temperate lakes (Kilham and Kilham 1990a, b) and developed a set of hypotheses. In the "endless summer" paper (Kilham and Kilham 1990a), we concluded that internal regeneration rates were greater than removal rates for phosphorus in tropical lakes as compared to temperate lakes, especially in larger lakes where epilimnetic mixing exceeded 50 m. Biological control of the elemental cycles dominates in tropical lakes year round, whereas nutrient cycles in temperate lakes are dominated by physical processes for a large part of the year which results in differences in the fundamental mechanisms of nutrient regeneration and their relationships to morphometric features of lakes in the two regions. In the "first law of limnology" paper (Kilham and Kilham 1990b), we questioned the limnological paradigm that small lakes are more productive than large lakes. Fee (p.414, 1979) called this "one of the oldest biological 'laws' of limnology." In

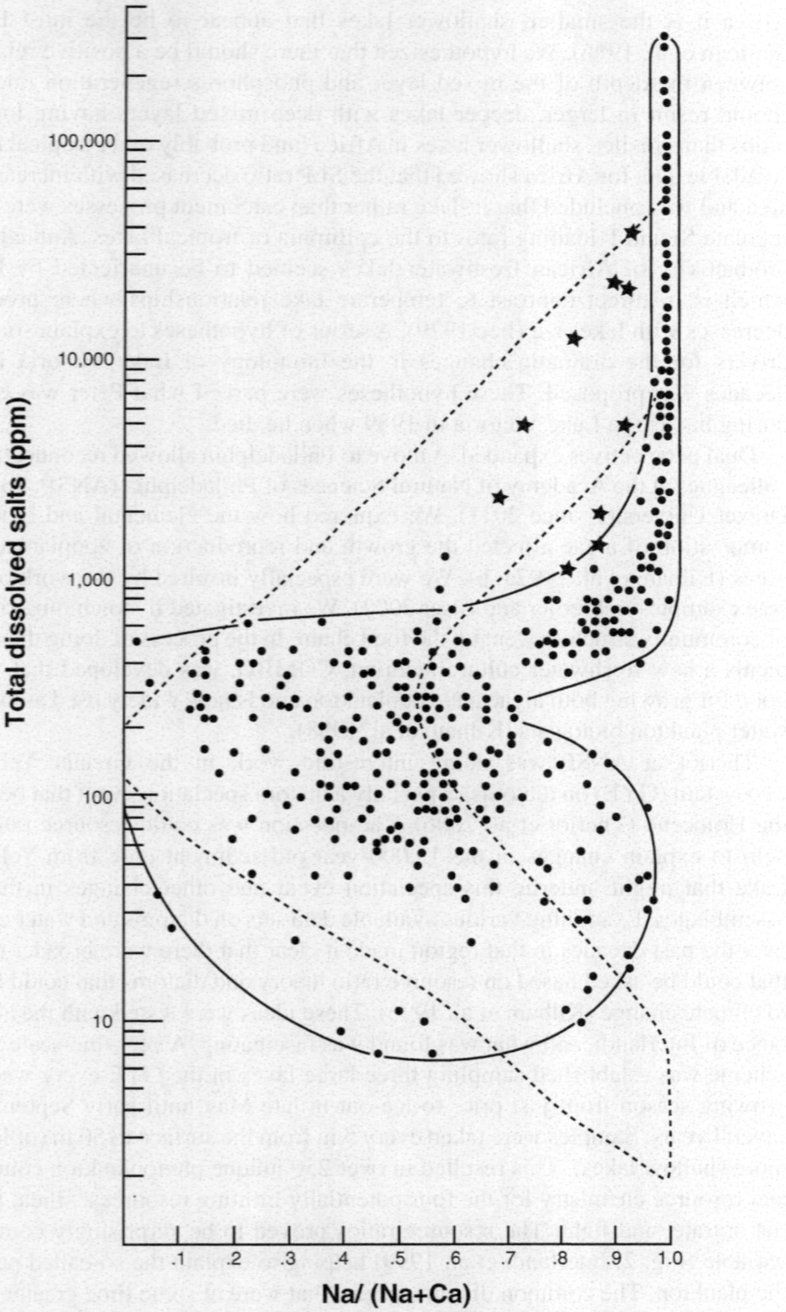

Fig. 1 Geochemical processes controlling water chemistry in African lakes and rivers (Fig. 1 in Kilham 1990b). Variation of the weight ratio of $Na^{+}/(Na^{+}=Ca^{2+})$ as a function of the total dissolved salts of African surface waters. Reproduced with permission of Limnology and Oceanography

Africa it is the smaller, shallower lakes that appear to be the most P-limited (Kilham et al. 1986). We hypothesized that there should be a positive relationship between the depth of the mixed layer and phosphorus regeneration rates which should result in larger, deeper lakes with deep mixed layers having lower Si:P ratios than smaller, shallower lakes in Africa (and probably other tropical regions). Available data for Africa showed that the Si:P ratio decreased with increasing lake area and we concluded that in-lake rather than catchment processes were likely to regulate Si and P loading rates to the epilimnia of tropical lakes. Annual primary productivity of African freshwater lakes seemed to be unaffected by lake size which is in direct contrast to temperate lake relationships where productivity decreases with lake size (Fee 1979). A series of hypotheses to explain some of the drivers for the dramatic changes in the limnology of Lake Victoria in recent decades was proposed. These hypotheses were part of what Peter was exploring during his trip to Lake Victoria in 1989 when he died.

Dual perspectives expanded. A move to Philadelphia allowed reconnections with colleagues at the Academy of Natural Sciences of Philadelphia (ANSP; now part of Drexel University since 2011). We explored how the elemental and biochemical composition of algae affected the growth and reproduction of zooplankton herbivores (Kilham et al. 1997a, b). We were especially inspired by the work of Sterner (see examples in Sterner and Elser 2002). We investigated if stoichiometric control of community structure went up the food chain. In the process of doing these experiments a new freshwater culture medium, COMBO, was developed that was very good for growing both algae and zooplankton and is now widely used among freshwater plankton biologists (Kilham et al. 1998).

Theriot at ANSP was doing interesting work in the Greater Yellowstone Ecosystem (GYE) on diatoms, especially a diatom speciation event that occurred in the Holocene (Theriot et al. 2006). The question was could resource ratio theory help to explain changes in the 13,000-year-old sediment core from Yellowstone Lake that might underlie this speciation event and other changes in the diatom assemblages. Examining various available data sets on diatoms and water chemistry over the past decades in that region made it clear that there were broader questions that could be asked based on resource ratio theory and diatoms that could be linked to climate change (Kilham et al. 1996). These ideas were tested with the able assistance of Interlandi, and what was found was fascinating. A very fine-scale sampling scheme was established sampling three large lakes in the GYE every week of the growing season from just prior to ice-out in late May until early September over several years. Samples were taken every 5 m from the surface to 50 m (or less in the more shallow lakes). This resulted in over 250 unique phytoplankton communities and resource chemistry for the four potentially limiting resources: silica, phosphorus, nitrate, and light. The resource ratios proved to be surprisingly complex and variable (Fig. 2; Interlandi et al. 1999) helping to explain the so-called paradox of the plankton. The common diatom species that were at some time greater than 5 % by biovolume of the total abundance (eight species) were ranked along the various resource ratio gradients and it became clear that diatom community structure could be largely explained by each species' unique 4D resource signature. The point of maximum abundance of any one species never overlapped that of any other species

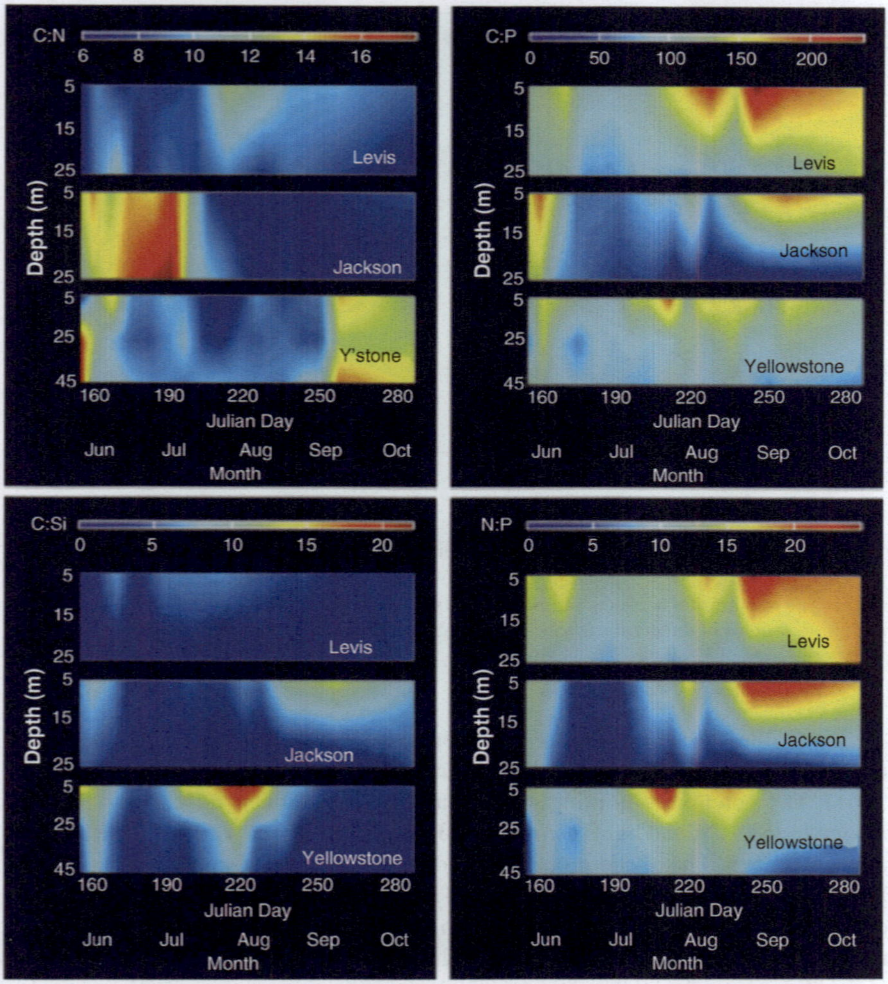

Fig. 2 Elemental particulate (phytoplankton) molar ratios versus depth and time of year of C:N, C:P, C:Si, and N:P in three Yellowstone lakes (Fig. 5 in Interlandi et al. 1999). Reproduced with permission of Limnology and Oceanography

during the growing season, supporting resource niche theory. These conclusions were supported by physiology experiments (Kilham and Taylor 2002) and bioassay experiments (Kilham and Interlandi 2005) of diatom species isolated from lakes in the GYE. One of those observations was that the endemic species *Stephanodiscus yellowstonensis* had a much lower requirement for N for growth than did the more widespread species from which it evolved, *Stephanodiscus niagarae*. The speciation event was perhaps influenced by changes in the nutrient cycling of N and P in Yellowstone Lake as it developed during the Holocene.

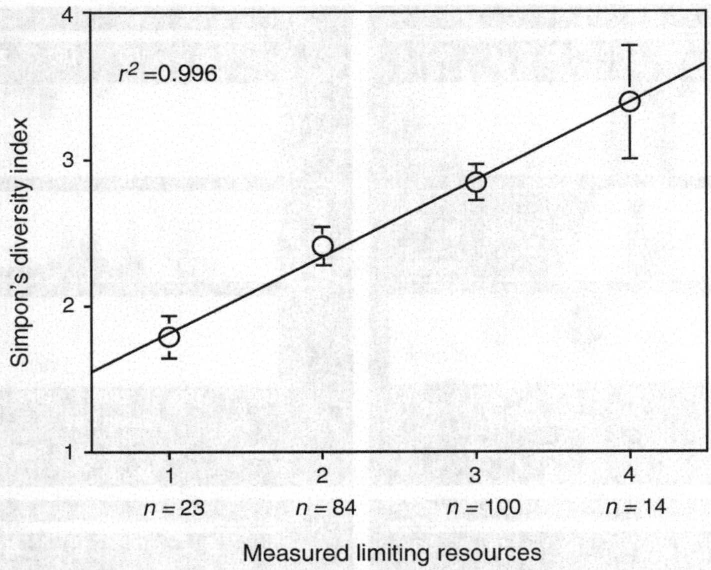

Fig. 3 The relationship between Simpson's diversity for phytoplankton species and the number of limiting resources in three Yellowstone lakes (Fig. 6 in Interlandi and Kilham 2001). Reproduced with permission of Ecology

One of the classic paradigms in ecology was also tested, namely does species diversity increase with the number of limiting resources (Interlandi and Kilham 2001). The potentially limiting concentrations were defined for each of the four resources (Si, P, N, and light) using two independent methods which were in remarkable agreement. This made it possible to define how many resources were limiting for each unique phytoplankton sample. The species diversity of each sample was calculated and compared to the number of limiting resources for that sample. The result was astonishing—a linear relationship with an r^2 of 0.996 (Fig. 3).

Implications about the impact of climate change on the GYE were also explored (Interlandi and Kilham 1998, Interlandi et al. 2003). Nitrogen loading to largely N-limited Yellowstone Lake is dominated by the spring melt of the accumulated snow pack. In El Niño years the snow pack is significantly decreased and in La Niña years it is increased compared to the long-term mean. This variability changes all the resource ratios with respect to N and the diatom species composition changes in a predictable manner. It is amazing that climate events in the southern Pacific Ocean affect the species composition and primary productivity of Yellowstone Lake in Wyoming.

Collaborations go well largely because of separate though compatible perspectives. Wonderful mentors, colleagues, and graduate students inspire with their new perspectives. We each separately and together made significant contributions bringing microbial ecology and biogeochemistry together into what is now the richly expanding field of ecological stoichiometry.

Susan S. Kilham and Peter Kilham

Susan K. Soltau and Peter Kilham met in 1965 when they were graduate students at Duke University. They married in 1967. Susan did her undergraduate work at Eckerd College where her mentor was George K. Reid. Growing up in Florida was an important influence on her choice of marine science as a major interest. She went to Duke University for doctoral studies where she was an NSF Oceanographic Trainee and worked on calcification of deep sea bivalve mollusks under Orrin Pilkey. Peter did his undergraduate studies at Dartmouth College where his mentor was Gene Likens. Growing up in a family of naturalists and his experience as a teenager in Uganda greatly influenced his career choices. Peter worked with Daniel Livingstone at Duke. Susan and Peter both went to Woods Hole Oceanographic Institution for postdoctoral work where both were mentored by Robert R.L. Guillard. They both went to the University of Michigan in 1972. Establishing dual careers at UM was a challenge, especially for Susan who had more than a few unfortunate experiences with sexism. We were both fortunate to have wonderful graduate students, our first one being David Tilman. Highlights of those years were sabbaticals, especially the year in Germany in 1987–1988. Peter died on a research trip to Lake Victoria on March 20, 1989, in Kisumu, Kenya, of a ruptured ulcer he did not even know he had. He was only 45 years old. Susan moved to Drexel University in 1991 where she continues as a Professor of Environmental Science. In recent years she has expanded her horizons in work on tropical stream ecosystems. Both of us were abundantly blessed with wonderful mentors, colleagues, and students.

References

Fee EJ (1979) A relation between lake morphometry and primary productivity and its use in interpreting wholelake eutrophication. Limnol Oceanogr 24:401–416

Gasse F, Talling JF, Kilham P (1983) Diatom assemblages in East Africa: classification, distribution, and ecology. Rev Hydrobiol Trop 16:3–34

Hecky RE, Kilham P (1973) Diatoms in alkaline saline lakes: ecology and geochemical implications. Limnol Oceanogr 18:53–71

Hecky RE, Kilham P (1988) Nutrient limitation of phytoplankton in marine and freshwater environments: a review of recent evidence on the effects of enrichment. Limnol Oceanogr 33:796–822

Interlandi SJ, Kilham SS (1998) Assessing the effects of nitrogen deposition on mountain waters: a study of phytoplankton community dynamics. Water Sci Technol 38:139–146

Interlandi SJ, Kilham SS (2001) Limiting resources and the regulation of diversity in phytoplankton communities. Ecology 85:1270–1282

Interlandi SJ, Kilham SS, Theriot EC (1999) Responses of phytoplankton to varied resource availability in the large lakes of the Greater Yellowstone Ecosystem. Limnol Oceanogr 44:668–682

Interlandi SJ, Kilham SS, Theriot EC (2003) Diatom-chemistry relationships in Yellowstone Lake (WY) sediments: implications for climatic and aquatic processes research. Limnol Oceanogr 48:79–92

Kilham P (1971) Biogeochemistry of African lakes and rivers. Ph.D. dissertation, Duke University

Kilham P (1971b) A hypothesis concerning silica and the freshwater planktonic diatoms. Limnol Oceanogr 16:10–18

Kilham SS (1975) Kinetics of silicon-limited growth in the freshwater diatom *Asterionella formosa*. J Phycol 11:396–399

Kilham SS (1978) Nutrient kinetics of freshwater planktonic algae using batch and semicontinuous methods. Verh Int Ver Theor Angew Limnol 21:141–151

Kilham SS (1986) Dynamics of Lake Michigan natural phytoplankton communities in continuous cultures along a Si:P loading gradient. Can J Fish Aquat Sci 43:351–360

Kilham P (1990a) The ecology of *Melosira* species in the Great Lakes of Africa. In: Tilzer MM, Serruya C (eds) Large lakes: ecological structure and function. Springer, Berlin, pp 414–427

Kilham P (1990b) Mechanisms controlling the chemical composition of lakes and rivers: data from Africa. Limnol Oceanogr 35:80–83

Kilham P, Hecky RE (1988) Comparative ecology of marine and freshwater phytoplankton. Limnol Oceanogr 33:776–795

Kilham SS, Interlandi SJ (2005) Nutrient bioassays on diatoms from the large lakes of the Greater Yellowstone Ecosystem. Verh Int Ver Theor Angew Limnol 29:268–272

Kilham SS, Kilham P (1978) Natural community bioassays: predictions of results based on nutrient physiology and competition. Verh Int Ver Theor Angew Limnol 20:68–74

Kilham P, Kilham SS (1980) The evolutionary ecology of phytoplankton. In: Morris I (ed) The physiological ecology of phytoplankton. Studies in ecology, vol 7. Blackwell, Oxford, pp 571–597

Kilham P, Kilham SS (1990a) Endless summer: internal loading processes dominate nutrient cycling in tropical lakes. Freshwater Biol 23:379–389

Kilham SS, Kilham P (1990b) Tropical limnology: do African lakes violate the “first law” of limnology? Verh Int Ver Theor Angew Limnol 22:435–439

Kilham SS, Taylor SM (2002) Resource requirements and competitive interactions of two closely related *Stephanodiscus* species for silicon and nitrogen. Verh Int Ver Theor Angew Limnol 28:928–931

Kilham P, Kilham SS, Hecky RE (1986) Hypothesized resource relationships among African planktonic diatoms. Limnol Oceanogr 31:1167–1179

Kilham SS, Theriot EC, Fritz SC (1996) Linking planktonic diatoms and climate change using resource theory in the large lakes of the Yellowstone ecosystem. Limnol Oceanogr 41:1052–1062

Kilham SS, Kreeger DA, Goulden CE, Lynn S (1997a) Effects of algal food quality on fecundity and population growth rates of *Daphnia*. Freshwater Biol 38:639–647

Kilham SS, Kreeger DA, Goulden CE, Lynn S (1997b) Effects of nutrient limitation on biochemical constituents of *Ankistrodesmus falcatus*. Freshwater Biol 38:591–597

Kilham SS, Kreeger DA, Lynn SG, Goulden CE, Herrera L (1998) COMBO: a defined freshwater culture medium for algae and zooplankton. Hydrobiologia 377:147–159

Lynn SG, Kilham SS, Kreeger DA, Interlandi SJ (2000) Effects of nutrient availability on the biochemical and elemental stoichiometry in the freshwater diatom *Stephanodiscus minutulus* (Bacillariophyceae). J Phycol 36:510–522

Mechling JA, Kilham SS (1982) Temperature effects on silicon limited growth of the Lake Michigan diatom *Stephanodiscus minutus* (Bacillariophyceae). J Phycol 18:199–205

Sommer U, Kilham SS (1985) Phytoplankton natural community experiments: a reinterpretation. Limnol Oceanogr 30:438–442

Sterner RW, Elser J (2002) Ecological stoichiometry. The biology of elements from molecules to the biosphere. Princeton University Press, Princeton, NJ, p 464

Theriot EC, Fritz SC, Whitlock C, Conley D (2006) Late quaternary rapid morphological evolution of an endemic diatom in Yellowstone Lake, Wyoming. Paleobiology 35:35–54

Tilman D (1977) Resource competition between planktonic algae: an experimental and theoretical approach. Ecology 58:338–348

Tilman D (1982) Resource competition and community structure. Princeton University Press, Princeton, NJ

Tilman D, Kilham SS (1976) Phosphate and silicate growth and uptake kinetics of the diatoms *Asterionella formosa* and *Cyclotella meneghiniana* in batch and semicontinuous culture. J Phycol 12:375–383

Tilman D, Kilham SS, Kilham P (1982) Phytoplankton community ecology: the role of limiting nutrient. Annu Rev Ecol Syst 13:349–372
Tilman D, Kiesling R, Sterner R, Kilham SS, Johnson FA (1986) Green, bluegreen and diatom algae: taxonomic differences in competitive ability for phosphorus, silicon and nitrogen. Arch Hydrobiol 106:473–485
Titman D (1976) Ecological competition between algae: experimental confirmation of resource-based competition theory. Science 192:463–465
Titman D, Kilham SS, Kilham P (1976) Morphometric changes in *Asterionella formosa* colonies under phosphate and silicate limitation. Limnol Oceanogr 21:883–886
Van Donk E, Kilham SS (1990) Temperature effects on the silicon and phosphorous limited growth and competitive interactions among three diatom species. J Phycol 26:40–50

Exploring the Implications of the Stoichiometric Modulation of Planktonic Predation

Aditee Mitra and Kevin J. Flynn

Introduction

Understanding the dynamics of predator–prey systems is central to ecology. Traditional marine ecosystem models were based entirely on a single nutrient (typically nitrogen, N; Fasham et al. 1990). Such models assumed no effect on predation caused by changes in food quality – indeed, they assumed no changes in quality as they assumed a fixed C:N:P ratio in accordance with the Redfield ratio (Redfield 1958). In nature, C:N:P stoichiometry is anything but fixed, especially in primary producers (Geider and La Roche 2002). Sterner and Elser's book on ecological stoichiometry (Sterner and Elser 2002) served to highlight, for many in the scientific community (notably non-modellers), the importance of the differences in stoichiometry between different organisms for ecology. In reality 'ecological stoichiometry' is a direct and inescapable consequence of the non-conservative nature of C:N:P between organisms. Indeed, anyone who attempts to make and use multi-nutrient, variable stoichiometric computational descriptions of ecology cannot fail to realise the importance of variable C:N:P for the differential flow of elements through food webs.

The application of intraspecific variable elemental stoichiometry in its simplest form has broadly linear consequences for trophic dynamics. For example, under N limitation, the quality of the phototrophic prey deteriorates (e.g., organism N:C halves). Under these conditions, the predator no longer has to release such an excess of ingested N to match respired C in comparison with feeding on a N replete food item. Hence the predator retains a proportionately higher amount of the ingested prey-N. The important knock-on consequence is that less N is regenerated to support the next generations of phototrophs, and so the next cycle of food for the consumers is of even lower quality with respect to N content. This is a positive feedback loop, with potential for the establishment of a persistent, poorly consumed, phototrophic biomass.

A. Mitra (✉) • K.J. Flynn
Swansea University, Wallace Building, Swansea SA2 8PP, UK
e-mail: a.mitra@swansea.ac.uk

P.M. Glibert, T.M. Kana (eds.), *Aquatic Microbial Ecology and Biogeochemistry: A Dual Perspective*, DOI 10.1007/978-3-319-30259-1_7

The situation with P-limitation differs from that for N in two important ways. Firstly, P can be readily accumulated to vast excess in many microbes (as polyphosphate), thus acting as a cushion against the absence of external nutrient-P. In contrast, an excess of inorganic N as ammonia/ammonium is toxic and thus cannot be accumulated to any significant extent; any accumulated inorganic N can only support a fraction of a generation while polyphosphate accumulations can support many generations of growth (Rhee 1973; Watanabe et al. 1987). Secondly, the relationship between cellular P:C and growth rate is curvilinear so that once P-stress develops it takes several generations before phytoplankton growth rate declines significantly (Flynn 2008a). In contrast, the relationship between N:C and phytoplankton growth rate is essentially linear, and growth declines very rapidly in the absence of incoming N (Flynn 2008a). These differences between N and P have important implications for the aforementioned positive feedback loop in stoichiometric ecology, because P-stress has a greater scope for impact on consumers than on primary producers. And that is so before we consider the topics of noxious metabolite production by phototrophs, and the consequential de-selection of unpalatable prey by predators.

The implications of variable stoichiometry for understanding ecological processes, for designing experimental work, and for modelling are, therefore, clear. It is not sufficient to consider just N:C or P:C; all three elements are required by organisms and we ignore that at our peril. Indeed, depending on the applications, Si, and perhaps Fe, also need to be considered. The reason for the need for C:N:P rests upon the requirement to properly account for the fate of the non- or lesser-limiting nutrient (Flynn 2005). In addition to the variability in stoichiometry between predator and prey, it is noteworthy that the variability within a particular organism can also differ significantly. The typical assumption is that predators and bacteria have a fixed stoichiometric ratio. However, while variability in stoichiometry is greatest in phototrophs, there is also a need to appreciate that C:N:P is not constant in the predators or in bacteria (Mitra and Flynn 2007; Polimene et al. 2015). However, in the past there has been something of an obsessive assumption of fixed Redfield ratios within marine foodweb models, which has in turn resulted in resistance to the incorporation of multi-element variable stoichiometric models (Flynn 2010).

Our contribution to the issue of stoichiometrics in trophic dynamics developed from 2005 (Mitra and Flynn 2005), based on observations that simple C:N:P 'ecological stoichiometry' is insufficient to account for observed changes in predator–prey interactions. There are several interesting lines of thought that has basis in our original observations; these we explore below.

Characterising the Predator–Prey Stoichiometric Link

When a consumer is presented with (encounters) food of suboptimal quality, there are three potential consequences (Fig. 1; Mitra et al. 2007). Feeding rates could remain as they are, and the stoichiometric consequences are passed on directly to the consumer; to explain crudely, if prey quality halves so predator growth halves. This is the outcome of simple 'ecological stoichiometry'. Alternately, feeding could be enhanced; the halving in prey quality could be countered by a doubling in prey consumption, thus maintaining immediate consumer growth, although the prey population is now diminished rapidly. The third option is that the consumption of the prey decreases due to de-selection, perhaps with feeding switched to an alternative food type which was formally of lower preference but is now of higher relative quality. We termed these alternatives as, respectively, zero, positive and negative stoichiometric modulation of predation (SMP; Fig. 1; Mitra and Flynn 2005).

The consequences of the third option, negative (−ve) SMP, are particularly profound for ecology for they lead to the potential development of an ungrazed, and ultimately ungrazable, population of phototrophic (or perhaps mixotrophic) prey. This occurs because, in the absence of grazing there is little effective recycling of the limiting nutrient, and hence the phototrophic prey become increasingly unpalatable. This event we proposed (Mitra and Flynn 2006b; Fig. 2) provides a mechanism for the formation of harmful algal blooms (HABs). More appropriately (escaping the often anthropocentric meaning of HABs), these blooms that form through a disruption of what may be considered as normal predator–prey activity would be better termed ecosystem disruptive algal blooms (EDABs). The term EDAB was coined by Sunda et al. (2006), but its meaning and importance becomes all the more apparent when placed in the context of −ve SMP.

The elucidation and modelling description of this mechanism of SMP required not only the use of multi-nutrient variable stoichiometric models of prey and predator (Mitra 2006), but the development of an alternative description of grazing kinetics and thence of prey selectivity. Prey selection is a common feature of predator behaviour, and yet traditional descriptions of grazing cannot describe de-selection or prey-switching (Mitra and Flynn 2006a). It was not possible to describe the dynamics of predator–prey interactions in multi-species scenarios without inclusion of these prey switching functions, which were linked to prey stoichiometry in a strongly nonlinear fashion (Mitra and Flynn 2006a, b). The implication is that chemicals that adversely affect palatability to consumers can rapidly accumulate in phototrophs during the onset of nutrient stress. Ecological stoichiometry, as typically described linked to elements (Sterner and Elser 2002), is insufficient (Mitra and Flynn 2005).

Models of trophic dynamics are not only sensitive to stoichiometry, but they are extremely sensitive to other factors affecting the kinetics of grazing and of digestion (Jones and Flynn 2005; Mitra et al. 2014a). Assimilation efficiency (AE) is notable in this context. Current ecosystem models do not typically consider variable AE at all; in those that do, the link to AE is to quantity (via changes in gut transit time; Mitra and Flynn 2007) rather than to quality per se (that response being associated primarily with de-selection of the poor prey type; Mitra and Flynn 2006a). It is well known that AE varies with the quality and also with the quantity of material ingested (e.g., Tirelli and Mayzaud 2005). We hypothesised how the relationship may work with simultaneous variations in quality and quantity (Mitra and Flynn 2007). Such variation in AE is important because, typically in nature, high quantities of food

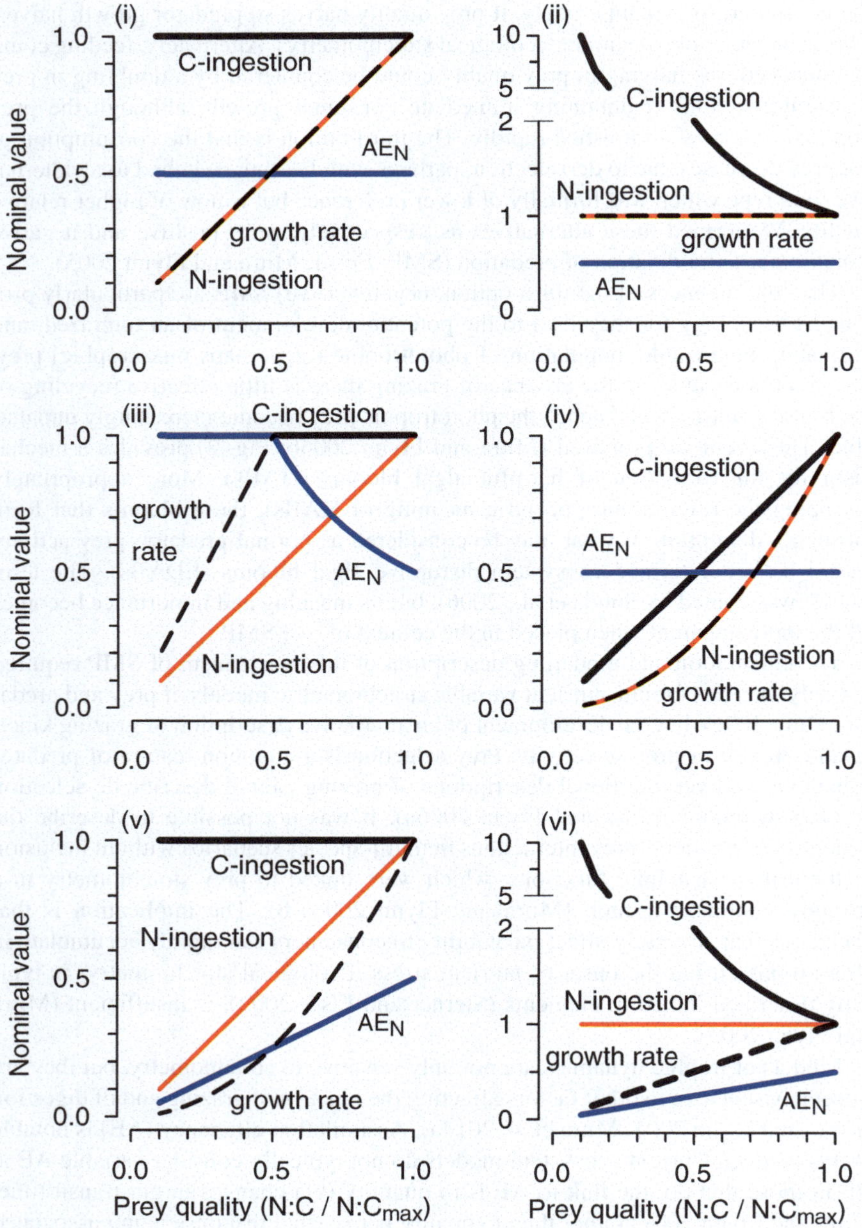

Fig. 1 Effects of prey quality on predator growth through neutral (0), negative (−ve) or positive (+ve) stoichiometric modulation of predation (SMP) acting at ingestion and/or assimilation. (*i*) Neutral stoichiometric modulation of predation (SMP) for ingestion and assimilation (0 SMP_{Ing} and 0 SMP_{AE}, respectively), with no modification of the ingestion behaviour or assimilation efficiency for the limiting nutrient (nitrogen; AE_N) in response to the presence of poor-quality prey. (*ii*) +ve SMP_{Ing} with 0 SMP_{AE}; ingestion of prey C is increased. (*iii*) 0 SMP_{Ing} with +ve SMP_{AE}; AE_N

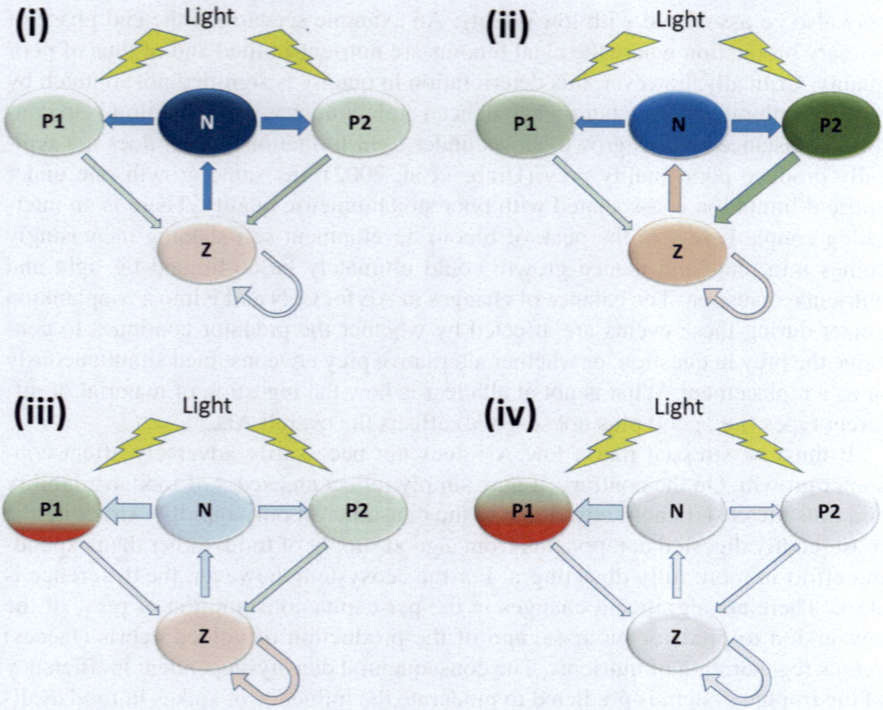

Fig. 2 Suggested route for the generation of an ecosystem disruptive algal bloom (EDAB) associated with negative stoichiometric modulation of predation (SMP; see Fig. 1). Two phytoplankton groups, of which P2 has a higher growth rate potential than does P1, consume nutrient (N) and are grazed by zooplankton (Z). Z regenerates nutrient, and is subject to intraguild cannibalism. At the start (*i*) there is an excess of N which supports phytoplankton growth; P2 grows faster. Both P1 and P2 are grazed upon by Z, whose population also increases. By stage (*ii*), most of N is consumed, and population P2 is dominant. Both P1 and P2 become nutrient stressed, and Z becomes subject to SMP. By stage (*iii*), however, the nutrient stress induced upon P1 has resulted in that species becoming unpalatable. Now Z graze only P2 and itself (cannibalism); there is a gradual transfer of nutrient into the ungrazed P1. The eventual state (*iv*) is thus a large ungrazed, and ungrazable, bloom of P1. Shown here, but not in the original (Mitra and Flynn 2006b), is a role of light; sufficient illumination both at the surface and penetrating the water column (a function of mixing depth and biomass density) is important for the deterioration of phytoplankton nutrient status

---

Fig. 1 (continued) is enhanced to a maximum of 100%, so initially maintaining N assimilation. (*iv*) −ve SMP_{Ing} with 0 SMP_{AE}; C ingestion decreases, decreasing N ingestion further. (*v*) 0 SMP_{Ing} with −ve SMP_{AE}; as panel (*i*) but now with AE_N decreasing. (*vi*) +ve SMP_{Ing} with −ve SMP_{AE}; as panel (*ii*) but now with AE_N decreasing. The original configuration of NPZ-type models for phytoplankton predation (which considers only N-biomass but describes N-deprivation and hence implicitly a decline in phytoplankton quality) accords with the pattern in panel (*ii*). *Y*-axis units are indicative of relative change with food quality only may also be associated with low quality. An example scenario is the end phase of primary production where the algal blooms are nutrient-limited and of thus of poor quality. Critically, however, this deterioration in quality is signified not so much by poor growth rates per se, but by unbalanced stoichiometry. This situation is in contrast to instances where growth slows under light limitation, which does not typically produce poor quality prey (Urabe et al. 2002); the same growth rate under nutrient limitation is associated with poor stoichiometric quality. There is an interesting couple here – at the peak of bloom development self-shading increasingly comes into play, and thence growth could ultimately be co-limited by light and nutrient exhaustion. The balance of changes in AE for C, N and P into a zooplankton grazer during these events are affected by whether the predator continues to consume the prey in question, or whether alternative prey are consumed simultaneously or as a replacement. What is not at all clear is how the ingestion of material of different types (i.e., good plus not so-good) affects the overall AE.

It must be stressed that a low AE does not necessarily adversely affect consumer growth. On the contrary, it may simply reflect an excess of food availability and thus the cost–benefit advantage to the consumer in only rapidly extracting the most readily digested components from a good supply of food, rather than expending effort in more fully digesting it. For the ecosystem, however, the difference is stark. There are significant changes in the per-capita consumption of prey, of the conversion to predator biomass, and of the production of voided debris (faeces) versus regeneration of nutrients. The consequential density-dependent inefficiency of the trophic system is predicted to moderate the influence of spikes in food availability (Mitra and Flynn 2007; Flynn 2009a). If the quality of components of feed had an additional impact upon AE then similar consequences could develop.

Elemental Stoichiometry and Commercial Microalgal Production

In the context of commercial microalgal culture, this whole predator–prey system takes on another perspective (Smith and McBride 2015). In microalgal crop production, the control of zooplanktonic pests is vital. Here, an understanding of the dynamics of EDAB development under nutrient limitation developing through –ve SMP (Fig. 2) is useful.

In production of biofuels from fatty acids of microalgae, nutrient exhaustion is required (Kenny and Flynn 2015); the balance of high crop production (per unit of area and of volume) together with nutrient exhaustion requires algal growth in systems that are optically shallow. Further, for commercial production of biofuels, a minor level of P-stress appears useful (Mayers et al. 2014) as it saves a valuable resource (P-fertiliser), does not significantly affect fatty acid synthesis, and yet is predicted to decrease palatability to a predator.

As noted in the Introduction, the linkage between N and light physiology is simple and broadly linear, while that involving P is quite different. Once the nutrient N:P ratio approaches and exceeds 64, and assuming light is not limiting and all nutrients are consumed (which is a function of N-supply, as well as optical depth and surface irradiance) then P-stress proper develops. This not only adversely affects fatty acid synthesis (and hence, perhaps, biofuels production) but on simple stoichiometric grounds predator–prey interactions will likely be severely compromised. Moreover, such levels of P-stress are also allied to the production of toxic secondary metabolites in some species (Granéli and Flynn 2006) that will have additional quality-related consequences limiting grazer activity.

From the above we can see that there is a critical window of opportunity for zooplanktonic pests in commercial microalgal cultivation, and that window relates to nutrient sufficiency rendering the crop as good quality, palatable, feed. There is another important facet to this topic though, and that is the development of genetically modified (GM) microalgae to enhance either the products for commercial exploitation (e.g., biofuels) or to directly make them pest-resistant. We explored this topic using models (Flynn et al. 2013b) and raised the spectre of the escape of fast growing, inedible GM microalgae into the wild; these would represent harmful algal bloom organisms par excellence.

Effects of Temperature, Ocean Acidification and Nutrient Excess

Changes in elemental (C:N:P) stoichiometry are not the same as biochemical (e.g., C:fatty acid) stoichiometry; while elements must be conserved within a system, biochemicals need not be, and typically they are not conserved except for compounds such as essential fatty acids. Minor changes in elemental stoichiometry may hide significant changes in biochemistry, some of which may have profound impacts on trophic dynamics. This was the basis of our original paper on ecological stoichiometry (Mitra and Flynn 2005) – that trophic interactions need not, indeed more often would not, relate in a linear fashion to elemental stoichiometry (likely explaining Jones and Flynn 2005).

Under anthropogenically driven climate change three factors of direct consequence to stoichiometry are of importance for marine plankton: increasing temperature, ocean acidification (OA) and changes in nutrient supply. These three factors inter-relate in various ways. Temperature and ocean acidification are directly associated with increased atmospheric levels of CO_2. Increased temperature increases water-column stability limiting nutrient exchange between photic and sub-photic zones (Doney et al. 2009). It also affects weather systems that alter rainfall patterns over land affecting agriculture and other aspects of human population growth and activity that in turn affect nutrient supply (concentrations and ratios) to coastal waters (Glibert et al. 2011). Within biological systems, temperature affects organism growth, and especially in predators affects processes fundamental to trophic dynamics that will not be altered pro rata; a notable example of the latter is variation in AE which is a function of processes outside of the direct sphere of influence of temperature on the predator (i.e., quality and quantity of prey). OA affects plankton in different ways, varying with size (affecting proximal pH gradients), activity, and differentially affecting succession through impacts of H^+ or CO_2 (Flynn et al. 2012, 2015). For phototrophs, the increase in dissolved CO_2 (with the increase in dissolved inorganic C, DIC, and also changes in DIC dissociation that raises the proportion of DIC as CO_2(aq)) has potential for raising primary production, hastening nutrient exhaustion and thence changes in elemental stoichiometry (Urabe et al. 2003).

Emphasis in most studies of phytoplankton physiology has been placed upon nutrient limitation and recovery from it. Far less has been researched concerning the consumption of different forms of non-limiting nutrients, an event which is of critical importance for production later in the season (Flynn 2005), and also for ecological stoichiometry. Changes in the availability of DIC, of different forms of N and of P all affect differential accumulation of carbohydrates and fatty acids. There is evidence of changes in biochemical composition of phytoplanktonic prey with growth under OA conditions that affect prey selection (Cripps et al. 2016). These interactions are of especial concern when set against observations that resilience by zooplankton against OA depends on a continued availability of good quality prey (Cripps et al. 2014).

Avoiding Predation

Prey selection is not solely a function of size across a narrow range (Hansen et al. 1997), indeed predators switch between prey depending on availability and nutritional status (quality) (Flynn et al. 1996; Mitra and Flynn 2006a, b). The best mechanism for handling your predator is not to kill it, for that benefits your competitors as well; rather it is to 'persuade' the predator to hunt others. For plankton that 'persuasion' can be achieved through production of noxious, rather than toxic, compounds (Flynn 2008b). The result is not only the removal of competitors but also the recycling of nutrients that are needed to support further growth (Mitra and Flynn 2006b). In ecological and evolutionary terms, and thence for modelling of such facets, this raises the question of which of two alternate evolutionary tactics is best:

1. For a species to grow quickly, to be a numerically dominant member of the prey field, but thus be a stronger driving factor in predator evolution.
2. To grow slowly, perhaps to be noxious, and thus to remain cryptic, so that predators will not have a strong selective advantage in evolving to consume such a prey, nor to neutralise or otherwise overcome any noxious compounds.

A traditional way of viewing these alternates would argue for a trade-off in allocation of resources towards growth versus the production of defences. There is no obvious overwhelming cost for phytoplankton in making secondary metabolites as these typically represent only a few % of cellular resources and are often synthesised during unbalanced growth (John and Flynn 2002). Thus such an argument based on trade-offs appears weak at best. There may be advantages in growing slowly, especially in 'slow' environments where nutrients are limiting (Flynn 2009b); being forced (so to speak) to grow slowly through being nutrient-stressed generates stress that is likely not good for overall cellular maintenance (Schaum and Collins 2014). If it were otherwise then one would expect all organisms to be capable of higher maximum growth rates than they are. Yet, toxins appear to be generated typically in nutrient-stressed conditions (Granéli and Flynn 2006). What we do not know is how many biochemicals are produced by these organisms under various conditions of stress that are unknown to research simply because they are not recognised as toxins by humans. A good example is the identity of the chemical that renders the widely used aquaculture species *Isochrysis* unpalatable to grazing when that species is deprived of N-nutrient (Flynn et al. 1996). This is an aspect of the intriguing issue of the role of chemical warfare in plankton ecology (Pohnert et al. 2007).

There thus appears something of a paradox; toxic species are typically not fast growing organisms and so may be expected to not be so likely to become nutrient-stressed, yet to become noxious they appear to need to become nutrient stressed. Explanations may rest for many species in the apparent conflict between heterotrophic versus phototrophic modes of growth in the one cell, for most HAB organisms are mixotrophic (Flynn et al. 2013a). Is it possible that while the heterotrophic core of these organisms is not stressed (explaining the low growth rates matched to 'slow' environments), the phototrophic component is *de facto* nutrient limited? This could be attained if the regulation of internal biochemical stoichiometry is not so well balanced as one may expect to occur in mixotrophs (Flynn and Mitra 2009), or indeed not so well balanced as one may expect from a consideration of solely elemental stoichiometry.

Stoichiometry and Mixotrophy

Hitherto we have considered stoichiometry and predator–prey interactions from the traditional point of view, with a clear separation between primary and secondary producers, between phototrophy and heterotrophy. Mixotrophic protists, now recognised as being near ubiquitous in many waters (Flynn et al. 2013a; Mitra et al. 2014a), combine these modes of nutrition in a single cell. As such they present an interesting subtopic for the study of stoichiometry and trophic interactions. The usual concept of mixotrophy is of mixing sources for nutrition, for example acquiring C from photosynthesis versus feeding. What actually seems to pass is that these organisms photosynthesise for C and energy and feed mainly to support acquisition of N, P and Fe (Mitra and Flynn 2010; Mitra et al. 2014b). Of course, in many instances, these other elements are acquired as inorganics during phototrophic physiology, while C is also acquired as a by-product of eating. Thus from a phagotrophic point of view, mixotrophy provides scope to minimise the otherwise certain loss of elements during anabolic and catabolic processing of food; it does this both via the provision of fixed C from photosynthesis that mitigates against the loss of C during catabolic respiration, and also through providing scope for the re-assimilation of regenerated nutrients de facto before they leave the cell. From a phototrophic point of view, mixotrophy provides scope to acquire elements that would otherwise be limiting if the organism depended solely on external inorganic sources (Mitra and Flynn 2010; Mitra et al. 2014b).

In theory, then, mixotrophy should provide scope for the maintenance of a heightened nutrient status, to enable continued good growth under conditions that would not be favourable to their non-mixotrophic counterparts. Mixotrophs may thus be expected to provide a superior food source for larger zooplankton, and indeed there is cause to suspect that this is so. On the other hand, as noted above, most HAB species are mixotrophic, which raises the alternative (or perhaps additional) role of phototrophy in providing secondary metabolites as toxins to deter grazers. And so we return to the issue of whether it is preferable from an evolutionary perspective to actually preserve some level of internal nutrient imbalance to render the cell as relatively unpalatable, and/or to remain cryptic, and hence not drive the evolution of grazers to develop biochemical counters to toxins. We do not as yet know, but for sure it will provide an interesting line in experimental and theoretical research.

Conclusions

As we noted at the outset, ecological stoichiometry is a natural and unavoidable consequence of interactions between organisms of different chemical composition. To model predator–prey interactions requires a recognition of the scope for variable composition, a recognition that has been surprisingly slow in coming; in large measure in marine modelling this has been justified by reference to the Redfield ratio. There are many facets of stoichiometry that remain elusive, including the nonlinear relationship between quality and prey selection, which is where we (authors) entered this arena.

Of one thing we can be sure, the whole topic of biochemical as well as variation in the bounds of elemental stoichiometry, when linked to prey (de)selection and trophic dynamics with prey quality and quantity, provide great scope for exploring the paradox of the plankton, why there are so many species of plankton when the simplistic views of trait trade-off often drive a vision of ecology dominated by the smallest cellular configuration.

Aditee Mitra and Kevin J. Flynn

"My organisms (zooplankton) eat yours!"-Aditee; "Yes, but my organisms (phytoplankton) may poison yours!"-Kevin. So what happened when we got together, courtesy of a grant, to "marry" our mathematical models of zooplankton and phytoplankton? We enjoyed debating the interactions of our favored plankton groups so much that we got married ourselves! And, several years on, we now have higher management to preside over our deliberations-our son, Rohan Mitra-Flynn. His preferred way of stalling our debates is to start an alternative discussion on giant squids or dinosaurs; change the topic, as per that classic senior management approach. Rohan entered the world of scientific debates at the Crete IMBER IMBIZO meeting, where he also had his first birthday party. He sat on a table next to his cake enjoying being the center of attention while a large crowd of marine scientists all sang "happy birthday." Rohan's party was even broadcast on Crete television, along with some mention about the conference, of course. Our first mixotroph meeting in Kalmar, Sweden, holds a special place in his memory bank because of the massive diggers which could be seen from the laboratory. Rohan aspires to become an engineer. An ambition of his is to design a deep sea submersible so that he can send (crazy) marine biologists to get videos of giant colossal squids. Meanwhile, Aditee and Kevin's level of banter along the lines of "zooplankton eating versus phytoplankton poisoning" has been contained by the more recent shared interest in mixotrophs, where primary and secondary productions live side by side within the same cell. Presently, Aditee is a lecturer in plant sciences and researcher in predator-prey trophic dynamics, while Kevin is a professor in marine microbiology. Both work at Swansea University, UK.

References

Cripps G, Lindeque P, Flynn KJ (2014) Have we been underestimating the effects of ocean acidification in zooplankton? Glob Chang Biol. doi:10.1111/gcb.12582

Cripps G, Flynn KJ, Lindeque PK (2016) Ocean acidification affects the phyto-zoo plankton trophic transfer efficiency. PLoS One. doi:10.1371/journal.pone.0151739

Doney SC, Fabrey VJ, Feely RA, Kleypas JA (2009) Ocean acidification: the other CO_2 problem. Ann Rev Mar Sci 1:189–192

Fasham MJR, Ducklow HW, McKelvie SM (1990) A nitrogen-based model of plankton dynamics in the oceanic mixed layer. J Mar Res 48:591–639

Flynn KJ (2005) Modelling marine phytoplankton growth under eutrophic conditions. J Sea Res 54:92–103

Flynn KJ (2008a) The importance of the form of the quota curve and control of non-limiting nutrient transport in phytoplankton models. J Plankton Res 30:423–438. doi:10.1093/plankt/fbn007

Flynn KJ (2008b) Attack is not the best form of defense; lessons from harmful algal bloom dynamics. Harmful Algae 8:129–139. doi:10.1016/j.hal.2008.08.007

Flynn KJ (2009a) Food-density dependent inefficiency in animals with a gut as a stabilising mechanism in trophic dynamics. Proc Biol Sci 276:1147–1152. doi:10.1098/rspb.2008.1575

Flynn KJ (2009b) Going for the slow burn: why should possession of a low maximum growth rate be advantageous for microalgae? Plant Ecol Divers 2:179–189. doi:10.1080/17550870903207268

Flynn KJ (2010) Ecological modelling in a sea of variable stoichiometry; dysfunctionality and the legacy of Redfield and Monod. Prog Oceanogr 84:52–65. doi:10.1016/j.pocean.2009.09.006

Flynn KJ, Mitra A (2009) Building the "perfect beast": modelling mixotrophic plankton. J Plankton Res 31:965–992. doi:10.1093/plankt/fbp044

Flynn KJ, Davidson K, Cunningham A (1996) Prey selection and rejection by a microflagellate: implications for the study and operation of microbial food webs. J Exp Marine Biol Ecol 196:357–372

Flynn KJ, Blackford JC, Baird ME, Raven JA, Clark DR, Beardall J, Brownlee C, Fabian H, Wheeler GL (2012) Changes in pH at the exterior surface of plankton with ocean acidification. Nat Clim Change 2:510–513. doi:10.1038/NCLIMATE1489

Flynn KJ, Stoecker DK, Mitra A, Raven JA, Glibert PM, Hansen PJ, Granéli E, Burkholder JM (2013a) Misuse of the phytoplankton-zooplankton dichotomy: the need to assign organisms as mixotrophs within plankton functional types. J Plankton Res 35:3–11. doi:10.1093/plankt/fbs062

Flynn KJ, Mitra A, Greenwell HC, Sui J (2013b) Monster potential meets potential monster; the pros and cons of deploying genetically modified microalgae for biofuels production. Interface Focus 3(1):20120037. doi:10.1098/rsfs.2012.0037

Flynn KJ, Clark DR, Mitra A, Fabian H, Hansen PJ, Glibert PM, Wheeler GL, Stoecker DK, Blackford JC, Brownlee C (2015) Ocean acidification with (de)eutrophication will alter future phytoplankton growth and succession. Proc R Soc B 282:20142604. doi:10.1098/rspb.2014.2604

Geider RJ, La Roche J (2002) Redfield revisited: variability of C:N:P in marine microalgae and its biochemical basis. Eur J Phycol 37:1–17

Glibert PM, Fullerton D, Burkholder JM, Cornwell J, Kana TM (2011) Ecological stoichiometry, biogeochemical cycling, invasive species and aquatic food webs: San Francisco Estuary and comparative systems. Rev Fish Sci 19:358–417

Granéli E, Flynn KJ (2006) Chemical And Physical Factors Influencing Toxin Production. In: Granéli E, Turner JT, Ecological Studies 189, (eds) Ecology of Harmful Algae. Springer, Berlin, pp 229–241

Hansen PJ, Bjornsen PK, Hansen BW (1997) Zooplankton grazing and growth: scaling within the 2–2,000-μm body size range. Limnol Oceanogr 42:687–704

John EH, Flynn KJ (2002) Modelling changes in paralytic shellfish toxin content of dinoflagellates in response to nitrogen and phosphorus supply. Mar Ecol Prog Ser 225:147–160

Jones RH, Flynn KJ (2005) Nutritional status and diet composition affect the value of diatoms as copepod prey. Science 307:1457–1459

Kenny P, Flynn KJ (2015) In silico optimization for production of biomass and biofuel feedstocks from microalgae. J Appl Phycol 27:33–48. doi:10.1007/s10811-014-0342-2

Mayers JJ, Flynn KJ, Shields RJ (2014) Influence of the N:P supply ratio on biomass productivity and time-resolved changes in elemental and bulk biochemical composition of *Nannochloropsis* sp. Bioresour Technol 169:588–595. doi:10.1016/j.biortech.2014.07.048

Mitra A (2006) A multi-nutrient model for the description of stoichiometric modulation of predation (SMP) in micro- and mesozooplankton. J Plankton Res 28:597–611

Mitra A, Flynn KJ (2005) Predator–prey interactions: is "ecological stoichiometry" sufficient when good food goes bad? J Plankton Res 27:393–399

Mitra A, Flynn KJ (2006a) Accounting for variation in prey selectivity by zooplankton. Ecol Model 199:82–92

Mitra A, Flynn KJ (2006b) Promotion of harmful algal blooms by zooplankton predatory activity. Biol Lett 2:194–197

Mitra A, Flynn KJ (2007) Importance of interactions between food quality, quantity, and gut transit time on consumer feeding, growth, and trophic dynamics. Am Nat 169:632–646

Mitra A, Flynn KJ (2010) Modelling mixotrophy in harmful algal blooms: More or less the sum of the parts? J Mar Syst 83:158–169. doi:10.1016/j.jmarsys.2010.04.006

Mitra A, Flynn KJ, Fasham MJR (2007) Accounting correctly for grazing dynamics in Nutrient-Phytoplankton-Zooplankton models. Limnol Oceanogr 52:649–661

Mitra A, Castellani C, Gentleman WC, Jónasdóttir SH, Flynn KJ, Bode A, Halsband C, Kuhn P, Licandro P, Agersted MD, Calbet A, Lindeque PK, Koppelmann R, Møller EF, Gislason A, Nielsen TG, John MS (2014a) Bridging the gap between marine biogeochemical and fisheries sciences; configuring the zooplankton link. Prog Oceanogr 129B:176–199. doi:10.1016/j.pocean.2014.04.025

Mitra A, Flynn KJ, Burkholder JM, Berge T, Calbet A, Raven JA, Granéli E, Hansen PJ, Stoecker DK, Thingstad F, Tillmann U, Våge S, Wilken S, Glibert PM, Zubkov MV (2014b) The role of mixotrophic protists in the biological carbon pump. Biogeosciences 11:1–11. doi:10.5194/bg-11-995-2014

Pohnert G, Steinke M, Tollrian R (2007) Chemical cues, defence metabolites and the shaping of pelagic interspecific interactions. Trends Ecol Evol 22:198–204

Polimene L, Mitra A, Sailley SF, Ciavatta S, Widdicombe C, Atkinson A (2015) Decrease in diatom palatability contributes to bloom formation in the Western English Channel. Prog Oceanogr. doi:10.1016/j.pocean.2015.04.026

Redfield AC (1958) The biological control of chemical factors in the environment. Am Sci 46:205–221

Rhee G-Y (1973) A continuous culture study of phosphate uptake, growth rate and polyphosphate in *Scenedesmus* sp. J Phycol 9:495–506

Schaum CE, Collins S (2014) Plasticity predicts evolution in a marine alga. Proc R Soc B 281:20141486

Smith VH, McBride RC (2015) Key ecological challenges in sustainable algal biofuels. J Plankton Res 37:671–682. doi:10.1093/plankt/fbv053

Sterner RW, Elser JJ (2002) Ecological Stoichiometry: the Biology of Elements from Molecules to the Biosphere. Princeton University Press, Princeton, NJ
Sunda WG, Granéli E, Gobler CJ (2006) Positive feedback and the development and persistence of ecosystem disruptive algal blooms. J Phycol 42:963–974
Tirelli V, Mayzaud P (2005) Relationship between functional response and gut transit time in the calanoid copepod *Acartia clausi*: role of food quantity and quality. J Plankton Res 27:557–568
Urabe J, Kyle M, Makino W, Yoshida T, Andersen T, Elser J (2002) Reduced light increases herbivore production due to stoichiometric effects of light/nutrient balance. Ecology 83:619–627
Urabe J, Togari JUN, Elser JJ (2003) Stoichiometric impacts of increased carbon dioxide on a planktonic herbivore. Glob Chang Biol 9:818–825
Watanabe M, Kohata K, Kunugi M (1987) 31P nuclear-magnetic-resonance study of intracellular phosphate pools and polyphosphate metabolism in *Heterosigma akashiwo* (Hada) Hada (Raphidophyceae). J Phycol 23:54–62

Part III
Understanding the Mysteries of Light and Nitrogen

On Saturating Response Curves from the Dual Perspectives of Photosynthesis and Nitrogen Metabolism

Todd M. Kana and Patricia M. Glibert

Introduction

Biological processes often respond in a way that "saturates", or comes to a maximum rate despite having increasingly available resources or substrate. A saturating response separates two important phenomena—a region where the substrate influences the rate, and a region where the process is functionally at or near a maximal rate and substrate has no or little influence. A key idea is that as the substrate increases in concentration or magnitude, its influence on the process ultimately stops. Accordingly, the substrate is *controlling* at low concentrations and *noncontrolling* at high concentrations. This simple, nontrivial, and profound concept is at the heart of nearly all biological and ecological systems models and is commonly the basis for evaluating "limitation" by environmental factors, whether it is at the level of cell physiology or ecosystem response. Rao (2000) considered the saturating response curve so fundamental to biological processes that he termed it "a curve for all reasons."

Classical examples of saturation curves are the photosynthesis–irradiance curve (the PE curve) and nutrient uptake kinetics (the V *vs* S curve), typically parameterized similarly, but not identically, as

$$P = P_{max}\,(1 - e(-\alpha E_o \,/\, P_{max})) \tag{1}$$

$$V = V_{max}\,[S \,/\, (K_s + S)] \tag{2}$$

where P is the photosynthesis rate, P_{max} is the maximal rate of photosynthesis, E_o is the light intensity, α is the initial slope (Jassby and Platt 1976; Smith 1936), V is the

T.M. Kana (✉) • P.M. Glibert
Horn Point Laboratory, University of Maryland Center for Environmental Science, Cambridge, MD 21613, USA
e-mail: kana@umces.edu

P.M. Glibert, T.M. Kana (eds.), *Aquatic Microbial Ecology and Biogeochemistry: A Dual Perspective*, DOI 10.1007/978-3-319-30259-1_8

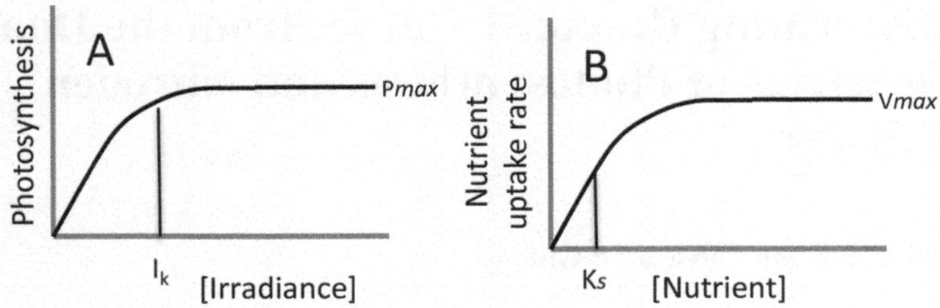

Fig. 1 Examples of saturating hyperbolic relationships and their application to (**a**) the rate of photosynthesis as a function of irradiance; (**b**) the rate of uptake of nutrients as a function of their concentration in the water column

specific uptake rate, V_{max} is the maximum uptake rate, S is the substrate concentration, and K_s is the half-saturation constant for uptake (Menten and Michaelis 1913; Fig. 1). Blackman and Tansley (1905) were the first to describe the saturating response as it relates to photosynthesis. "*…[T]hat photosynthesis in Nature is proportional to light intensity is only true up to a limit set by the amount of CO_2 that can reach the plastids by diffusion.*" Blackman and Tansley recognized that photosynthesis ceded control from irradiance to CO_2 supply. Application of the saturation curve to the concept of limitation is perhaps one of the most well-known concepts in phytoplankton ecology, with the half-saturation constant as the break-point wherein nutrient limitation is considered to be overcome. For photosynthesis, the break-point for light limitation is typically evaluated using the parameter, E_k ($=P_{max}/\alpha$), the light intensity at which photosynthesis begins to saturate and has been used to characterize organisms according to their "sun" vs. "shade" response (e.g. Prézelin 1981).

For organisms, biochemical and physiological processes are generally constrained by a maximum rate of reaction and saturating responses are the basis of most models describing organism responses to the environment. At larger scales, saturation responses have been applied to ecosystem management of nutrient loading responses; there is a region of rapid response where systems are changing and a region of slow response where systems have considerable resilience to change to either increasing nutrient loading or nutrient reductions (Glibert et al. 2010). Across these scales, the processes of photosynthesis and nutrient acquisition are central to understanding environmental limitation on growth and ecology of phytoplankton and this brings together our dual perspectives on the biological regulation of saturating responses and implications for understanding ecological behavior.

Static vs. Dynamic Behavior

In classical phytoplankton physiology, the kinetics of nutrient uptake and growth (analogous to the relationship describing enzyme–substrate kinetics) utilizes a half-saturation "constant" (K_s) calculated from a curve fit, and that parameter is used to assess the degree of limitation of that nutrient for growth. Often, this parameter is assumed to be characteristic or fixed and the literature is replete with comparisons of K_s and V_{max} for different species or different water bodies (e.g. Kudela et al. 2008). The use of "fixed kinetics" continues to persist in spite of the recognition decades ago that there is considerable variation, even within a given organism, in such relationships (e.g. Goldman and Glibert 1983 and references therein). As with nutrient kinetics, photosynthesis kinetics vis-à-vis irradiance effects also exhibit a well-characterized saturation response (we ignore photoinhibition) and curve fit equations are plentiful (Jassby and Platt 1976) and the literature is replete with species comparisons of PE parameters. The importance of those parameters that define saturating curves cannot be overstated as they are the input to mathematical models formulated to describe both cellular and ecological behavior of more complex systems.

Implicit in a saturating response curve is the notion that the cell follows the response as the independent factor changes. Photosynthesis follows a PE curve as irradiance changes. Likewise, N uptake follows an uptake curve as N concentration changes. At one level, the cell must follow the curve—it is an empirical result. At this most simple conceptual level, we can think of a cell's physiology as "running up and down" the saturation curve as substrate availability changes. But it has also been long known that saturating response curves depend on physiological state of the cell and/or the manner in which the experiment is performed (e.g. Harris 1978; Goldman and Glibert 1983 and references therein), so any specific curve has some arbitrariness related to the measurement protocol. This dynamic is nicely exemplified by "rapid light curves" (i.e. PE curves measured in short duration) using variable fluorescence methods (White and Critchley 1999). Rapid light curves are strongly dependent on the investigator-determined duration of irradiance at each step, which affects the physiological state and resulting response curve. Similarly, nitrogen uptake curves have long been known to be dependent on the length of time an experiment is performed (e.g. Wheeler et al. 1982). Therefore, it can be argued that a cell does not "run up and down the curve", but simply responds instantaneously to its environment based on its physiological state, which is dependent on its history.

The challenge is to relate measured saturating curves (implicitly static) to the activity of cells that undergo regulation during or around the process of measurement. A common approach has been to obtain "catalogs" of curves for processes, as was done for many years with PE measurements in efforts to understand species responses to environmental factors. Here we develop an alternative perspective whereby the curves are used to inform us about mechanisms of regulation and from an understanding of those mechanisms, we develop a perspective on how the organism "sees" the world and manifests saturation curves under experimentation.

Gradient Signals and Dynamics of Response Curves

Thirty years ago we conducted experiments with the recently-discovered marine cyanobacterium, *Synechococcus*, with an interest in growth and photosynthesis capability over a broad irradiance range (it had been considered a low-light adapted organism; Kana and Glibert 1987a, b). Using cultures from eight growth irradiances spanning a saturating growth rate curve, we measured PE responses. And, relating carbon uptake rates to four different basis units (cell number, cell carbon, chlorophyll, and phycoerythrin), the data provided us with a suite of 32 unique PE curves for this one species. The *Synechococcus* experiments clearly demonstrated that under steady-state conditions rates of photosynthesis and growth were linked, but that potential short-term photosynthesis at irradiances different from the growth irradiance (i.e. across a PE response) exhibited varied, but highly regulated rates. This finding paralleled earlier observations for N uptake, which clearly demonstrated that short-term rates of N uptake and growth were uncoupled, but were also highly regulated and followed similar patterns (McCarthy and Goldman 1979; Goldman and Glibert 1983).

In both cases, only under conditions of maximal growth rate (μ_{max}) did the maximal rate of photosynthesis (P_{max}) or nutrient uptake (V_{max}) balance the growth demand. At light limitation for growth, there was an excess capacity of photosynthesis that was not utilized under the growth conditions. When comparisons were made of the rates of P_{max} relative to P_i (photosynthesis at the growth irradiance) in relation to the ambient growth rate (μ_i) relative to μ_{max} (i.e. μ_i:μ_{max}, or the relative growth rate), one finds that excess photosynthetic capacity diminishes as relative growth rate approaches 1, and similarly for nitrogen uptake, V_{max} exceeds V_i (ambient uptake rate) under nitrogen limitation and the ratio V_{max}:V_i diminishes toward μ_{max} (McCarthy and Goldman 1979; Goldman and Glibert 1982, 1983 and references therein; Kana and Glibert 1987b; Fig. 2).

The parallels between nutrient- and light-dependent responses to relative growth rate imply that the uptake *capacities* for the major resources (nutrients and photons) regulate to a balance point that satisfies μ_{max}. An alternative way of looking at this is that μ_{max} provides the rate constraints (i.e. slow steps) for regulation of light and nutrient harvesting. An important implication of this is that growth rate per se can be used as a "grand integrator" of metabolism. Growth rate links all processes related to nutrient and energy acquisition, a conclusion that would be drawn a priori from mass balance considerations (Shuter 1979).

Early work on photoacclimation (termed photoadaptation prior to the mid-1980s), which is ubiquitous among plants and algae, was generally in the context of "sun vs. shade" or "high vs. low" irradiance acclimation. In that experimental context, species appeared to sort themselves out in terms of two or more "strategies" depending on patterns of change in α and/or P_{max} (Prézelin 1981; Richardson et al. 1983). However, the *Synechococcus* light gradient study (Kana and Glibert 1987b) demonstrated that all of the strategies previously described existed in one organism when observed over a growth irradiance range that encompassed limiting and saturating irradiances. This implied that there must be a single mechanism for photoacclimation rather than multiple strategies. Subsequently, it was demonstrated that the "light meter" for photoacclimation resided in the electron transport chain and was related to the reduction state of the plastoquinone (PQ) pool (Escoubas et al. 1995; Maxwell et al. 1995) whereby a shift in reduction state

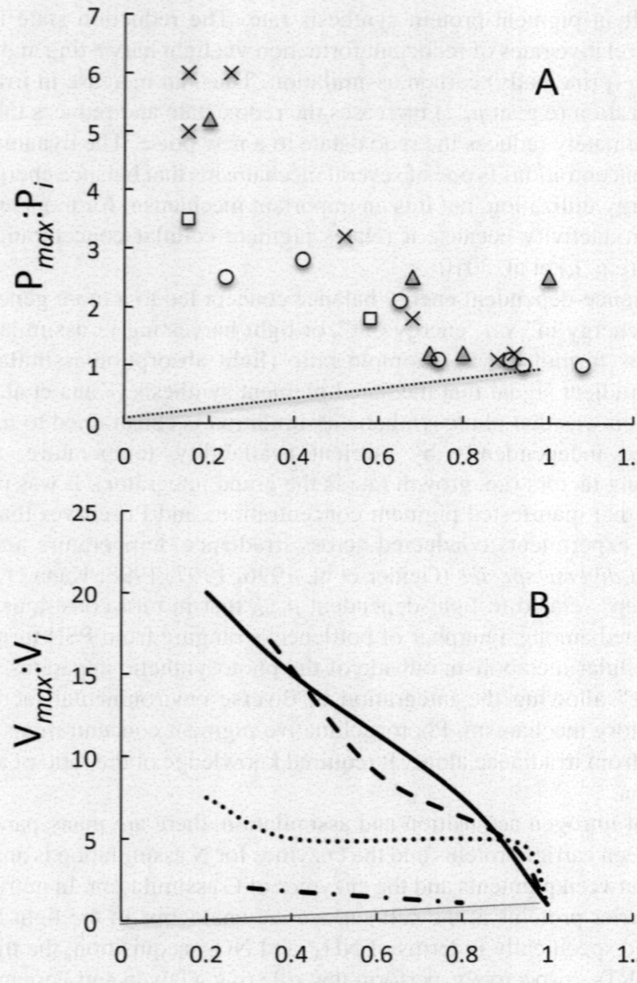

Fig. 2 Comparison of the relationships between $P_{max}:P_i$ (**a**) and $\mu_i:\mu_{max}$ and $V_{max}:V_i$ and $\mu_i:\mu_{max}$ (**b**). In panel (**a**) four species are compared: *circle*, *Synechococcus* WH7803; *triangle*, *Microcystis aeruginosa*; *crosses*, *Phaeodactylum tricornutum*; and *squares*, *Alexandrium tamarensis*. Figure redrawn from, and original data sources given in, Kana and Glibert 1987b. In panel (**b**) four species are compared: *solid line*, *Phaeodactylum tricornutum*; *dashed line*, *Thalassiosira weissflogii*; *dotted line*, *Chaetoceros simplex*; and *dot-dash line*, *Dunaliella teriolecta*. Figure redrawn from, and original data derived from Goldman and Glibert (1982). The fine *dashed lines* in both panels represent the equivalency of the rates

caused a shift in pigment-protein synthesis rate. The reduction state is directly related to the relative rates of reductant formation via light harvesting and reductant utilization via (principally) carbon assimilation. Thus, an increase in irradiance at constant utilization (e.g. at μ_{max}) increases the redox state and reduces the pigmentation that ultimately reduces the redox state to a new poise. The dynamic balance of pigment concentrations is one of several mechanisms that balance energy absorption with energy utilization, but it is an important mechanism for modelers of phytoplankton productivity because it relates pigment cellular concentrations to the environment (e.g. Li et al. 2010).

This irradiance-dependent energy balance concept led to a more general analysis relating "energy in" vs. "energy out", or light harvesting *vs* assimilation. This "balance" was formulated as a simple ratio (light absorption/assimilation) that served as a gradient signal that mediated pigment synthesis (Kana et al. 1997). A key assumption was that photosynthetic assimilation is constrained to μ_{max}, which is constrained independently by nutrient availability, temperature, and other growth-limiting factors (i.e. growth rate is the grand integrator). It was possible to model a cell that manifested pigment concentrations and PE curves that behaved according to experiments conducted across irradiance, temperature and nutrient gradients *and diverse species* (Geider et al. 1996, 1997, 1998; Kana et al. 1997). The "slow step" related to light-dependent μ_{max}, that in turn constrains P_{max}, and could be located among a number of bottlenecks ranging from PSII turnover time to rates of cellular metabolism outside of the photosynthetic apparatus. This provided a "rule" allowing the integration of diverse environmental factors into a single regulatory mechanism. Photoacclimative pigment concentrations could not be predicted from irradiance alone. It required knowledge of the ratio of absorption to assimilation.

In terms of nitrogen acquisition and assimilation, there are many parallels. The balance between carrier proteins and the enzymes for N assimilation is analogous to the balance between pigments and the enzymes of C assimilation. In nutrient acquisition, the carrier proteins at the cell surface are analogous to the light harvesting apparatus, and specifically in terms of NH_4^+ and NO_3^- acquisition, the transporters AMTs and NRTs, respectively, perform that role (e.g. Galván and Fernandez 2001; Rogato et al. 2015).

However, in terms of nutrient acquisition, the parallels with light acquisition are more complicated because not all nutrient substrates follow the same kinetics. In fact, even within the inorganic nitrogen forms, there are key differences in the nutrient uptake response as a function of variable substrate supply (e.g. Glibert et al. 2016). In general, NO_3^- transporters are induced by the presence of their substrate (NO_3^-), whereas NH_4^+ transporters are induced by the absence or deficiency of their substrates, or repressed by increased availability of their substrate, NH_4^+ (Glibert et al. 2016 and references therein). Thus, increasing concentrations of NO_3^- yield more NRTs, whereas increasing concentrations of NH_4^+ yield fewer AMTs. In this regard, the regulation of NH_4^+ acquisition is more similar to light acquisition in that absence leads to up-regulation of the acquisition pathways. Such a phenomenon has been well documented in both culture and field experiments, where N limitation results in uptake rates that far exceed the nutrient that would be required to balance growth (e.g. Conway et al. 1976; McCarthy and Goldman 1979; Glibert and Goldman 1981; Fig. 2). This rapid or "surge uptake" is, in concept, the same as the excess PS capacity relative to balanced growth shown in low light grown cultures (Kana and Glibert 1987b; Fig. 2). Due to the differing nature of the regulation of NO_3^- vs NH_4^+ transporters, vis-à-vis what signals their up-regulation when substrate is limiting, rapid or surge uptake of NH_4^+ is more likely than that of NO_3^-. It, like light harvesting antennae, is "primed and ready" to respond to any increase in substrate availability, whereas the up-regulation of NO_3^- transporters in most cases require time to respond, the so-called "shift-up" response (e.g. Berges et al. 2004).

If the "light meter" for photosynthetic regulation is the energy pressure and state of the PQ pool, what is the "nutrient meter" and how does it sense a state of sufficiency or saturation? All nitrogen forms are ultimately reduced to NH_4^+ before assimilation into amino acids and proteins. Assimilation of NH_4^+, either derived from direct uptake or from reduction of NO_3^- or NO_2^-, occurs via a series of reactions involving (for most algal species) the enzymes glutamine (Gln) synthetase (GS) and glutamate (Glu) synthase (GOGAT; also known as glutamine-2-oxoglutarate amidotransferase). This pathway yields Glu, the product of Gln and oxoglutarate (2-OG) (Scanlan and Post 2008). The availability of Gln and the Gln/Glu ratio govern the NO_3^- reducing capacity in the cell; when Gln levels are low, and when NO_3^- is available, nitrate reductase (NR) is up-regulated. Alternatively, when Gln levels are high, NR activity levels are dialed back (Flynn et al. 1994; Campbell 1999). As the supply of NH_4^+ becomes insufficient to maintain a high internal N-status, indicated by a decline in internal Gln:Glu ratios (Flynn et al 1989, 1994), then the ability to transport and use NO_3^- is up-regulated. AMTs in some species are up-regulated by the depletion of NO_3^-, but the inverse relationship does not appear to be the case; that is they are not down-regulated by the prevalence of NO_3^- (e.g. Hockin et al. 2012; Glibert et al. 2016). Thus, the "nutrient meter" for all forms of N acquisition is the GS-GOGAT state and therefore the ability to assimilate NH_4^+ (directly or from reduction of NO_3^- or NO_2^-) relative to the ability to use that nitrogen downstream. In all, the cells have tuning knobs, the PQ redox state and the relative levels of Gln:Glu, that serve as the signals to up-regulate or down-regulate acquisition to meet the needs of the cells when resources are low, and to switch off acquisition when resources are sufficient.

Overall Perspective on Dynamic Kinetics

The ratio of light absorption to assimilation capacity turned out to be a robust modeling parameter that "self-regulated" cellular pigments under diverse environmental factors (Geider et al. 1996, 1997, 1998; Kana et al. 1997). Energy input was related to irradiance and light harvesting (pigment content), whereas assimilation capacity was regulated by cellular limits determined by factors such as temperature and nutrient limitation. Effectively, these models scaled pigment responses to growth rate as the "grand rate setter" and by doing so, it was possible to coalesce diverse phytoplankton species variation for pigment regulation into a single regulatory structure for multiple environmental effects. This result also supports the notion described above that the cell does not "ride up and down" a PE curve but rather the PE curve is merely a consequence of the nature of light harvesting efficiency and maximum rate constraints at any given time—both of which lead to a saturating response curve. By the same token, a cell does not "ride up and down" a nutrient kinetic curve; the shape of the curve at any given time is a consequence of the nature of the nutrient acquisition machinery and the rate constants of the suite of nitrogen assimilating enzymes at any given point in time.

As is the case with photoacclimation, which is recognized to depend on the dynamic balance of energy flow through the entire photosynthetic apparatus and cell, nutrient assimilation should be recognized to depend on the balance of nutrient acquisition at the cell surface and the maximal rate at which these nutrients can be assimilated within the cell, a balance between surface uptake sites and internal enzymes (Smith et al. 2009). This dynamic balance approach recognizes that even at the level of saturation the cell continues to regulate its nutrient metabolism through processes of internal feedbacks and controls. Such a suite of feedbacks may result in considerable adjustment of nutrient uptake in the region where nutrients are normally considered "saturating". Such adjustments may lead to short-term uptake curves showing continued increase in uptake, leading to biphasic kinetics or even inhibition (Glibert et al. 2016). In fact, kinetic relationships should be viewed as continually varying within the bounds of a response surface and deviations from a single, classically defined kinetic relationship, should be viewed as the norm rather than the exception (Goldman and Glibert 1983; Smith et al. 2009; Glibert et al. 2013, 2016).

One approach that is showing promise in capturing dynamic regulation of nutrient kinetics is that of optimal kinetics (Aksnes and Egge 1991; Smith et al. 2009). This approach recognizes that the ability of the cell to up- or down-regulate nutrient uptake is a function of the potential maximum uptake sites, internal enzymes and rates of assimilation. Instead of a half-saturation constant, this approach calculates an affinity uptake rate:

$$V_{\text{aff}} = [(V_{\text{max}} S) / ((V_{\text{max}} / A) + S] \tag{3}$$

wherein the relationship substitutes the more classic half-saturation constant (K_s) with an affinity constant, V_{max} ratioed to A, the affinity. In such a formulation, both the affinity and V_{max} may vary with cellular physiology. Thus, as with the photosynthetic "regulatory term", light harvesting/assimilation, a ratio provides a more robust measure of the relative abilities of all species to compete for nutrients (Smith et al. 2009). In essence, optimal kinetics assumes that the cells dynamically balance the efficiency of nutrient acquisition at the cell surface and the maximal rate at which these nutrients can be assimilated within the cell, a balance between surface uptake sites and internal enzymes (Smith et al. 2009; Pahlow and Oschlies 2013). Bonachela et al (2012) have also proposed a dynamic formulation of nutrient uptake in which a model cell allows for dynamic regulation of cell transport proteins, leading to flexibility in the maximal uptake rate as well as the limiting portion of the curve. Others (e.g. Klausmeier et al. 2007) have addressed model formulations that allow for flexibility in the uptake of more than one nutrient resource. Such an approach allows for dynamic changes in uptake and in allocation strategies of the different nutrients.

The implications of a dynamically varying, rather than fixed kinetic model are important. On the one hand, nutrient stress can develop before nutrient availability declines below the conventionally defined half saturation value (and bearing in mind how poorly this value is typically known), while on the other hand, regulation of nutrient uptake does not cease when availability of nutrient reaches values defined as "saturating" (Glibert et al. 2013). Thus, application of fixed kinetics to the concept of nutrient limitation fails to recognize the complexity of regulation that occurs across the entire range of substrate availability. Furthermore, regulation of nutrient uptake along this continuum may differ for different nutrients or for different forms of the same nutrient. Importantly not only are there differences in cellular nutrient content between taxa, but within taxa at any given time there are differences in the plasticity or flexibility in nutrient content. Such regulation is fine-tuned and balanced at steady state. However, natural communities are rarely growing at steady state under single nutrient sources and fixed concentrations and are composed typically of numerous taxa. Conceptualizing the relationships between physiological processes and growth as dynamic rather than as fixed kinetic relationships, and understanding how this regulation may differ for different nutrients, has further implications for understanding cell properties and ultimately for ecosystem metabolism.

The brief review above has emphasized that with both light acquisition and CO_2 assimilation as well as with nutrient acquisition and assimilation, there is strong biological regulation between the demand side (getting what is needed when the supply is low) and the assimilation of the resource (the maximal rate always set by biochemical reactions and their constants) for each set of environmental conditions. It is time to incorporate these tuning knobs in models rather than fixed half saturation or photosynthetic constants. Progress is being made. More needs to be done-both experimentally and computationally to move forward from our fixed and invariant notions of kinetics to better representation of the dynamics of biology.

Acknowledgements This is contribution number 5127 from the University of Maryland Center for Environmental Science.

Todd M. Kana and Patricia M. Glibert

Ask each of us when we met and you will get a different answer (clearly one of us made a better impression on the other when we first met!). We do agree that we met in college, but we were attending different colleges. After several years of a long-distance relationship while we pursued our master's degrees and first job opportunities, we both applied to Harvard for our Ph.D.—and remarkably it was the only place that accepted us! We were married later that year, much to the surprise of many faculty and students who thought we had just met. Pat studied N cycling, but Todd was not aquatically oriented and studied plant autecology. Pat finished first and took a postdoc at Woods Hole Oceanographic Institution (WHOI). The writing was on the wall: we were going to be located at a marine lab, but WHOI at that time forbade spouses from both being hired. Nor were there opportunities for Todd's expertise. Todd learned that hanging out with oceanographers could be fun and he gravitated back to his interests in basic photosynthetic processes, but now in the context of the newly discovered marine *Synechococcus*. Pat stayed focused on how and why phytoplankton could cope with vanishingly low nitrogen in the oceans as well as becoming a first time mom (WHOI's first female scientist to give birth). A move to the Horn Point Laboratory in Maryland was welcome when positions for both of them were offered and when HPL was developing a core group of plankton ecologists and recognized the advantages of spousal hiring. Two more children, many challenges, wonderful colleagues, good students, and fun travels have filled our lives over the past 30+ years. Todd now spends more time on instrumentation and applications development and less time on photophysiology. Pat focuses on how and why phytoplankton can cope with all the excess nitrogen that has eutrophified our estuaries and coasts. The questions have changed; our work has evolved, but the excitement of the science has not.

References

Aksnes DL, Egge JK (1991) A theoretical model for nutrient uptake in phytoplankton. Mar Ecol Prog Ser 70:65–72

Berges JA, Gibson CE, Stewart BM (2004) Physiological responses of phytoplankton communities in the Irish Sea to simulated upwelling. Hydrobiologia 517:121–132

Blackman FF, Tansley AG (1905) Ecology in its physiological and phytotopographical aspects. A review. New Phytol 4:232–253

Bonachela JA, Raghib M, Levin SA (2012) Dynamic model of flexible phytoplankton nutrient uptake. Proc Natl Acad Sci 108(51):20633–20638. doi:10.1073/pnas.1118012108

Campbell WH (1999) Nitrate reductase structure, function and regulation: bridging the gap between biochemistry and physiology. Annu Rev Plant Physiol Plant Mol Biol 50:277–303

Conway HL, Harrison PJ, Davis CO (1976) Marine diatoms grown in chemostats under silicate or ammonium limitation. II. Transient response of *Skeletonema costatum* to a single addition of the limiting nutrient. Mar Biol 35:187–199

Escoubas JM, Lomas M, La Roche J, Falkowski PG (1995) Light-intensity regulation of CAB gene transcription is signaled by the redox state of the plastoquinone pool. Proc Natl Acad Sci 92:10237–10241

Flynn KJ, Dickson DMJ, Al-Amoundi OA (1989) The ratio of glutamine:glutamate in microalgae: a biomarker for N-status suitable for use at natural densities. J Plankt Res 11:165–170
Flynn K, Franco JM, Fernández P, Reguera B, Zepata M, Wood G, Flynn KJ (1994) Changes in toxin content, biomass and pigments of the dinoflagellate *Alexandrium minutum* during nitrogen refeeding and growth into nitrogen and phosphorus stress. Mar Ecol Prog Ser 111: 99–109
Galván A, Fernandez E (2001) Eukaryotic nitrate and nitrite transporters. Cell Mol Life Sci 58:225–233
Geider RJ, MacIntyre HL, Kana TM (1996) A dynamic model of photoadaptation in phytoplankton. Limnol Oceanogr 41:1–15
Geider RJ, MacIntyre HL, Kana TM (1997) Dynamic model of phytoplankton growth and acclimation: responses of the balanced growth rate and chlorophyll *a*:carbon ratio to light, nutrient-limitation and temperature. Mar Ecol Prog Ser 148:187–200
Geider RJ, MacIntyre HL, Kana TM (1998) A dynamic regulatory model of phytoplankton acclimation to light, nutrients and temperature. Limnol Oceanogr 43:679–694
Glibert PM, Goldman JC (1981) Rapid ammonium uptake by marine phytoplankton. Mar Biol Lett 2:25–31
Glibert PM, Boyer J, Heil C, Madden C, Sturgis B, Wazniak C (2010) Blooms in lagoons: different from those of river-dominated estuaries. In: Kennish M, Paerl H (eds) Coastal lagoons: critical habitats of environmental change. Taylor and Francis, Boca Raton, FL, pp 91–113
Glibert PM, Kana TM, Brown K (2013) From limitation to excess: consequences of substrate excess and stoichiometry for phytoplankton physiology, trophodynamics and biogeochemistry, and implications for modeling. J Mar Syst 125:14–28
Glibert PM, Wilkerson FP, Dugdale RC, Raven JA, Dupont C, Leavitt PR, Parker AE, Burkholder JM, Kana TM (2016) Pluses and minuses of ammonium and nitrate uptake and assimilation by phytoplankton and implications for productivity and community composition, with emphasis on nitrogen-enriched conditions. Limnol Oceanogr. 61:165–197. doi:10.1002/lno.10203
Goldman JC, Glibert PM (1982) Comparative rapid ammonium uptake by four species of marine phytoplankton. Limnol Oceanogr 27:814–827
Goldman JC, Glibert PM (1983) Kinetics of inorganic nitrogen uptake. In: Carpenter EJ, Capone DG (eds) Nitrogen in the marine environment. Academic, New York, pp 233–274
Harris G (1978) Photosynthesis, productivity and growth: The physiological ecology of phytoplankton. Archiv für Hydrobiol IV. 171
Hockin NL, Mock T, Mulholland F, Kopriva S, Malin G (2012) The response of diatom central carbon metabolism to nitrogen starvation is different from that of green algae and higher plants. Plant Physiol 158:299–312. doi:10.1104/pp.111.184333
Jassby AD, Platt T (1976) Mathematical formulation of the relationship between photosynthesis and light for phytoplankton. Limnol Oceanogr 21:540–547
Kana TM, Glibert PM (1987a) Effect of irradiances up to 2000 μE m^{-2} sec^{-1} on marine *Synechococcus* WH7803: II. Photosynthetic responses and mechanisms. Deep-Sea Res 34: 497–516
Kana TM, Glibert PM (1987b) Effect of irradiances up to 2000 μE m^{-2} sec^{-1} on marine *Synechococcus* WH7803: I. Growth, pigmentation, and cell composition. Deep-Sea Res 34:479–495
Kana TM, Geider RJ, Critchley C (1997) Photosynthetic pigment regulation in microalgae by multiple environmental factors: a dynamic balance hypothesis. New Phytol 137:629–638
Klausmeier CA, Litchman E, Levin SA (2007) A model of flexible uptake of two essential resources. J Theor Biol 246:278–289
Kudela RM, Lane JQ, Cochlan WP (2008) The potential role of anthropogenically derived nitrogen in the growth of harmful algae in California, USA. Harmful Algae 8:103–110
Li QP, Franks PJS, Landry MR, Goericke R, Taylor AG (2010) Modeling phytoplankton growth rates and chlorophyll to carbon ratios in California coastal and pelagic ecosystems. J Geophys Res 115, G04003. doi:10.1029/2009JG001111

Maxwell DP, Laudenbach DA, Huner HPA (1995) Redox regulation of the light harvesting complex II and cab mRNA abundance in *Dunaliella salina*. Physiol Psychol 109:787–795

McCarthy JJ, Goldman JC (1979) Nitrogenous nutrition of marine phytoplankton in nutrient-depleted waters. Science 203:670–672

Menten L, Michaelis MI (1913) Die kinetik der invertinwirkung. Biochem Zeitung 49:333–369

Pahlow M, Oschlies A (2013) Optimal allocation backs Droop's cell-quota model. Mar Ecol Prog Ser 473:1–5

Prézelin BB (1981) Light reactions in photosynthesis. In: Platt T (ed) Physiological bases of phytoplankton ecology. Canadian Bulletin of Fisheries and Aquatic Sciences, Ottawa, pp 1–43, no. 210

Rao TR (2000) A curve for all reasons: the rectangular hyperbola in biology. Resonance 85–90

Richardson K, Beardall J, Raven JA (1983) Adaptation of unicellular algae to irradiance: and analysis of strategies. New Phytol 93:157–191

Rogato A, Amato D, Iudicone M, Chiurazzi MIF, d'Alcalà MR (2015) The diatom molecular toolkit to handle nitrogen uptake. Mar Genomics 24(Pt 1):95–108. doi:10.1016/j.margen.2015.05.018

Scanlan DJ, Post AF (2008) Aspects of marine cyanobacterial nitrogen physiology and connection to the nitrogen cycle. In: Capone DG, Bronk DA, Mulholland MR, Carpenter EJ (eds) Nitrogen in the marine environment. Elsevier, Burlington, MA, pp 1073–1096

Shuter B (1979) A model of physiological adaptation in unicellular algae. J Theor Biol 78:519–552

Smith EL (1936) Photosynthesis in relation to light and carbon dioxide. Proc Natl Acad Sci 22:504–511

Smith S, Yamanaka Y, Pahlow M, Oschlies A (2009) Optimal uptake kinetics: physiological acclimation explains the patterns of nitrate uptake by phytoplankton in the ocean. Mar Ecol Prog Ser 384:1–12

Wheeler PA, Glibert PM, McCarthy JJ (1982) Ammonium uptake and incorporation by Chesapeake Bay phytoplankton: short-term uptake kinetics. Limnol Oceanogr 27:1113–1128

White AJ, Critchley C (1999) Rapid light curves: a new fluorescence method to assess the state of the photosynthetic apparatus. Photosyn Res 59:63–72

Nitrate Reductase: A Nexus of Disciplines, Organisms, and Metabolism

Erica B. Young and John A. Berges

Introduction

Aquatic ecosystems present unique challenges to microbial ecologists because of the difficulty in representing dynamic field conditions in the laboratory, as well as the logistical difficulties of working in the field, especially in ocean and large lake ecosystems. This has motivated development and application of physiological ecology (or ecophysiology), especially exploring use of biochemical indices of cell function and processes to gain insight into an organism's physiology and roles in ecosystem processes. While there is a wide array of biochemical aspects of organisms which could be used (e.g. cell composition, adjustments in cell microenvironments; see Hochachka and Somero 1984), responses in functional proteins are among the most interesting and enzyme activities in particular have proven useful (Newsholme and Crabtree 1986). Enzymes represent the critical mediators of microbial nutrient transformations and thus biogeochemical cycles. In this article, we use the enzyme *nitrate reductase* to illustrate how examination of enzymes can lead to broader understanding of organism and ecosystem functions.

Why Nitrate Reductase?

Nitrogen plays a central role in biogeochemistry on earth, and its multiple oxidation states, multiple reservoirs on the planet and common roles as limiting nutrient and culprit in eutrophication make it one of the most fascinating elements to aquatic biologists (see Mulholland and Lomas 2008). In terms of acquisition of nitrogen by

---

E.B. Young (✉) • J.A. Berges
Department of Biological Sciences and School of Freshwater Sciences, University of Wisconsin-Milwaukee, Milwaukee, WI 53211, USA
e-mail: ebyoung@uwm.edu

P.M. Glibert, T.M. Kana (eds.), *Aquatic Microbial Ecology and Biogeochemistry: A Dual Perspective*, DOI 10.1007/978-3-319-30259-1_9

autotrophs, assimilatory nitrate reductase (for eukaryotes usually classified under Enzyme Commission 1.7.1.1 and 1.7.1.2 and Gene Ontology 0009703 and 0050463, and for cyanobacteria EC 1.7.7.2, GO 0047889; Berges and Mulholland 2008) functions at a central point in algal nitrogen metabolism (Mulholland and Lomas 2008) and integrates responses to several environmental variables, including nitrogen and light availability, as well as internal carbon metabolism (Young and Beardall 2005; Young et al. 2007a) (see Fig. 1). Therefore, it is unsurprising that the nitrate reductase (NR) enzyme has received considerable attention in seminal studies ranging from laboratory algal physiology (e.g. Morris and Syrett 1963) through to oceanographic field work (e.g. Eppley et al. 1969), as well as applied sciences (e.g. Ntoko and Senwo 2012).

Many aspects of NR in aquatic organisms have been reviewed relatively recently (Berges and Mulholland 2008; Mulholland and Lomas 2008), thus our goal here is to selectively review established ideas and the most critical recent advances in our emerging understanding of the enzyme, and identify some of the more pressing needs for future advances.

In terms of a basic overview, NR enzymes are widespread among autotrophs (from phytoplankton to macroalgae, seagrasses and their epiphytes) and, although NR forms can differ in properties, their activities can generally be assayed in a straightforward manner and can often be quantitatively related to rates of nitrate incorporation (e.g. Berges and Harrison 1995). NR activity can be regulated by a variety of transcriptional and post-translational mechanisms (see Berges 1997), but in at least some algae, synthesis and degradation of the enzyme is a major factor on a timescale of hours to days (e.g. Vergara et al. 1998; Young et al. 2009). With some exceptions in specific taxa, NR activity responds to irradiance (e.g. Berges and Harrison 1995; Young et al. 2007a), nitrate supply (e.g. Young et al. 2009), presence of ammonium (e.g. Berges et al. 1995; Vergara et al. 1998; Young et al. 2007a, 2009), and temperature (e.g. Berges et al. 2002). While some phytoplankton taxa like diatoms are capable of storing considerable nitrate, it is the multicellular algae that are true specialists, developing substantial internal pools, to some extent uncoupling inorganic nitrogen assimilation and incorporation from ambient nitrogen availability (e.g. Young et al. 2007b). Macroalgae also show marked seasonality in NR activity (e.g. Young et al. 2007b) and there is evidence that NR is regulated differently depending on position in the intertidal zone (e.g. Young et al. 2007a).

Understanding That Has Emerged from Recent NR Measurements

Measuring NR can provide information about more than just nitrogen assimilation. Although unassimilated nitrate can be stored by cells, assimilation, including reduction by NR, requires committing energy and thus is subject to cellular regulation and integration with other metabolic processes. It has been noted previously that, in metabolic pathways, regardless of which step is limiting a particular moment, all enzymes in the

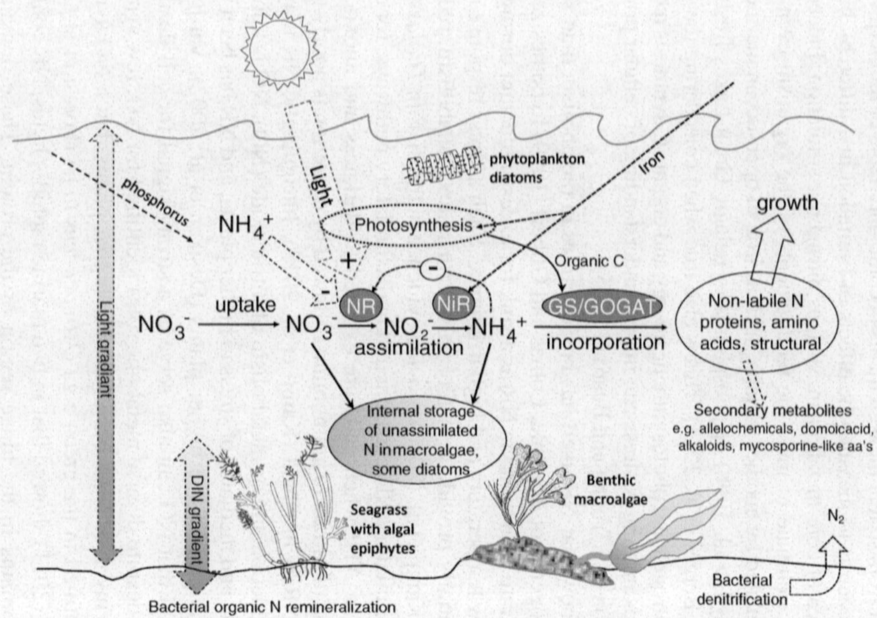

Fig. 1 Conceptual diagram illustrating the role of nitrate reductase (NR) in the context of nitrate uptake, assimilation and incorporation into biomass by photoautotrophs. Organisms covered in this review are shownalong with gradients in light and inorganic nitrogen availability and other relevant nutrient resources. Upregulation (+) or down-regulation (-) of NR activity is shown in relation to light and photosynthesis, availability ofmore reduced N forms (NH_4^+). Abbreviations: DIN – dissolved inorganic nitrogen, NR – nitrate reductase, NiR – nitrite reductase, GS – glutamine synthetase, GOGAT – glutamate synthase

pathway acclimate to the prevailing flux through the pathways (see Newsholme and Crabtree 1986). This appears to apply at the level of organisms as well; what advantage would an organism chronically limited for one element have in maintaining the capacity to assimilate non-limiting elements at rates greatly out of proportion to its composition? Such an effect can be exploited. For example, when examining cells or populations limited by low iron or phosphorus, it is tempting to focus directly on aspects of iron or phosphorus availability, uptake or metabolism. But our ability to measure iron or phosphorus at the relevant physiologically limiting concentrations is technically difficult and subject to large errors so that measurements become unreliable or uninterpretable. An alternative approach is to measure the effects of limitation in a different, but still strongly affected area of core metabolism, such as nitrogen assimilation. Indeed, NR activity appears to acclimate to meet the requirements of the organism even when growth is being regulated by another variable such as iron in open-ocean marine phytoplankton (e.g. Boyd et al. 1998) or phosphorus in pelagic Great Lakes freshwater phytoplankton (e.g. Berges et al. 2013). This is also true when considering the strong coupling between nitrogen and photosynthetic energy and organic C synthesis required for N assimilation; examining N stress can inform understanding of cellular priorities for carbon acquisition (e.g. Young and Beardall 2005).

Recent results show that NR itself has broader roles in metabolism than simply assimilation of nitrogen. For example, Lomas and Glibert (1999) hypothesized that diatoms can use nitrate reduction via NR as a sink for excess reductant during periods of growth at low temperature and high irradiance. Such a stress response mechanism would essentially provide a "safety valve" for redox metabolism, reducing oxidative damage. Kamp et al. (2013) provide evidence in the diatom *Thalassiosira weissflogii*, that assimilatory NR may function in dissimilatory pathways, for example, to help generate ATP when cells are exposed to darkness and anoxia. This short-term reduction of nitrate to ammonium followed by excretion is poorly understood, but clearly NR is involved (Kamp et al. 2013). Intriguingly, NR may also function in the production and/or metabolism of nitric oxide (NO). NO is known as a signaling molecule and mediator of stress in higher plants, and NR has been implicated in NO generation in some higher plants (Desikan et al. 2002). Vardi et al. (2008) demonstrated that NO can also serve as a strong signal for cell death in a diatom. Curiously, despite diverse metabolic roles identified for NO, few synthesis pathways or genes involved have been found in higher plants, but a NO synthase gene has been identified in the green alga *Ostreococcus tauri* (Foresi et al. 2015). Stewart and Coyne (2011) showed that in two marine raphidophytes, NR contained 2/2 hemoglobin domains in the hinge region of the protein. These domains are homologous to those in higher plants shown to interact with NO. Coupled with experimental evidence, this suggests that NR may play a role in converting NO to nitrate either as an assimilation or detoxification mechanism (Stewart and Coyne 2011). In *Chlamydomonas*, N uptake and assimilation including NR activity can be inhibited by NO (Sanz-Luque et al. 2013), NR is required for nitrite-dependent NO biosynthesis (Sakihama et al. 2002), and in the red macroalga *Gracilaria chilensis*, additions of NO-generating or scavenging molecules were found to modulate NR activity (Chow et al. 2013), suggesting strong relationships between NO and NR.

NR activity assays have provided insight into physiological function, but NR has also proven to be an important marker of organism diversity. Allen et al. (2005) noted considerable NR gene sequence diversity among marine phytoplankton, findings supported by more detailed phylogenetic analysis showing distinct diatom, rhodophyte, prasinophyte, and chlorophyte clades based on NR genes (Ghoshroy and Robertson 2015). Measurements of this diversity are also clearly possible in natural communities; Ward (2008) demonstrated the potential of functional microarrays based on *rbc*L and NR oligonucleotide probes to examine both phytoplankton composition and gene expression at relatively high resolution and throughput. At a much finer scale, Paerl et al. (2012) were able to exploit differences in assimilatory NR gene (*nar*B) sequences in marine *Synechococcus* to show seasonal dynamics in distinct clades. However, so far, the use of gene-based methods in natural ecosystems remains constrained somewhat by the restricted representation of algal genes (aside from rRNA and commonly used marker genes such as *rbc*L) in genetic databases, which limits taxonomic resolution for application to NR (Ward 2008).

In trying to understand NR and metabolism, divisions between "lab" and "field" biology or "ecologically relevant" organisms and "lab weeds" lead us astray. While compromises in sampling regime and experimental design often need to be made for logistic and practical purposes, there is great value in "intermediates" between lab and field studies. For example, in work on NR in Irish intertidal *Laminaria* and *Fucus*, large flow-through tanks of natural lough water were used for incubation of whole thalli, allowing for easier round-the-clock sampling access, but with ambient temperature, light and water nutrient conditions perturbed as little as possible from the field (Young et al. 2007a, 2009). This was complemented with in situ sampling on the seashore, including in the intertidal zone at midnight (Young et al. 2007a). In work in pelagic ecosystems, where vessel movement and water column mixing are issues, experiments involving NR measurements in contained whole-water samples (and sometimes ship-board incubators), though imperfect, provide valuable data that complement both culture experiments and in situ measurements (e.g. Berges et al. 1995; Boyd et al. 1998; Berges et al. 2013). Larger-scale mesocosms have been used in work on NR in sea grass systems (Burkholder et al. 1992).

Research on NR in laboratory model (often "weedy") organisms has attracted the criticism of doubtful "ecological relevance". While laboratory model organisms might not dominate in nature, they may still provide good experimental background data for generating hypotheses to test in the field (Ward 2008; Alexander et al. 2015). Moreover, we often lack good criteria for determining ecological relevance of taxa. For example, the question of why a species such as *Phaeodactylum tricornutum,* that rarely occurs as a dominant in nature, is capable of taking over a culture from just about any other species in the laboratory is surely relevant if the goal is to understand the mechanisms of an ecological process like competition. Furthermore, Ward (2008) used NR gene probes and regularly detected *P. tricornutum* in environments where it was seldom found using more conventional methods. We have also learned much from the *P. tricornutum* genome and many of the genes and their organization appear "typical" of diatoms and even larger taxonomic groups (Bowler et al. 2008).

Recent Advances and Emerging Challenges

Symbioses, communities, and communication. The idea of organisms functioning within communities rather than as individuals complicates understanding of physiology but, like the human microbiome "movement", this also acknowledges realistic and probably critical metabolic interactions. For example, bulk community measurements of NR activity make it difficult to partition N assimilation functions within complex communities, but genetic analysis can help (Alexander et al. 2015). In bacteria, similar concepts around organismal interactions have been examined for some time; for example the process of quorum sensing, whereby cell–cell communication is mediated by small molecules such as *N*-acyl homoserine lactone, allows populations to coordinate processes including biofilm formation and toxin production (Dobretsov et al. 2009).

Only relatively recently has the ability of eukaryotic organisms to sense, exploit, and even disrupt such prokaryotic signaling for their own benefit become clear (e.g. Joint et al. 2002). The degree to which marine eukaryotes communicate within populations or across the eukaryote–prokaryote boundary is poorly understood, but there are significant physiological and ecological implications (Joint et al. 2002). As one illustration of tight interactions between eukaryotes and prokaryotes, diatoms can have both or either extracellular and intracellular symbiotic cyanobacteria. The cyanobacteria, *Richelia*, is an N_2-fixing *endo*symbiont of diatoms, showing genome reduction including loss of NR genes, relinquishing these nitrogen assimilation functions to the host algal cell, but the closely related cyanobacterial genus *Calothrix* is an diatom *ecto*symbiont but shows no similar genomic reduction (Hilton et al. 2013). In tight associations between diatoms and heterotrophic bacteria, cells may exchange ammonium and amino acids, as well as N-containing signaling molecules (Amin et al. 2015), so NR regulation in both cells may be influenced by this association.

Similarly, cell–cell biochemical interactions within algal-epiphyte communities and possibilities of extracellular regulation was an idea (Joint et al. 2002) that stimulated our work on the potential of microbial populations and communities to cycle nutrients sequentially (Lee et al. 2015). Using molecular approaches incorporating community taxonomic analysis and both metagenome and metatranscriptome measurements, we are now exploring nitrogen assimilation and transformations, including diversity and expression of NR genes, to examine metabolic partitioning and collaborative metabolism within complex natural algal-bacterial communities (Zulkifly et al. 2012; Young 2015).

In a broader conceptualization of associations between organisms, Lima-Mendez et al. (2015) examined oceanic plankton samples using metagenomics and targeted amplifications to develop interactive networks of symbiotic and parasitic associations between prokaryotes, eukaryotic cells, metazoans, and viruses across geographical spatial scales. Developing such "interactome" networks is facilitated by large research directives such as the Tara project (Lima-Mendez et al. 2015), and provide exciting opportunities to examine how organismal interactions are constrained and promoted by environmental variables. Their finding that nitrite concentration was one of the most important global abiotic factors influencing network patterns (Lima-Mendez et al. 2015), suggests an important central role for planktonic NR in mediating or responding to environmental conditions.

After decades of reducing from organism to gene, we are now reassembling. Earlier work which focused on the NR protein and enzyme activity, often with an emphasis on cells cultured under controlled laboratory conditions, has provided a foundation for understanding NR function. But emerging technology and techniques in genomics and community metagenomes and metatranscriptomes are now providing new and powerful perspectives and opportunities. The combination of transcriptomic and proteomics has proven powerful in addressing questions in individual species. For example, Hockin et al. (2012) were able to provide evidence for post-transcriptional regulation of NR in *Thalassiosira pseudonana*, and show that, unlike the case in many higher plants, nitrogen starvation caused an increase in glycolytic and tricarboxylic acid pathways, perhaps to provide C-skeletons in preparation for anticipated N assimilation, or more carbon for fatty acid biosynthesis to support increases in lipid content. Applying this understanding, Levitan et al. (2015) showed that knocking down NR expression 40–50% in *Phaeodactylum tricornutum* produced a comparable increase in lipid content, a finding with important implications for biofuels production.

Although some gene regulatory regions have been identified in algal NR genes (e.g. Cannons and Cannon 2002), a remaining critical gap in integrating knowledge of NR is a real lack of understanding of general transcription factors in algal genomes, and in particular for NR gene regulation. A recent review of higher plant NR transcription factors emphasizes the importance of nodule-inception-like proteins (NLPs) and nitrate-responsive elements (NREs) (Yanagisawa 2014); there is almost nothing known and little comparable work about such proteins in aquatic organisms.

The potential contribution of emerging genome, transcriptome, and proteome information to our understanding of organismal and community ecophysiology depends on an understanding of what organisms do (functional physiology), and therefore how they might interact within communities, and the potential constraints of nutrients and other resources in different ecosystems. A physiological understanding of how NR is regulated in response to light and external and internal nitrate and ammonium availability (e.g. Dortch 1990; Vergara et al. 1998; Young et al. 2007a, b, 2009) is an important background for designing experiments utilizing powerful gene-based approaches. And in mixed communities, an understanding of how NR differs and is regulated across key groups of organisms (e.g. diatoms, chlorophytes; Ghoshroy and Robertson 2015) can help us decide where to focus attention in metagenomes and metatranscriptomes (Alexander et al. 2015).

Genomic and transcriptomic data are also being complemented by proteomic approaches, providing new possibilities for more complex integration of genetic regulation, environmental responses, and functional diversity. Information on gene function is now facilitating examination of not only individual taxa but also population level changes in proteins responding to environmental dynamics. For example, Saito et al. (2014) recently used proteomics to track the nitrogen response proteins NtcA and P-II and other peptides in *Prochlorococcus* populations in relation to nutrient availability across Pacific Ocean gyres. This approach offers unprecedented insights into specific nutrient stress responses of cells across broad spatial scales and thus environmental gradients. Our understanding of the central regulatory role of NR in nitrogen assimilation makes it an excellent target for similar proteomics investigations. Expanding NR gene information for diverse taxa (e.g. Allen et al. 2005; Ward 2008; Ghoshroy and Robertson 2015) will provide tools to increase resolution of NR protein functional analysis in natural communities.

The next critical step will be to integrate genomic, transcriptomic and proteomic data, as well as biochemical measurements. This so-called "metabolomics" (which is surely a synonym for "physiology") represents a return to requirements for more organism-based, functional understanding. In the same way that early genetic data came from a very few species and clones, and forced us to recognize the difficulties in generalizing to broader taxonomic groups, we foresee a new approach requiring experts in different methodologies to contribute to a more holistic picture of organismal function. While traditional physiologists often study single "model" organisms, aquatic ecophysiologists have rarely been able to afford such restrictions because they must be able to generalize across (functional) groups of taxa, and so they may have a key role to play in helping to manage this transition.

Practical applications. There are also much more practical applications of knowledge related to NR. For example, development of the NR system as a bioreporter involves genetically modifying an organism to show an observable response when NR transcription is induced, allowing it to be used as a sensor. Bioreporters have been used with some success, but these require knowledge about specific gene regulation, which we have for NR in only a limited number of algal taxa. However, it also takes considerable work to transform the reporter organism and achieve stable and reliable expression, e.g. diatoms transformed with NR regulatory regions and GFP reporter (Poulsen and Kröger 2005), and *Volvox carteri* transformed with NR gene promoter driving a luciferase reporter (von der Heyde et al. 2015). Fortunately, recent advances have been made in transformation of groups such as diatoms (Karas et al. 2015). Furthermore, reporter organisms are often chosen because they can be readily genetically transformed, and not for their ecological suitability, so choices need to be made carefully to provide useful information about how natural communities respond to nutrient stress (Bullerjahn et al. 2010). An application of an NR-based bioreporter could be to indicate changes in presence of nitrate and ammonium in biotechnology culturing or wastewater environments where switches between ammonium and nitrate may indicate changing balance of bacterial nitrogen transformation processes.

Another application involves use of NR to mitigate excess nitrate in soils or waters (e.g. Ntoko and Senwo 2012), a goal which has been pursued for decades commercially (e.g. the Nitrate Elimination Company, www.nitrate.com/about_us). Such efforts have largely used bacterial and higher plant enzymes and have mostly resulted in products like test kits for nitrate rather than actual remediation protocols. But interest continues and NR within or from aquatic organisms, coupled with nitrite treatment options, would be a logical target for addressing excess nitrate in aquatic ecosystems.

Conclusion

Much of our collaborative research has taken place at a time when we perceive that organismal biology and ecophysiology has been under-valued and under-represented in strategic research. Nonetheless, there have also been very exciting developments both in technology and in understanding. With the rise of "metabolomics" we can even anticipate a new "golden age" for integrated physiological exploration and understanding.

Erica B. Young and John A. Berges

Erica and John met at a European Phycological Congress in Germany. Seeing her conversing in Swedish, John concluded Erica was a Swede and later complimented her on her excellent, though Australian-accented, English! Returning to different hemispheres, they corresponded about science and life for 4 years before finally meeting again on the shores of Strangford Lough, Northern Ireland. A period of kelp nitrogen metabolism, Irish music, amateur theater, and Irish pubs followed. In 2002 Erica was offered a faculty position (and John a spousal accommodation) at the University of Wisconsin-Milwaukee on the shores of Lake Michigan, where they started to explore freshwaters. The unexpectedly early arrival of their son Theo in 2003, with a rare genetic syndrome (www.m-cm.net/), changed their lives dramatically. Erica and John found their personal and professional partnership challenged by the demands of understanding and advocating for medical and support services. Against a backdrop of several years of little sleep, patchy child care, and regular hospital visits, they continued to balance research programs and teaching to meet tenure requirements. But both the family and professional successes have been all the sweeter for the struggles. One family contribution to science not noted on their CVs is a medical breakthrough clarifying the genetic basis of Theo's syndrome in which the family's genes featured anonymously (Rivière et al. 2012).

References

Alexander H, Jenkins BD, Rynearson TA, Dyhrman ST (2015) Metatranscriptome analyses indicate resource partitioning between diatoms in the field. Proc Natl Acad Sci U S A 112:E2182–E2190

Allen AE, Ward BB, Song B (2005) Characterization of diatom (Bacillariophyceae) nitrate reductase genes and their detection in marine phytoplankton communities. J Phycol 41:95–104

Amin SA, Hmelo LR, van Tol HM, Durham BP, Carlson LT, Heal KR, Morales RL, Berthiaume CT, Parker MS, Djunaedi B, Ingalls AE, Parsek MR, Moran MA, Armbrust EV (2015) Interaction and signalling between a cosmopolitan phytoplankton and associated bacteria. Nature 522:98–101

Berges JA (1997) Algal nitrate reductases. Eur J Phycol 32:3–8

Berges JA, Harrison PJ (1995) Nitrate reductase activity quantitatively predicts the rate of nitrate incorporation under steady-state light limitation: a revised assay and characterization of the enzyme in three species of marine phytoplankton. Limnol Oceanogr 40:82–93

Berges JA, Cochlan WP, Harrison PJ (1995) Laboratory and field responses of algal nitrate reductase to diel periodicity in irradiance, nitrate exhaustion, and the presence of ammonium. Mar Ecol Prog Ser 124:259–269

Berges JA, Mulholland MR (2008) Enzymes and nitrogen cycling. In: Capone DG, Bronk DA, Mulholland MR, Carpenter EJ (eds) Nitrogen in the marine environment. Elsevier, New York, pp 1385–1444

Berges JA, Varela DE, Harrison PJ (2002) Effects of temperature on growth rate, composition and nitrogen metabolism in the marine diatom *Thalassiosira pseudonana* (Bacillariophyceae). Mar Ecol Prog Ser 225:139–146

Berges JA, Jiang Y, Sterner RW, Bullerjahn GS, Ivanikova NA, McKay RML (2013) Identification of factors constraining nitrate assimilation in Lake Superior, Laurentian Great Lakes. Hydrobiologia 731:81–94

Bowler C, Allen AE, Badger JH, Grimwood J, Jabbari K, Kuo A, Maheswari U, Martens C, Maumus F, Otillar RP, Rayko E, Salamov A, Vandepoele K, Beszteri B, Gruber A, Heijde M, Katinka M, Mock T, Valentin K, Verret F, Berges JA, Brownlee C, Cadoret J-P, Chiovitti A, Choi CJ, Coesel S, De Martino A, Detter JC, Durkin C, Falciatore A, Fournet J, Haruta M, Huysman MJJ, Jenkins BD, Jiroutova K, Jorgensen RE, Joubert Y, Kaplan A, Kroger N, Kroth PG, La Roche J, Lindquist E, Lommer M, Martin-Jezequel V, Lopez PJ, Lucas S, Mangogna M, McGinnis K, Medlin LK, Montsant A, Oudot-Le Secq M-P, Napoli C, Obornik M, Parker MS, Petit J-L, Porcel BM, Poulsen N, Robison M, Rychlewski L, Rynearson ATA, Schmutz J, Shapiro H, Siaut M, Stanley M, Sussman MR, Taylor AR, Vardi A, von Dassow P, Vyverman W, Willis A, Wyrwicz LS, Rokhsar DS, Weissenbach J, Armbrust EV, Green BR, Van de Peer Y, Grigoriev IV (2008) The *Phaeodactylum* genome reveals the evolutionary history of diatom genomes. Nature 456:239–244

Boyd PW, Berges JA, Harrison PJ (1998) In vitro iron enrichment experiments at iron-rich and -poor sites in the NE subarctic Pacific. J Exp Mar Biol Ecol 227:133–151

Bullerjahn GS, Boyanpalli R, Rozmarynowycz MJ, McKay RML (2010) Cyanobacterial bioreporters as sensors of nutrient availability. Adv Biochem Eng Biotechnol 118:165–188

Burkholder JM, Mason KM, Glasgow HB (1992) Water-column nitrate enrichment promotes decline of eelgrass *Zostera marina*—evidence from seasonal mesocosm experiments. Mar Ecol Prog Ser 81:163–178

Cannons AC, Cannon J (2002) The stability of the *Chlorella* nitrate reductase mRNA is determined by the secondary structure of the 5′-UTR: implications for post-transcriptional regulation of nitrate reductase. Planta 214:488–491

Chow F, Pedersen M, Oliveira MC (2013) Modulation of nitrate reductase activity by photosynthetic electron transport chain and nitric oxide balance in the red macroalga *Gracilaria chilensis* (Gracilariales, Rhodophyta). J Appl Phycol 25(6):1847–1853

Desikan R, Griffiths R, Hancock J, Neill S (2002) A new role for an old enzyme: nitrate reductase-mediated nitric oxide generation is required for abscisic acid-induced stomatal closure in *Arabidopsis thaliana*. Proc Natl Acad Sci U S A 99(25):16314–16318

Dobretsov S, Teplitski M, Paul V (2009) Mini-review: quorum sensing in the marine environment and its relationship to biofouling. Biofouling 25:413–427

Dortch Q (1990) The interaction between ammonium and nitrate uptake in phytoplankton. Mar Ecol Prog Ser 61:183–201

Eppley RW, Coatsworth JL, Solorzano L (1969) Studies of nitrate reductase in marine phytoplankton. Limnol Oceanogr 14:194–205

Foresi N, Mayta ML, Lodeyro AF, Scuffi D, Correa-Aragunde N, Garcia-Mata C, Casalongue C, Carrillo N, Lamattina L (2015) Expression of the tetrahydrofolate-dependent nitric oxide synthase from the green alga *Ostreococcus tauri* increases tolerance to abiotic stresses and influences stomatal development in Arabidopsis. Plant J 82(5):806–821

Ghoshroy S, Robertson DL (2015) Molecular evolution of nitrogen assimilatory enzymes in marine prasinophytes. J Mol Evol 80:65–80

Hilton JA, Foster RA, Tripp HJ, Carter BJ, Zehr JP, Villareal TA (2013) Genomic deletions disrupt nitrogen metabolism pathways of a cyanobacterial diatom symbiont. Nat Commun 4:1767

Hochachka PW, Somero GN (1984) Biochemical adaptation. Princeton University Press, Princeton

Hockin NL, Mock T, Mulholland F, Kopriva S, Malin G (2012) The response of diatom central carbon metabolism to nitrogen starvation is different from that of green algae and higher plants. Plant Physiol 158:299–312

Joint I, Tait K, Callow ME, Callow JA, Milton D, Williams P, Camara M (2002) Cell-to-cell communication across the prokaryote-eukaryote boundary. Science 298:1207

Kamp A, Stief P, Knappe J, de Beer D (2013) Response of the ubiquitous pelagic diatom *Thalassiosira weissflogii* to darkness and anoxia. PLoS One 8(12), e82605

Karas BJ, Diner RE, Lefebvre SC, McQuaid J, Phillips APR, Noddings CM, Brunson JK, Valas RE, Deerinck TJ, Jablanovic J, Gillard JTF, Beeri K, Ellisman MH, Glass JI, Hutchison CA III, Smith HO, Venter JC, Allen AE, Dupont CL, Weyman PD (2015) Designer diatom episomes delivered by bacterial conjugation. Nat Commun 6:6925. doi:10.1038/ncomms7925

Lee PO, McLellan SL, Graham LK, Young EB (2015) Effects of Dreissenid mussel invasion on benthic bacterial communities in nearshore Lake Michigan. FEMS Microbiol Ecol 91:1–12

Levitan O, Dinamarca J, Zelzion E, Lun DS, Guerra LT, Kim MK, Kim J, Van Mooy BAS, Bhattacharya D, Falkowski PG (2015) Remodeling of intermediate metabolism in the diatom *Phaeodactylum tricornutum* under nitrogen stress. Proc Natl Acad Sci USA 112:412–417

Lima-Mendez G, Faust K, Henry N, Decelle J, Colin S, Carcillo F, Chaffron S, Ignacio-Espinosa JC, Roux S, Vincent F, Bittner L, Darzi Y, Wang J, Audic S, Berline L, Bontempi G, Cabello AM, Coppola L, Cornejo-Castillo FM, d'Ovidio F, De Meester L, Ferrera I, Garet-Delmas M-J, Guidi L, Lara E, Pesant S, Royo-Llonch M, Salazar G, Sánchez P, Sebastian M, Souffreau C, Dimier C, Picheral M, Searson S, Kandels-Lewis S, Gorsky G, Not F, Ogata H, Speich S, Stemmann L, Weissenbach J, Wincker P, Acinas SG, Sunagawa S, Bork P, Sullivan MB, Karsenti E, Bowler C, de Vargas C, Raes J (2015) Determinants of community structure in the global plankton interactome. Science 348(6237):1262073. doi:10.1126/science.1262073

Lomas MW, Glibert PM (1999) Temperature regulation of nitrate uptake: a novel hypothesis about nitrate uptake and reduction in cool-water diatoms. Limnol Oceanogr 44:556–572

Morris I, Syrett PJ (1963) The development of nitrate reductase in *Chlorella* and its repression by ammonium. Arch Mikrobiol 47:32–41

Mulholland MR, Lomas MW (2008) Nitrogen uptake and assimilation. In: Capone DG, Bronk DA, Mulholland MR, Carpenter EJ (eds) Nitrogen in the marine environment. Elsevier, San Diego, pp 303–384

Newsholme EA, Crabtree B (1986) Maximum catalytic activity of some key enzymes in provision of physiologically useful information about metabolic fluxes. J Exp Zool 239:159–167

Ntoko FA, Senwo ZN (2012) Nitrate reduction by commercially available nitrate reductases: Biocatalytic potentials and enzymatic activities in the presence of metals ions. J Environ Sci Health A Tox Hazard Subst Environ Eng 47:2028–2034

Paerl RW, Turk KA, Beinart RA, Chavez FP, Zehr JP (2012) Seasonal change in the abundance of *Synechococcus* and multiple distinct phylotypes in Monterey Bay determined by rbcL and narB quantitative PCR. Environ Microbiol 14:580–593

Poulsen N, Kröger N (2005) A new molecular tool for transgenic diatoms: control of mRNA and protein biosynthesis by an inducible promoter-terminator cassette. FEBS J 272:3413–3423

Rivière J-B, Mirzaa GM, O'Roak BJ, Beddaoui M, Alcantara D, Conway RL, St-Onge J, Schwartzentruber JA, Gripp KW, Nikkel SM, Worthylake T, Sullivan CT, Ward TR, Butler HE, Kramer NA, Albrecht B, Armour CM, Armstrong L, Caluseriu O, Cytrynbaum C, Drolet BA, Innes AM, Lauzon JL, Lin AE, Mancini GMS, Meschino WS, Reggin JD, Saggar AK, Lerman-Sagie T, Uyanik G, Weksberg R, Zirn B, Beaulieu CL, Majewski J, Bulman DE, O'Driscoll M, Shendure J, Graham JM, Boycott KM, Dobyns WB (2012) De novo germline and postzygotic mutations in AKT3, PIK3R2 and PIK3CA cause a spectrum of related megalencephaly syndromes. Nat Genet 44:934–940

Saito MA, McIlvin MR, Moran DM, Goepfert TJ, DiTullio GR, Post AF, Lamborg CH (2014) Multiple nutrient stresses at intersecting Pacific Ocean biomes detected by protein markers. Science 345:1173–1177

Sakihama Y, Nakamura S, Yamasaki H (2002) Nitric oxide production mediated by nitrate reductase in the green alga *Chlamydomonas reinhardtii*: an alternative NO production pathway in photosynthetic organisms. Plant Cell Physiol 43:290–297

Sanz-Luque E, Ocaña-Calahorro F, Llamas A, Galvan A, Fernandez E (2013) Nitric oxide controls nitrate and ammonium assimilation in *Chlamydomonas reinhardtii*. J Exp Bot 64:3373–3383

Stewart JJ, Coyne KJ (2011) Analysis of raphidophyte assimilatory nitrate reductase reveals unique domain architecture incorporating a 2/2 hemoglobin. Plant Mol Biol 77:565–575

Vardi A, Bidle KD, Kwityn C, Hirsh DJ, Thompson SM, Callow JA, Falkowski P, Bowler C (2008) A diatom gene regulating nitric-oxide signaling and susceptibility to diatom-derived aldehydes. Curr Biol 18:895–899

Vergara JJ, Berges JA, Falkowski PG (1998) Diel periodicity of nitrate reductase activity and protein levels in the marine diatom *Thalassiosira weissflogii* (Bacillariophyceae). J Phycol 34:952–961

von der Heyde EL, Klein B, Abram L, Hallmann A (2015) The inducible nitA promoter provides a powerful molecular switch for transgene expression in *Volvox carteri*. BMC Biotechnol 15:5. doi:10.1186/s12896-015-0122-3

Ward BB (2008) Phytoplankton community composition and gene expression of functional genes involved in carbon and nitrogen assimilation. J Phycol 44:1490–1503

Yanagisawa S (2014) Transcription factors involved in controlling the expression of nitrate reductase genes in higher plants. Plant Sci 229:167–171

Young EB (2015) Metagenome and metatranscriptome of complex algal communities growing in wastewater: bioremediation, nutrient transformation, carbon sequestration. Community Science Program (CSP) plans for fiscal year 2015. Joint Genome Institute. http://jgi.doe.gov/our-projects/csp-plans/fy-2015-csp-plans/

Young EB, Beardall J (2005) Modulation of photosynthesis and inorganic carbon acquisition in a marine microalga by nitrogen, iron and light availability. Can J Bot 83:917–928

Young EB, Dring MJ, Berges JA (2007a) Distinct patterns of nitrate reductase activity in brown algae: light and ammonium sensitivity in *Laminaria digitata* is absent in *Fucus* species. J Phycol 43:1200–1208

Young EB, Dring MJ, Savidge G, Birkett DA, Berges JA (2007b) Seasonal variations in nitrate reductase activity and internal N pools in intertidal brown algae are correlated with ambient nitrate concentrations. Plant Cell Environ 30:764–774

Young EB, Berges JA, Dring MJ (2009) Physiological responses of intertidal marine brown algae to nitrogen deprivation and resupply of nitrate and ammonium. Physiol Plant 135:400–411

Zulkifly SB, Hanshew A, Young EB, Lee PO, Graham ME, Graham LE (2012) The epiphytic microbiota of the globally widespread macroalga *Cladophora* (Chlorophyta, Cladophorales). Am J Bot 99:1542–1553

The Ammonium Paradox of an Urban High-Nutrient Low-Growth Estuary

Frances Wilkerson and Richard Dugdale

High-Nutrient Low-Growth Estuaries and Oligotrophication

Not all estuaries with high concentrations of nutrients exhibit elevated rates of primary production and phytoplankton biomass, symptomatic of cultural eutrophication. Some systems show the opposite symptoms with lower productivity and biomass than would be expected based upon nutrient input. For example, the San Francisco Bay-Delta (Bay-Delta) (Fig. 1) has been described as nutrient replete (Cloern and Jassby 2012) supplied largely by anthropogenic nutrient sources, especially wastewater effluent from multiple treatment plants (e.g., Jassby 2008; Parker et al. 2012a) and agricultural runoff. However, since the late 1980s there has been persistently low phytoplankton biomass (e.g., Cloern 1996) that is considered to have contributed to a decline in pelagic fishes (the Pelagic Organism Decline, POD, Sommer et al. 2007). Measurements of annual primary production, especially in the northern estuary (e.g., Kimmerer et al. 2012; Parker et al. 2012c; Wilkerson et al. 2015) are among the lowest rates of estuarine-coastal ecosystems of the world (Cloern et al. 2014) and place the estuary in the lower end of the oligotrophic category developed by Nixon (1995), i.e., <100 g C m^{-2} $year^{-1}$. As such it is a high-nutrient estuary with low chlorophyll variously called high-nutrient low-chlorophyll (HNLC, Cloern 2001) or high-nutrient low-growth (HNLG, Sharp 2001). Here, we prefer to use the latter term to differentiate it from the more typical open ocean HNLC condition.

F. Wilkerson (✉) • R. Dugdale
Romberg Tiburon Center of Environmental Studies, San Francisco State University, Tiburon, CA 94920, USA
e-mail: fwilkers@sfsu.edu

P.M. Glibert, T.M. Kana (eds.), *Aquatic Microbial Ecology and Biogeochemistry: A Dual Perspective*, DOI 10.1007/978-3-319-30259-1_10

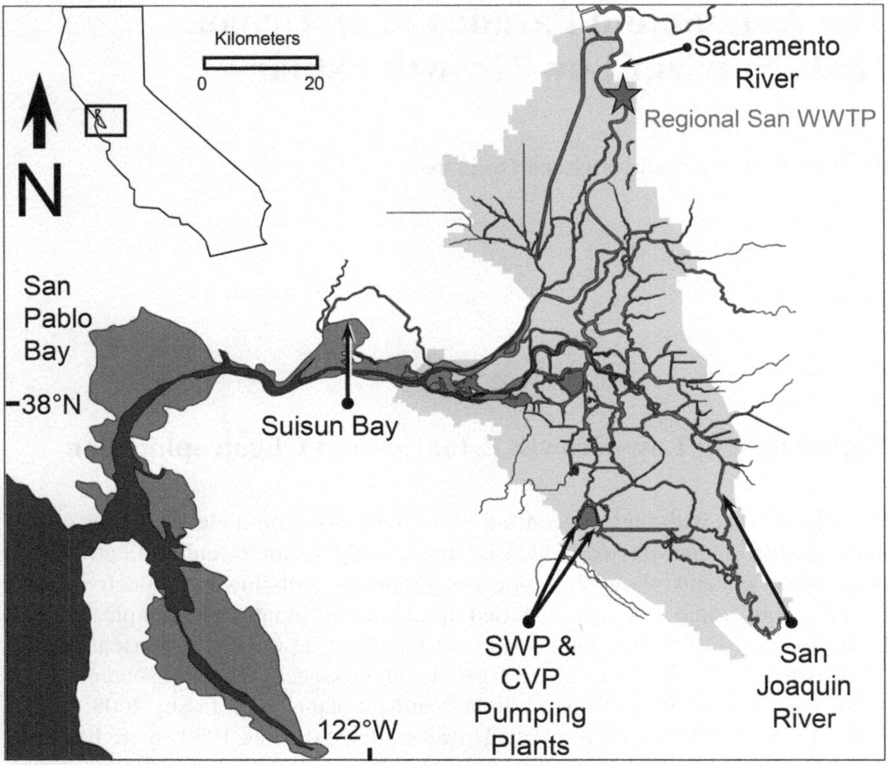

Fig. 1 Map of the San Francisco Bay-Delta showing Suisun Bay (northern San Francisco Bay) where the data in Fig. 2 were collected, and location of the major wastewater treatment plant, the Sacramento Regional Sanitation District Wastewater Treatment Plant (Regional San) that discharges into the Sacramento River that enters Suisun Bay

Prior studies in the Bay-Delta attributed the low standing stock of phytoplankton and low primary productivity to light limitation (Cole and Cloern 1984) resulting from high turbidity (Cloern 1987) and benthic grazing (Nichols and Thompson 1985) especially by the invasive Asian clam, *Potamocorbula amurensis* (Alpine and Cloern 1992). But, *P. amurensis* is unlikely to control spring phytoplankton blooms in the northern San Francisco Bay (that have been sporadically observed, Wilkerson et al. 2006; Dugdale et al. 2012) since it is abundant only in late summer and early fall (Greene et al. 2011). Generally, nutrients have been dismissed as a factor in the low productivity condition on the basis that concentrations never reach limiting or exhausted levels. However, paradoxically, elevated levels of NH_4 may contribute to the HNLG condition of the Bay-Delta in ways not fully appreciated.

Observation of an Ammonium Paradox

The traditional paradigm is that nutrient enrichment leads to eutrophication (the production of organic matter) but this is not the response in all estuaries (Sharp 2001). For example, Cox et al. (2009) observed low algal biomass in the Scheldt Estuary that was "contrary to expectations from the classical eutrophication response". Similarly, Nixon (2009) described a low productivity response to elevated nutrients in Narragansett Bay and termed it oligotrophication. Working with nearly 40 years of estuarine data for the Delaware Estuary, Yoshiyama and Sharp (2006) observed primary productivity to decline exponentially with increasing NH_4 concentration (>10 μmol L^{-1} NH_4^+) and attributed this effect to a shift in the form of DIN being used and leading to lower growth. They recognized that not all chemical forms of DIN result in the same outcome. The different metabolic response by phytoplankton to the different chemical forms of nitrogen (i.e., NO_3^- versus NH_4^+) is the underpinning of the concept of new and regenerated production (Dugdale and Goering 1967). Elevated NH_4^+ (from wastewater effluent) was also implicated in depressed primary production (and lower uptake of nitrate, NO_3^-) along the California coast (MacIsaac et al. 1979), in the Saronikos Gulf of Greece (Dugdale and Hopkins 1972) and Wascana Creek, Canada (Waiser et al. 2010).

In the Bay-Delta, observations suggest a similar scenario with spring bloom suppression accompanied by historic increased NH_4^+ loading from sewage effluent. Studies in the Bay-Delta during the 1960s and 1970s documented regular blooms that were associated with low NH_4^+ concentrations and almost complete nutrient (NH_4^+ and NO_3^-) drawdown by the phytoplankton (Ball and Arthur 1979). However the Bay-Delta is now HNLG although when favorable irradiance conditions occur, rare episodic blooms (e.g., Wilkerson et al. 2006; Dugdale et al. 2012) with elevated chlorophyll have been documented. The reported chlorophyll accumulation was accompanied by high rates of NO_3^- uptake and low NH_4^+ conditions (Dugdale et al. 2007; Wilkerson et al. 2015). The proposed mechanism for such bloom development is that the well-known metabolic NH_4 repression of NO_3^- uptake (e.g., Dortch 1990) is alleviated and the phytoplankton can access the larger component of the DIN pool in the Bay-Delta which is NO_3^-.

As shown in Fig. 2, to drive a sustained bloom of >25 μg L^{-1} as observed in the past (Ball and Arthur 1979) and in recent episodic events (e.g., Dugdale et al. 2012), the NO_3^- needs to be accessed in addition to NH_4^+. To make 1 μg L^{-1} of chlorophyll typically requires 1 μmol L^{-1} N (for discussion and citations, see Dugdale et al. 2007, 2012), so using only ambient NH_4^+ (typically peak values are ~10 μmol L^{-1}) will yield chlorophyll concentrations of only <10 μg L^{-1}. Conversion from nutrient drawdown to phytoplankton biomass calculated in this way results in biomass accumulation only with suitable physical conditions, such as low flow and favorable irradiance. Additionally, use of NO_3^- could also result in twice the chlorophyll that can be produced from using only NH_4^+. In low-light treated enclosures containing water from the Bay-Delta and enriched with N, the ratio of chlorophyll increase to NO_3^- drawdown was 2:1 compared to 1:1 with NH_4 (Glibert et al. 2014b).

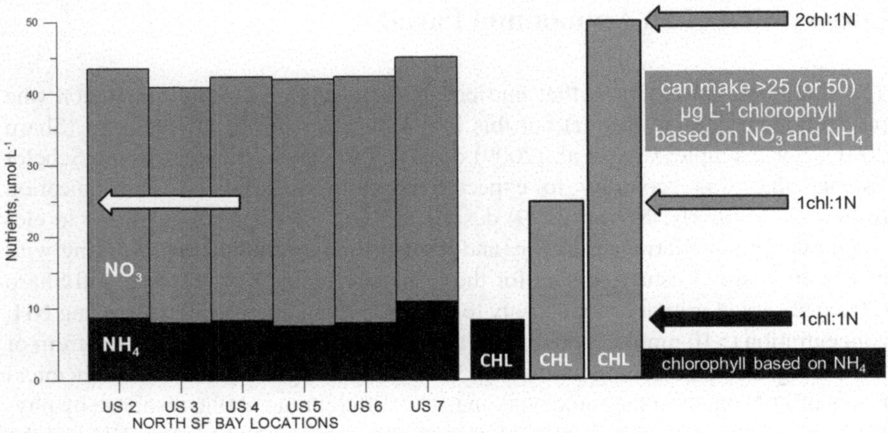

Fig. 2 Relative concentrations of NH_4^+ (*black*) and NO_3^- (*gray*) making up the dissolved inorganic nitrogen pool in the northern San Francisco Bay, shown to demonstrate how more chlorophyll can be made when the ambient NO_3^- (if it is available to the phytoplankton) is used compared to that using only ambient NH_4^+

The concept of NH_4^+ repressing NO_3^- uptake in the Bay-Delta phytoplankton was explained as the "Ammonium Paradox" by Dugdale et al. (2012), similar to the characterization of NH_4^+ as a "paradoxical" nutrient repressing growth in higher plants by Britto and Kronzucker (2002). In the Bay-Delta scenario, NH_4^+ prevents phytoplankton from being able to access NO_3^-, resulting in persistent low chlorophyll, lack of blooms and the export of unused NO_3^- to the coastal ocean. Loading of NH_4^+ into the Bay-Delta by the largest wastewater treatment plant has been increasing from ~5 tons NH_4^+-N in the 1970s to present-day levels of ~15 tons NH_4^+-N day^{-1} (Jassby 2008) with a concurrent increase in the proportion of NH_4^+:NO_3^-. This NH_4^+ adds to the DIN pool but is likely to yield less chlorophyll as it will restrict the NO_3^- from being used.

If NH_4^+ repression is relieved by a decrease in NH_4^+ concentration to below some physiological threshold (e.g., by phytoplankton uptake, nitrification, or dilution), the phytoplankton can accelerate their NO_3^- uptake in response, and build biomass (i.e., chlorophyll) at a rate that outpaces grazing losses in a repeatable sequence. This sequence was observed in experimental enclosures (Dugdale et al. 2007; Parker et al. 2012b), and at stations in the Bay-Delta during 2010 when NH_4^+ loading was reduced and NO_3^- drawdown occurred with a resultant spring bloom (Dugdale et al. 2012). NO_3^- drawdown and uptake by the phytoplankton (as measured with ^{15}N tracers) did not start until NH_4^+ concentrations were below a threshold of ~ 4 µmol L^{-1}. Then NO_3^- uptake rates accelerated due to metabolic "shift-up" in NO_3^- metabolism (e.g., Berges et al. 2004) and carbon uptake and chlorophyll concentrations increased. Wilkerson et al. (2015) recently showed the same predictable sequence to occur at a shallow shoal station in the Bay-Delta. In the field, the sequence to access the available NO_3^- is modulated by river flow conditions and residence time as well as continual inputs of NH_4^+ from the wastewater treatment plants (Dugdale et al. 2012, 2013).

NH_4^+ repression of NO_3^- uptake may not be the only factor suppressing growth, as there may be additional negative physiological aspects of NH_4^+, e.g., failure of electron pathways to balance metabolism and redox (Glibert et al. 2016). Also, in estuaries or situations, where NH_4^+ is the major source of N, the Ammonium Paradox does not apply. With sufficient residence time, a bloom based on growth on NH_4^+ would be expected to occur (e.g., Esparza et al. 2014).

Ammonium: The Gatekeeper Controlling Access to Nitrate

NH_4^+ is proposed to act as a gatekeeper controlling the flux of NO_3^- into the phytoplankton thereby driving phytoplankton growth rate and allowing biomass accumulation. Through the metabolic repression of NO_3^- uptake, NH_4^+ impedes access by the phytoplankton to the more abundant pool of DIN (NO_3^-). Any factors that influence NH_4^+ concentrations or the drawdown of NH_4^+ by phytoplankton will regulate the nutrient uptake sequence allowing NO_3^- use and alleviation of the HNLG condition. We have developed a conceptual model applying a holistic approach in which abiotic and biotic variables control the NH_4^+ concentration through changes in mass balance and uptake (Fig. 3), which in turn controls access of the phytoplankton to NO_3^- depicted as a spring valve (although it is actually a kinetic effect, repression of NO_3^- uptake by NH_4^+).

Parameters that would close the gate in the Bay-Delta are numerous and just some examples are considered here. High freshwater (or river) flow could supply more nutrients and make the system more sensitive to source NH_4^+ concentrations by decreasing residence time and shortening the time available for phytoplankton to reduce ambient NH_4^+. Excessive flow would ultimately "wash out" the phytoplankton population. Grazing will reduce the phytoplankton standing stock and limit their ability to absorb inflowing NH_4^+ allowing NH_4^+ concentrations to increase. Any factors that disrupt or change phytoplankton metabolism and impact NH_4^+ uptake capability (e.g., unfavorable irradiance, toxic contaminants, availability of other nutrients such as phosphorus, silicate, iron, etc.) will also close the gate, as elevated ambient NH_4^+ will not be drawn down to below threshold values.

Figure 3a shows how available irradiance may act as a valve, with either too little light decreasing phytoplankton NH_4^+ uptake or possibly too high being photoinhibitory. A representative NH_4^+ uptake versus irradiance curve measured in the Bay-Delta (inset Fig. 3a) shows that either low or high irradiances could restrict the uptake of NH_4^+ tipping the mass balance of NH_4^+ and forcing the gate to close, stopping the access to NO_3^- (Fig. 3a, white arrow). Similarly, a toxic contaminant could negatively impact phytoplankton metabolism (e.g., herbicides, Blaser et al. 2011) and decrease NH_4^+ uptake capacity, such that inflowing NH_4^+ would be unused, closing the access gate to NO_3^-.

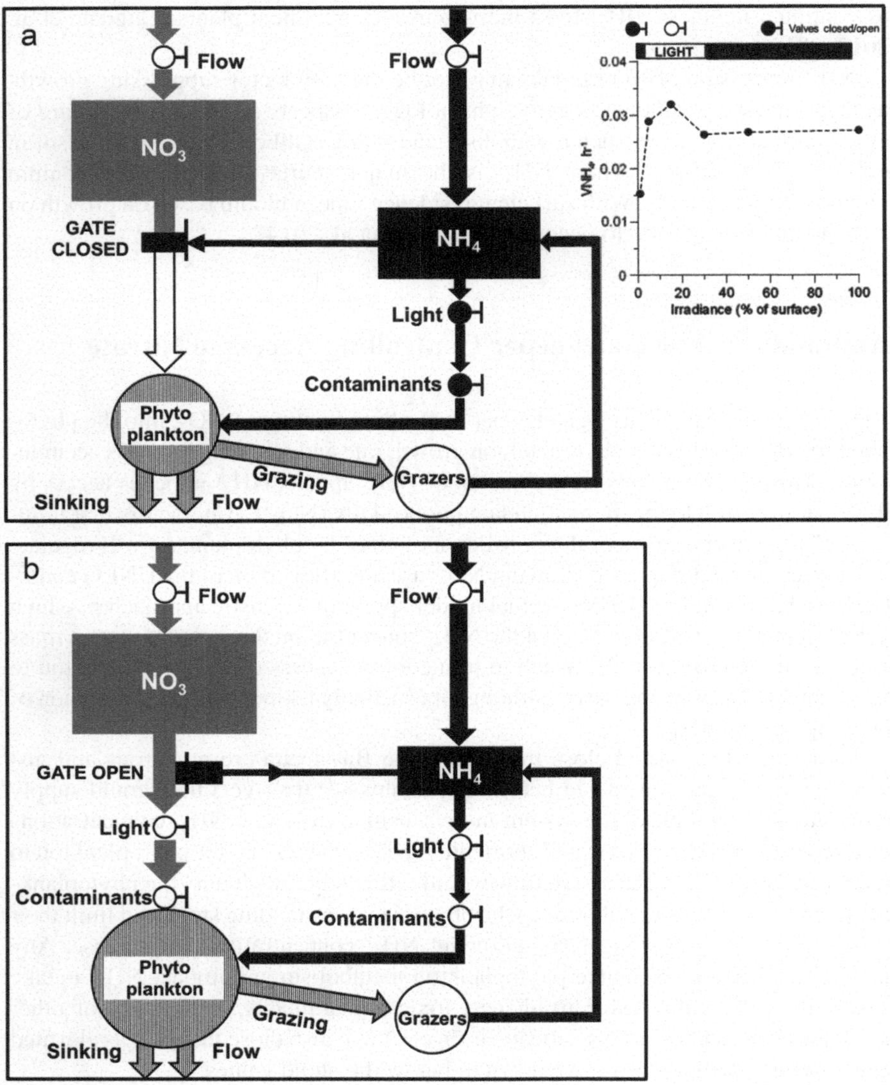

Fig. 3 Conceptual model showing how NH_4^+ acts as a gatekeeper, controlling phytoplankton access to NO_3^-. (**a**) NH_4^+ gate is closed; i.e., flow of NO_3^- to the phytoplankton is blocked (*white arrow*). High flow (*open white valves*) ensures there are elevated nutrients (pools—*filled boxes* of NO_3^- and NH_4^+). Any one of a variety of valves (e.g., low/high light, contaminant toxicity) can close (*black solid valve*), shutting down phytoplankton uptake of NH_4^+ thereby maintaining an inhibitory NH_4^+ pool—keeping the gate closed. Inset shows biomass specific NH_4^+ uptake versus irradiance in northern San Francisco Bay; i.e., how low and high light will decrease NH_4^+ uptake, result in a closed "light" valve and close the gate to access NO_3^-. (**b**) NH_4^+ gate is open; i.e., phytoplankton can access NO_3^- (*solid gray arrow*) as all valves are open (*white*), i.e., there are favorable conditions for lowered NH_4^+ concentration (e.g., flow, light, contaminants shown here). The gate (*spring valve*) is pulled back and NO_3^- can be taken up

To open the NH_4^+ gate and access NO_3^- (Fig. 3b), all parameters (flow, light, toxic contaminants among others) need to be favorable. The open valves will ensure that NH_4^+ will be drawn down to below threshold concentrations for NO_3^- uptake and assimilation. With the gate open and access to NO_3^-, enhanced NO_3^- uptake results in increased biomass that overcomes grazing, sinking, and washout losses. Grazers return some NH_4^+ to the NH_4^+ pool via excretion and regeneration. The uptake of NO_3^- by the phytoplankton will be similarly influenced by the same parameters as for NH_4^+ uptake (e.g., light, contaminants, temperature, etc.) but since these were favorable for NH_4^+ uptake, they should also allow maximal NO_3^- uptake. Increased microbial nitrification (NH_4^+ oxidation) is thought to have lowered the NH_4^+ levels in the drought spring of 2014 that opened the gate, enabling NO_3^- access and a spring bloom (Glibert et al. 2014a).

It is important to distinguish between direct effects on the gate brought about by anthropogenic loading which directly influences the ambient NH_4^+ concentration, and the indirect effects that alter the NH_4^+ concentration brought about by environmental variables such as those described above (light, contaminants temperature, etc.). When NH_4^+ loading exceeds the ability of the phytoplankton to absorb it, NH_4^+ increases and the gate closes. We suggest that in the Bay-Delta HNLG system, NH_4^+ concentration acts as an indicator of ecosystem condition, with tractable management actions; e.g., NH_4^+ loading could be lowered through regulation, unlike other challenging options such as grazing control or turbidity reduction. For example, in 2010, practices resulted (by serendipity) in both lowered effluent NH_4^+ concentrations and flow changes in the Bay-Delta that enabled a spring bloom (Dugdale et al. 2012). Reduced NH_4^+ discharge by the major wastewater treatment plant, due to more advanced effluent treatment, is planned for the future (in 2021) to meet various water quality requirements.

Elevated NH_4^+ and a closed gate for access to NO_3^- indicate the potential for reduced phytoplankton biomass and productivity, but do not necessarily mean that NH_4^+ inputs are the cause. A successful search for the proximate cause of the elevated NH_4^+ could allow an assessment of whether possible mitigation actions might exist. By understanding the role of NH_4^+ as gatekeeper to the use of NO_3^-, and the likelihood of phytoplankton blooms, it may be possible to manage the ecosystem to minimize the HNLG condition and improve production. Changed chlorophyll conditions could be beneficial in ameliorating the POD situation in the Bay-Delta.

Nutrient pollution and enrichment, long recognized as stressors for urbanized estuaries and coasts, are on the rise with increased population growth. The classic eutrophication paradigm is that excess nutrients will result in elevated phytoplankton biomass, low oxygen, occurrence of HABs and other deleterious effects. Alternatively, with NH_4^+ enrichment, oligotrophication may occur resulting in a HNLG condition. This chapter provides one mechanistic explanation for the HNLG condition in the Bay-Delta and relevant for other urban estuaries. The take-home messages are (1) ambient NH_4^+ at elevated concentrations saturating for phytoplankton uptake may result in decreased phytoplankton production. (2) That DIN should not be considered in management decisions as a single nutrient pool that combines NH_4^+ and NO_3^-, but that the reduced and oxidized chemical forms should be considered separately as they affect algal physiology differently. (3) There are interactions between nutrients, e.g., repression by NH_4^+ of NO_3^- uptake. (4) That some HNLG estuaries may in part be a consequence of the Ammonium Paradox with NH_4^+ acting as a gatekeeper, controlling access to the greater DIN pool (of NO_3^-) to be used for growth, and environmental factors combining to set the gate open or closed. Here, the urban HNLG system has been simplified to some extent (i.e., an NH_4^+-centric view) to purposely focus on one parameter that can be regulated or managed—a knob that can be turned to alleviate such a condition if it was shown to be beneficial. In the future, more data collection is needed along with appropriate nutrient parameterization in simulation models to offer a more comprehensive and holistic view of the HNLG condition in urban estuaries, with anthropogenic influence.

Acknowledgements This builds on research supported by San Francisco Bay and Central Valley Regional Water Quality Control Boards, Delta Stewardship Council, the Interagency Ecological Program and the State and Federal Contractors Water Agency. We thank Dr. P. Glibert for the invitation to contribute. Especially, we wish to thank Dr. A.E Parker for many years of collaboration and discussion of the Ammonium Paradox, and to all the students and technicians who helped collect and analyze the data that led to this conceptual model (A. Marchi, V. Hogue, A. Pimenta, S. Blaser, E. Antell, J. Fuller, A. Kleckner, E. Kress, C. Buck, J. Lee, A. Johnson, N. Travis, S. Strong, and T. Lee).

Frances Wilkerson and Richard Dugdale

We first met in 1983 when Frances was giving a seminar at the University of Southern California (USC) on symbiotic jellyfish from Palau. Dick was intrigued that these jellyfish were acting like large dinoflagellates migrating and using ammonium, and Frances wanted to ask him why she had linear nutrient uptake kinetics in her giant clam incubations. He suggested it was surge uptake, which was reported in the subsequent Marine Biology paper. Since Frances knew about jellyfish and Dick knew almost nothing about jellyfish, Dick then asked if she would like to go to Greece in his place and give an invited paper at the United Nations Environmental Program on jellyfish, to which she agreed. So the next few weeks were spent putting together a paper on how the extensive jellyfish blooms observed in Greece might be part of a food web based on sewage derived nutrients and Frances headed off to Athens to present it! On her return she started as a Research Associate in his coastal upwelling research group at USC and learnt how to use N-15 isotopic tracers. Later Dick took her to a dinner/lecture by Helen Kazantzakis (widow of Nikos Kazantzakis—the author of Zorba the Greek) and the next year we went to Greece, first to work as consultants on the new Athens sewage outfall and then to visit the island of Paros. Next was a proposal (of marriage not science) in Venice, a Greek wedding in Pasadena, then a son Nicholas to accompany stepson Alexis, plus numerous papers usually based on nitrogen except for those on the silicate pump. We still visit Paros every summer where we write proposals and papers and have hosted an AGU Chapman conference. Otherwise we are based near San Francisco as a team working still with phytoplankton and nutrients, both sewage related ammonium impacts and nitrate driven upwelling productivity.

References

Alpine AE, Cloern JE (1992) Trophic interactions and direct physical effects control phytoplankton biomass and production in an estuary. Limnol Oceanogr 37:946–955

Ball MD, Arthur JF (1979) Planktonic chlorophyll dynamics in the Northern San Francisco Bay and Delta. In: Conomos J (ed) San Francisco Bay: the urbanized estuary investigations into the natural history of San Francisco Bay and delta with reference to the influence of man. Pacific Division, AAAS, San Francisco, pp 265–286

Berges JA, Gibson CE, Stewart BM (2004) Physiological responses of phytoplankton communities in the Irish Sea to simulated upwelling. Hydrobiologia 517:121–132

Blaser S, Parker AE, Wilkerson FP (2011) Diuron and imazapyr herbicides impact estuarine phytoplankton carbon assimilation: evidence from an experimental study. Interagency Ecol Prog San Francisco Estuary Newsl 24(3):3–11

Britto DT, Kronzucker HJ (2002) NH_4^+ toxicity in higher plants: a critical review. J Plant Physiol 159:567–584

Cloern JE (1987) Turbidity as a control on phytoplankton biomass and productivity in estuaries. Cont Shelf Res 7(11):1367–1381

Cloern JE (1996) Phytoplankton bloom dynamics in coastal ecosystems: a review with some general lessons from sustained investigation of San Francisco Bay, California. Rev Geophys 34(2):127–168

Cloern JE (2001) Our evolving conceptual model of the coastal eutrophication problem. Mar Ecol Prog Ser 210:223–253

Cloern JE, Jassby AD (2012) Drivers of change in estuarine-coastal ecosystems: discoveries from four decades of study in San Francisco Bay. Rev Geophys 50(4):RG4001. doi:10.10292012RG000397

Cloern JE, Foster SQ, Kleckner AE (2014) Phytoplankton primary production in the worlds estuarine ecosystems. Biogeosciences 11:2477–2501

Cole BE, Cloern JE (1984) Significance of biomass and light availability to phytoplankton productivity in San Francisco Bay. Mar Ecol Prog Ser 17:15–24

Cox T, Maris T, Soetaert K, Conley D, Van Damme S, Meire P, Middelburg J, Vos M, Struyf E (2009) A macro-tidal freshwater ecosystem recovering from hypereutrophication; the Schelde case study. Biogeosciences 6:2935–2948

Dortch Q (1990) The interaction between ammonium and nitrate uptake in phytoplankton. Mar Ecol Prog Ser 61:183–201

Dugdale RC, Goering JJ (1967) Uptake of new and regenerated forms of nitrogen in primary productivity. Limnol Oceanogr 12:196–206

Dugdale RC, Hopkins TS (1972) Predicting the structure and dynamics of a pollution-driven marine ecosystem embedded in an oligotrophic sea. Thalass Jugosl 14(1-2):107–126

Dugdale RC, Wilkerson FP, Hogue VE, Marchi A (2007) The role of ammonium and nitrate in spring bloom development in San Francisco Bay. Estuar Coast Shelf Sci 73:17–29

Dugdale RC, Wilkerson FP, Parker AE, Marchi A, Taberski K (2012) River flow and ammonium discharge determine spring phytoplankton blooms in an urbanized estuary. Estuar Coast Shelf Sci 115:187–199

Dugdale RC, Wilkerson FP, Parker AE (2013) A biogeochemical model of phytoplankton productivity in an urban estuary: the importance of ammonium and freshwater flow. Ecol Model 263:291–307

Esparza ML, Farrell AE, Craig DJ, Swanson C, Dhaliwal BS, Berg GM (2014) Impact of atypical ammonium concentrations on phytoplankton abundance and composition in fresh vs estuarine waters. Aquat Biol 21:191–204

Glibert PM, Dugdale RC, Wilkerson F, Parker AE, Alexander JA, Antell E, Blaser S, Johnson A, Lee J, Lee T, Murasko S, Strong S (2014a) Major—but rare—spring blooms in 2014 in San Francisco Bay Delta, California, a result of the long-term drought, increased residence time, and altered nutrient loads and forms. J Exp Mar Biol Ecol 460:8–18. doi:10.1016/j.jembe.2014.06.001

Glibert PM, Wilkerson FP, Dugdale RC, Parker AE, Alexander JA, Blaser S, Murasko S (2014b) Microbial communities from San Francisco Bay Delta respond differently to oxidized and reduced nitrogen substrates—even under conditions that would otherwise suggest nitrogen sufficiency. Front Mar Sci. doi:10.3389/fmars.2014.00017

Glibert PM, Wilkerson FP, Dugdale RC, Raven JA, Dupont C, Leavitt PR, Parker AE, Burkholder JM, Kana TM (2016) Pluses and minuses of ammonium and nitrate uptake and assimilation by phytoplankton and implications for productivity and community composition, with emphasis on nitrogen-enriched conditions. Limnol Oceanogr. doi:10.1002/lno.10203

Greene V, Sullivan LJ, Thompson JK, Kimmerer WJ (2011) Grazing impact of the invasive clam *Corbula amurensis* on the microplankton assemblage of the northern San Francisco Estuary. Mar Ecol Prog Ser 431:183–193

Jassby A (2008) Phytoplankton in the upper San Francisco Estuary: recent biomass trends, their causes, and their trophic significance. San Francisco Estuary Watershed Sci 6(1). jmie_sfews_10995. Retrieved from: http://eprints.cdlib.org/uc/item/71h077r1

Kimmerer WJ, Parker AE, Lidstrom U, Carpenter EJ (2012) Short-term and inter-annual variability in primary productivity in the low-salinity zone of the San Francisco Estuary. Estuar Coasts 35(4):913–929

MacIsaac JJ, Dugdale RC, Huntsman SA, Conway HL (1979) The effect of sewage on uptake of inorganic nitrogen and carbon by natural populations of marine phytoplankton. J Mar Sci 37:51–66

Nichols FH, Thompson JK (1985) Time scales of change in the San Francisco Bay benthos. In: Cloern JE, Nicholas FH (eds) Temporal dynamics of an estuary: San Francisco Bay. Springer, Dordrecht, pp 121–138

Nixon SW (2009) Eutrophication and the macroscope. In: Andersen JH, Conley DJ (eds) Eutrophication in coastal ecosystems. Developments in hydrobiology, 207:5–19

Nixon SW (1995) Coastal marine eutrophication—a definition, social causes and future concerns. Ophelia 41:199–219

Parker AE, Dugdale RC, Wilkerson FP (2012a) Elevated ammonium concentrations from wastewater discharge depress primary productivity in the Sacramento River and the northern San Francisco Estuary. Mar Pollut Bull 64:574–586

Parker AE, Hogue VE, Wilkerson FP, Dugdale RC (2012b) The effect of inorganic nitrogen speciation on primary production in the San Francisco Estuary. Estuar Coast Shelf Sci 104:91–101

Parker AE, Kimmerer WJ, Lidstrom U (2012c) Re-evaluating the generality of empirical models for light-limited primary production in the San Francisco Estuary. Estuar Coasts 35(4):930–942

Sharp JH (2001) Marine and aquatic communities, stress from eutrophication. Encycl Biodivers 4:1–11

Sommer TR, Armor C, Baxter R, Breuer R, Brown L, Chotkowski M, Culberson S, Feyrer F, Gingas M, Herbold B, Kimmerer W, Müller-Solger A, Nobriga M, Souza K (2007) The collapse of the pelagic fishes in the upper San Francisco Estuary. Fisheries 32:270–276

Waiser MJ, Tumber V, Holm J (2010) Effluent-dominated streams. Part 1: Presence and effects of excess nitrogen and phosphorus in Wascana Creek, Saskatchewan, Canada. Environ Toxic Chem 30:496–507

Wilkerson FP, Dugdale RC, Hogue VE, Marchi A (2006) Phytoplankton blooms and nitrogen productivity in the San Francisco Bay. Estuar Coasts 29:401–416

Wilkerson FP, Dugdale RC, Parker AE, Blaser S, Pimenta A (2015) Spring phytoplankton blooms in Suisun Bay: intensive observations of nutrients and productivity rates from 2010-2012. Aquat Ecol 49(2):211–233

Yoshiyama K, Sharp JH (2006) Phytoplankton response to nutrient enrichment in an urbanized estuary: apparent inhibition of primary production by overeutrophication. Limnol Oceanogr 51:424–434

Why Is Planktonic Nitrogen Fixation So Rare in Coastal Marine Ecosystems? Insights from a Cross-Systems Approach

Roxanne Marino and Robert W. Howarth

Whether primary production in an aquatic ecosystem is limited more by nitrogen (N) or phosphorus (P) is a critical issue for management, and understanding the controls that lead to N versus P limitation is also a fundamental ecological and biogeochemical question. The microbially mediated process of di-nitrogen (N_2) fixation brings new N into an ecosystem, and as such it is critical to understand the regulation and constraints on this process in the context of nutrient limitation, eutrophication, and the anthropogenic acceleration of the N cycle. We have long been intrigued by the paradox of how N limitation can occur and persist in some aquatic ecosystems (particularly coastal marine ecosystems) despite the fact that N_2-fixing cyanobacteria taxa are widespread and diverse, and therefore should have a substantial competitive advantage when the availability of N is low relative to P and the cellular needs of plankton (Vitousek and Howarth 1991). In this paper we briefly review a body of research we have published in collaboration with several colleagues, investigating the mechanisms of control on planktonic N_2 fixation in lakes and estuaries using a cross-systems comparative approach and considering several hierarchical levels of control on N_2 fixation from cellular and physiological to the ecosystem scale (Vitousek et al. 2002).

Nitrogen fixation by planktonic cyanobacteria is a ubiquitous process in many lakes and tropical and subtropical ocean waters, yet it is largely absent as an ecosystem N input to estuaries and coastal marine ecosystems when the salinity is greater than 10 ppt (Howarth et al. 1988a; Marino et al. 2002; Paerl 1996). In aquatic ecosystems where it occurs, N_2 fixation is often an important process, for instance helping to maintain P limitation or co-limitation of P and N in many lakes and in tropical and subtropical open oceans, as originally suggested by Alfred Redfield (Schindler 1977; Howarth and Marino 2006; Karl et al 2002). Many estuaries and coastal marine ecosystems are

R. Marino (✉) • R.W. Howarth
Department of Ecology and Evolutionary Biology, Cornell University,
Ithaca, NY 14853, USA
e-mail: rmm3@cornell.edu

P.M. Glibert, T.M. Kana (eds.), *Aquatic Microbial Ecology and Biogeochemistry: A Dual Perspective*, DOI 10.1007/978-3-319-30259-1_11

moderately to strongly N-limited, making the lack of planktonic N_2 fixation and the general absence of heterocystous cyanobacteria in these systems all the more striking, and paradoxical in contrast to the often extensive blooms by N_2-fixing heterocystous cyanobacteria observed in nutrient-enriched lakes (Howarth 1988; Paerl 1996; Howarth et al. 2011; Howarth and Marino 2006). Several heterocystous taxa typical of freshwater blooms can grow in culture and fix N_2 at salinities that span the range found in estuaries from oligohaline to full salinity seawater (Howarth et al. 1988a; Paerl and Zehr 2000; Marino 2001), yet blooms are rarely observed in these systems despite favorable conditions of N depletion, ample available P, and a regular source of these organisms from freshwater and littoral areas to estuaries via rivers (Marino et al. 2006).

In the early 1980s, we began to investigate whether a low availability of the transition metal molybdenum (Mo) in seawater may contribute to the lack of planktonic N_2 fixation in estuaries (Howarth and Cole 1985). Very low Mo concentrations had been hypothesized as a factor limiting productivity in oligotrophic, N-deficient Castle Lake by Goldman (1964), and Mo deficiency had long been recognized as a constraint on production in agricultural systems. Molybdenum is required in the conventional form of nitrogenase, the enzyme that catalyzes N fixation and can be synthesized by all N-fixing organisms. Vanadium (V)- or iron (Fe)-only alternative nitrogenases are possible but can only be synthesized by some N-fixing organisms under highly Mo-deficient or reducing conditions, and have a lower specific activity and less efficient N_2 fixation (Eady 1996). While alternative (non-Mo) nitrogenase activity has been demonstrated by some free-living terrestrial diazotrophs, these enzymes are highly unlikely to be important in marine waters, where Mo is an abundant trace metal (Bellenger et al. 2014). Our interest in molybdenum biogeochemistry grew out of our background in measuring sulfate reduction in salt marsh and estuarine sediments (Howarth and Teal 1979), where molybdate (at millimolar concentrations) is commonly used as a control to stop sulfate reduction. Molybdate (MoO_4^{2-}) is the major form of Mo in oxic waters, and the stereochemistry of molybdate and sulfate (SO_4^{2-}) are very similar; hence we asked if the large amount of sulfate in seawater might inhibit the assimilation of molybdate by cyanobacteria, and thus constrain their ability to fix N_2 and form the large blooms typically observed in freshwater lakes under N-deficient conditions. Previous studies had demonstrated an interaction between sulfate and Mo uptake in tomato plants, through the guts of animals, and in pure cultures of *Clostridium* (Howarth et al. 1988b and references therein).

Ecosystem-scale, biogeochemical evidence indeed strongly indicates that the bioavailability of Mo is low in seawater compared to other trace metals, or compared to Mo availability in freshwaters. The concentration of dissolved Mo in seawater is the highest for any trace metal, while the amount of Mo in plankton and seston is the lowest of the biologically important trace metals (Howarth et al. 1988b). The ratio of particulate (seston) to dissolved Mo in seawater is 3.5×10^{-5}, compared to ratios of 0.2 to 20 or more for bio-reactive metals such as iron, manganese, zinc, copper, nickel, and cadmium. On the other hand, the ratio for Mo in freshwaters ranges from 0.4 to 1.6, indicating a high bioavailability (Howarth et al. 1988b). In seawater, sulfate is the second most abundant anion after chloride, and sulfate concentrations are five orders of magnitude higher than for Mo: 28 mM vs. 0.11 μM for full-salinity water, whereas sulfate concentrations are far lower in freshwaters, typically ranging from 0.05 to 0.3 mM (Marino et al. 1990).

While the concentrations of sulfate and Mo are conservative with salinity in oxic seawater, and so the sulfate to Mo ratio is essentially constant, salt lakes have a wider range of sulfate concentrations and sulfate to Mo ratios. In an empirical study of 13 salt lakes in Canada where sulfate concentrations and dissolved sulfate to Mo ratios ranged from typical freshwater to nearly ten times full salinity seawater, we found that the only significant variable which predicted the abundance of planktonic cyanobacteria and heterocyst numbers was the ratio of sulfate to Mo (Marino et al. 1990).

In the late 1980s, we performed a series of short-term Mo uptake studies and demonstrated a highly significant inhibition of assimilation of ^{99}Mo-molybdate by sulfate in pure cultures of cyanobacteria and in ambient populations of plankton in both lakes and low-salinity Baltic Sea water (Howarth and Cole 1985; Cole et al. 1986, 1993). Originally, we attributed the results of these experiments to competitive inhibition, and accordingly designed a series of experiments to test our hypothesis, assuming that the effect of the inhibitor (sulfate) would be fully reversible by the addition of enough substrate (molybdate). We first conducted a mesocosm experiment in collaboration with colleagues at the Institute for Ecosystem Studies and the University of Rhode Island, using Narragansett Bay water (salinity 29–32 ppt) to which we added molybdate at levels we expected would largely offset the inhibiting effect of sulfate. Because planktonic, N_2-fixing cyanobacteria have not been observed in Narragansett Bay or other such saline estuaries, we added a small "seed" source of filamentous, heterocystous cyanobacteria from a seawater macroalgal nutrient "scrubber" system (Marino et al. 2002, 2006). The effect of the added Mo on cyanobacteria growth and N_2 fixation was far less than we had predicted based on our kinetic studies (Marino et al. 2003).

In a follow-up microcosm-scale experiment, we added sulfate and molybdate at seawater concentrations to P-enriched pond water with heterocystous cyanobacteria that were actively growing and fixing N_2; after 8 days of incubation under natural light conditions we found that sulfate inhibited N_2 fixation, while chloride addition controls for ionic strength and salinity showed little effect. However, Mo at very high levels (10× seawater and 200× the ambient pond water concentration) did not fully reverse this inhibition (Marino et al. 2003). Our conclusion from these experiments, and from a re-analysis of our previous kinetic data, was that the sulfate effect was not one of competitive inhibition of a single uptake enzyme, which would result in a common maximum uptake rate of Mo across a range of concentrations of the inhibitor (Fig. 1). Rather, we postulated that Mo uptake by these planktonic cyanobacteria might involve two or more uptake systems, activated in response to the environmental conditions. That is, when sulfate concentrations are low, a fast but low-specificity enzyme can be used for Mo assimilation, but at higher sulfate concentrations, a more highly selective uptake system is needed to discriminate between molybdate and sulfate. This results in a higher energetic cost that can slow the growth rate of the cyanobacteria (Marino et al. 2003).

Cellular and physiological constraints on growth rate would not alone prevent cyanobacteria from accumulating biomass and fixing ecologically significant amounts of nitrogen. Using a simulation model, we explored how growth rates may interact with mortality from grazing by zooplankton to control heterocystous cyanobacteria populations and rates of N_2 fixation in lakes and coastal marine ecosystems of the

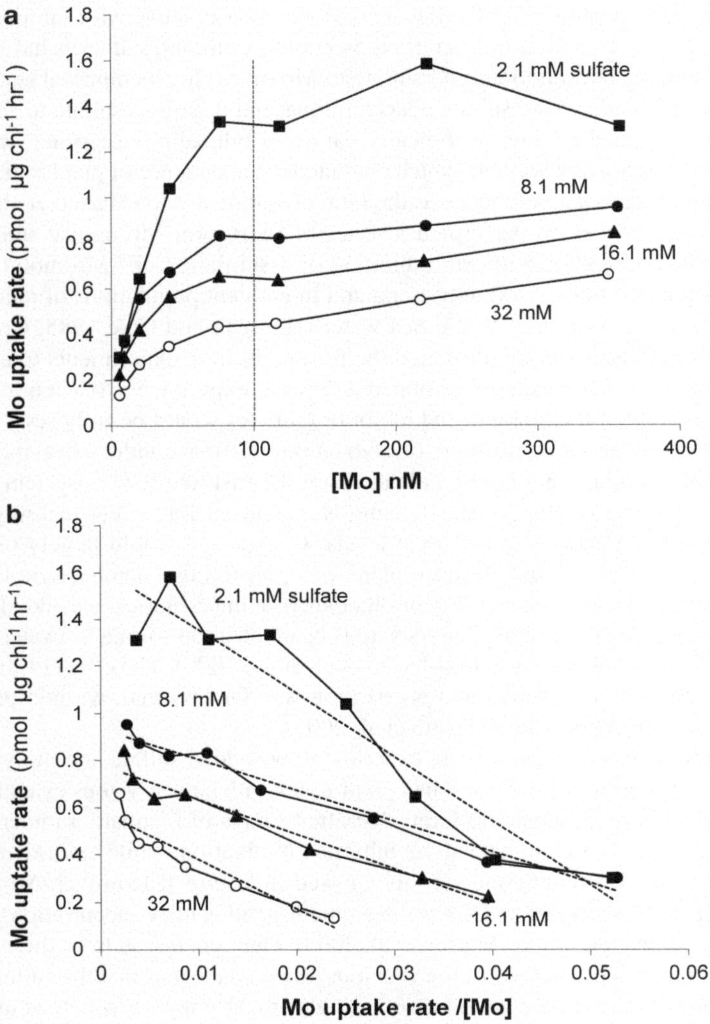

Fig. 1 (**a**) Uptake of molybdate, normalized to chlorophyll, as a function of molybdate concentration for a culture of *Anabaena cylindrica*, in the presence of 4 levels of sulfate ranging from oligohaline to full salinity seawater concentrations. *Vertical line* indicates Mo concentration in seawater. (**b**) Eadie–Hofstee plot of the data in **a**; R^2 values for linear regressions at each sulfate level: 2.1 mM = 0.91, 8.1 mM = 0.97, 16.1 mM = 0.95, 32 mM = 0.90. Competitive inhibition of Mo uptake by sulfate would be indicated by the regression lines all having the same y-intercept, or maximum rate of Mo uptake at some saturating level of Mo. Modified from Marino et al. 2003

Temperate Zone, in the absence of limitation by essential elements other than N and Mo (Howarth et al. 1999). Earlier, together with colleagues, we had demonstrated that high levels of zooplankton grazing can reduce rates of planktonic N_2 fixation in freshwater ponds (Schaffner et al. 1994). Our model well-captured those freshwater pond results, and further predicted that even a modest level of grazing by zooplankton

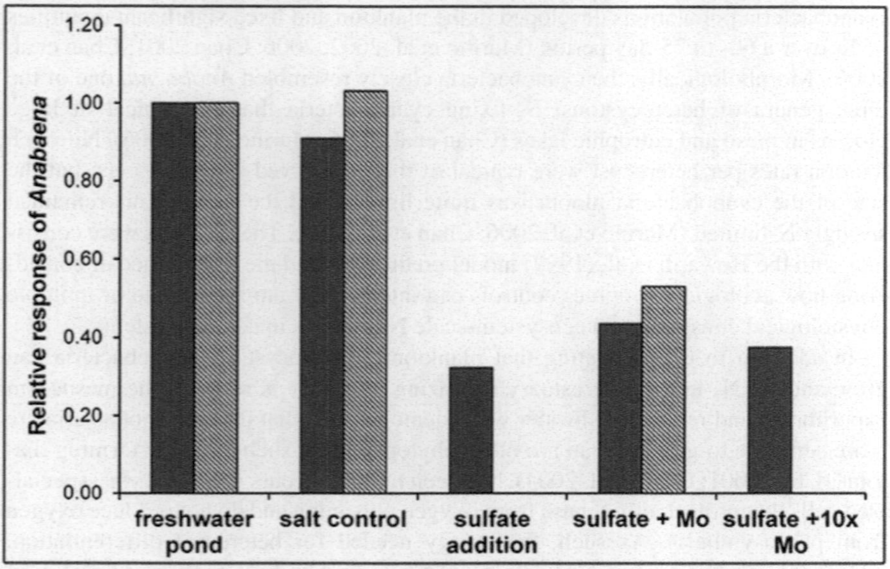

Fig. 2 Relative nitrogen fixation (*dark bars*) and cell growth (*striped bars*) responses of *Anabaena* spp. in a P-enriched, N-limited freshwater pond microcosm experiment to additions of either sulfate alone at seawater concentration (28 mM) or sulfate + Mo to seawater (0.1 μM Mo) and 10× seawater (1 μM Mo) concentrations. Bars are scaled to ambient freshwater pond (=1; 0.02 mM sulfate, 5.2 nM Mo). Salt Control maintained total equivalents of anion charge and total moles of cations constant with the sulfate addition treatments using chloride salts. The effect of sulfate on reducing N_2 fixation (assayed by acetylene reduction) was highly significant, as was the effect on cell growth rate, relative to the freshwater or salt control treatments. The addition of Mo to either seawater or 10× seawater levels only partially reversed the inhibition by sulfate. Further details are presented in Marino et al. (2003)

could suppress blooms of N_2-fixing cyanobacteria over an approximately 2.5 month growing season, if intrinsic growth rates were significantly slower than those observed in the freshwater experiments (Howarth et al. 1999). In our freshwater microcosm experiments discussed above, we observed a two- to threefold lower rate of both N_2 fixation and cyanobacteria growth as a result of adding seawater levels of sulfate to pond water with actively growing cyanobacteria (Fig. 2), in good agreement with our kinetic data which predict a 25–65 % reduction in Mo availability under seawater vs. typical freshwater conditions (Marino et al. 2003).

Returning to mesocosms on the shore of Narragansett Bay, we experimentally tested this prediction that growth rate and grazing interact to highly constrain development of populations of planktonic N_2 fixers in estuaries, and so limit this mechanism for alleviating N limitation. As for the previously described mesocosm experiment, a seed source of heterocystous cyanobacteria was added and nutrient conditions were highly favorable for planktonic N_2 fixation (Marino et al. 2006). When macrozooplankton were present in numbers typical of Narragansett Bay, few if any N_2-fixing cyanobacteria were present. However, in treatments where we added zooplanktivorous fish to keep the zooplankton population very low, heterocystous cyanobacteria populations developed in the plankton and fixed significant quantities of N_2 over a 60- to 75-day period (Marino et al. 2002, 2006; Chan 2001; Chan et al. 2006). Morphologically, the cyanobacteria closely resembled *Anabaena*, one of the three genera of heterocystous, N_2-fixing cyanobacteria that are typical of large blooms in meso and eutrophic lakes (Chan et al. 2006; Marino et al. 2006). Nitrogen fixation rates per heterocyst were typical of those observed in freshwaters, but the size of the cyanobacteria bloom was quite limited and the mesocosms remained strongly N-limited (Marino et al. 2006; Chan et al. 2006). These results were consistent with the Howarth et al. (1999) model predictions and the importance of considering how ecological (trophic) controls can interact and amplify single or multiple physiological constraints on ecosystem-scale N_2 fixation in aquatic systems.

In addition to demonstrating that planktonic, heterocystous cyanobacteria can grow and fix N_2 in a saline estuary if grazing mortality is relaxed, the mesocosm experiments and related freshwater work demonstrated that these cyanobacteria are more sensitive to grazing than are other phytoplankton, such as chain-forming diatoms (Chan 2001; Chan et al. 2004). Nitrogen fixation occurs in heterocysts, specialized cells that protect nitrogenase from oxygen poisoning and do not produce oxygen from photosynthesis. As such, the energy needed for heterocyst differentiation, nitrogenase synthesis, and N_2 fixation must be supplied by a sufficient number of photosynthetic, vegetative cells in a cyanobacteria filament (Wolk et al. 1994). On average for the cyanobacteria filaments in our mesocosms, 16 vegetative cells were needed to support the production and activity of one heterocyst (Chan et al. 2006), and the net growth rate was closely tied to overall mean filament length. *Anabaena* populations in the mesocosms where grazers were present had short filaments (average <16 cells) and so very few heterocysts; note that exogenous N to support cell growth was kept very low. Consequently both N_2 fixation and the growth of the population in these treatments were highly suppressed relative to the no-grazer treatments, where filaments were significantly longer. Grazing by generalist zooplankton such as the *Acartia tonsa* present in our experiments results in short, clipped filaments and thus has a disproportionate negative effect on further growth and N_2 fixation by the cyanobacteria. Of interest, *Acartia* showed no preference for grazing diatoms over cyanobacteria, and consumed similar-sized chains of both equally (Chan et al. 2006). A revision of our 1999 model using our experimental and literature survey data on estuarine and freshwater growth rates further indicated a sharp nonlinearity in the response of filamentous cyanobacteria abundance to increased zooplankton biomass, as N_2 fixation to support rapid cell growth was strongly limited by short filaments and so insufficient numbers of heterocysts (Marino et al. 2002; Fig. 3). It is important to note that we are not postulating a systematic difference in generalized grazing pressure across freshwater and saline estuarine systems, as our model and experiments used typical macro zooplankton densities for both.

Our Narragansett Bay mesocosms were enriched with P at loading rates similar to both those used for whole-lake experiments on ecosystem N_2 fixation at the Experimental Lakes Area (ELA) in Canada, and also to the loading rates we used in some freshwater mesocosm (tanks identical to the seawater mesocosms) and

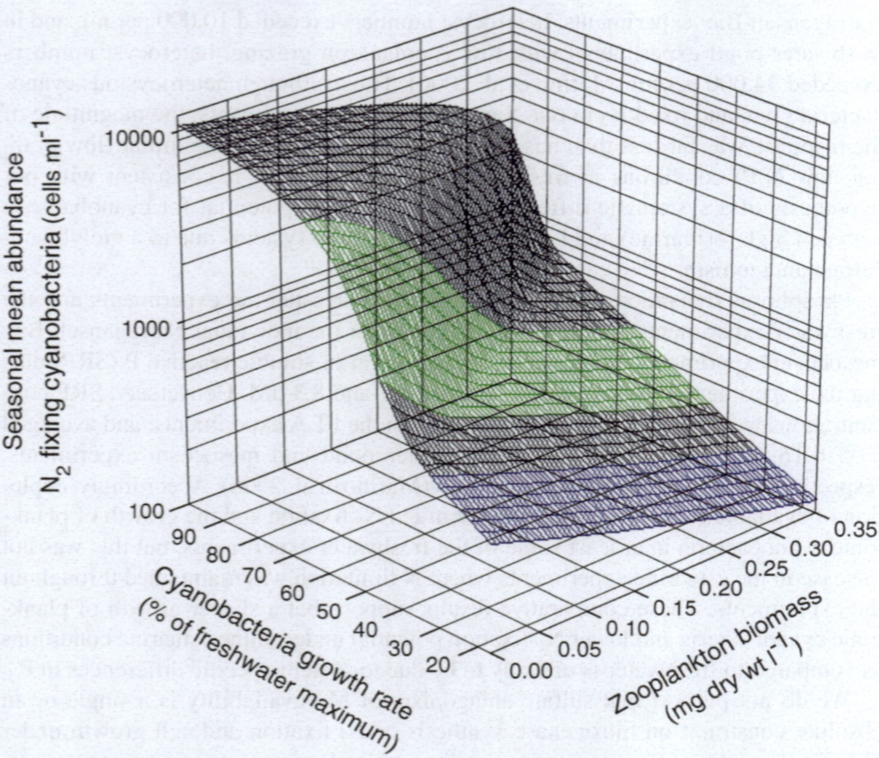

Fig. 3 Modeled sensitivity of heterocystic cyanobacteria abundance to interacting controls by growth rate (indicated as a percentage of freshwater maximum) and zooplankton biomass; *green shading* indicates freshwater and *blue shading* indicates estuarine system conditions. Model structure is presented elsewhere (Howarth et al. 1999). Actual magnitude of seasonal mean abundance (*z* axis) is sensitive to initial conditions of DIP, DIN, and cyanobacteria cell numbers for a given model run (1 μM, 10 μM, and 240 cells ml^{-1} respectively for plot shown here), however the pronounced nonlinearity of the response surface is unaffected. Modified from Marino et al. 2002

freshwater pond experiments (see Table 2 in Marino et al. 2006, and references therein). In all of these experiments, the N:P ratio of inputs was well below the molar Redfield ratio of 16:1 and thus provided conditions that are considered conducive for N_2 fixation. These similarities in experimental design allow for some interesting comparisons in the responses of the freshwater vs. saline systems (Marino 2001; Marino et al. 2006). In the estuarine mesocosms, heterocyst numbers varied between 50 and 295 per ml when zooplankton populations were suppressed, and were very low (0–11 per ml) when zooplankton were present at densities typical of the natural system. Heterocyst numbers were far higher in the ELA experiments, which had natural lake populations of zooplankton: 1800 to 3470 per ml. In freshwater mesocosms where zooplankton populations were suppressed as in the

Narragansett Bay experiments, heterocyst numbers exceeded 10,000 per ml, and in freshwater pond experiments with low zooplankton grazing, heterocyst numbers exceeded 34,000 per ml (Marino et al. 2006). Thus, although heterocystous cyanobacteria grew and fixed N_2 in our Narragansett Bay experiments, the magnitude of the response was far less than has been observed to occur under similar (low grazing, low N:P) conditions in freshwater experiments. This is consistent with our hypothesis of a systematic difference in gross growth potential for cyanobacteria between high (estuarine) and low (freshwater) sulfate systems due to a molybdate-sulfate antagonism.

Phosphorus dynamics also differed between our estuarine experiments and the freshwater experiments (Marino et al. 2006). Over the individual Narragansett Bay mesocosm experiments, the average concentration of soluble reactive P (SRP) during the experimental period ranged between 1.7 and 8.3 μM. Conversely, SRP concentrations were below the limit of detection in the ELA experiments, and averaged 0.05–0.16 μM and 1.1 μM in the freshwater pond and mesocosm experiments, respectively, despite similar P loading rates (Marino et al. 2006). Accordingly, depletion of available P may have eventually limited N_2 fixation and the growth of planktonic cyanobacteria in at least some of the freshwater experiments, but this was not the case in the estuarine experiments where N limitation was maintained throughout the experiments. These comparative results support that a slower growth of planktonic cyanobacteria and lower N_2-fixation potential under saline estuarine conditions as compared to freshwater is unlikely to be due to system-specific differences in P.

We do not purport that sulfate antagonism of Mo availability is a single or an absolute constraint on nitrogenase synthesis or N_2 fixation and cell growth under N-limited conditions in marine waters, but rather that it can be an important constraint on gross growth rate in saline (i.e. high sulfate) waters, which then can interact with other factors to limit pelagic cyanobacteria blooms (Howarth et al. 1988b; Marino et al. 2006). Blooms of heterocystous cyanobacteria are not uncommon in the oligohaline areas of the Baltic Sea (<10 ppt). At these lower salinities (and so sulfate concentrations), the inhibition of Mo assimilation is less severe, and cyanobacteria are able to grow fast enough to withstand moderate grazing by zooplankton (Howarth et al. 1999). In the very few estuaries where cyanobacteria blooms have been reported to persist at higher salinities such as the Peel-Harvey in Australia, often pronounced seasonal differences in precipitation and freshwater inputs result in wide fluctuations in salinity, approaching 0 at the time of peak freshwater input but much higher at other times of the season. Cyanobacteria blooms in these estuaries can initiate when the salinity is low, which in addition to favoring higher Mo availability is often a time of low grazing, since the freshwater excludes estuarine and marine grazers. Blooms can persist for a while as the salinity rises, but have not been observed to commence when the salinity is high (Lukatelich and McComb 1986; Jones et al. 1994).

Along with Mo, iron (Fe) is an essential micronutrient in the nitrogenase complex and so can constrain cyanobacteria growth, particularly under N-limited conditions. Early in our work, we recognized the importance of Fe (Howarth et al. 1988b) but chose to focus on the role of Mo in part because we viewed the latter as more novel, with the possible sulfate-Mo antagonism as an intriguing systematic difference on growth potential across freshwater and estuarine systems. Many factors affect Fe availability in natural waters, including inputs from land, oxidation level, complexation by dissolved organic matter, and salinity. The latter likely affects Mo and Fe similarly, decreasing availability with increasing salinity in estuaries and coastal marine ecosystems (Howarth et al. 1988b). High dissolved organic matter concentrations and inputs of Fe from the watershed in many nutrient-impacted aquatic systems, including the Baltic, would likely lessen constraints on growth from Fe limitation.

Although our conceptual model focuses on the interaction of physiological and ecological constraints on the development of sufficient biomass and so N_2 fixation by planktonic cyanobacteria in estuaries, aspects of this construct are generally applicable to benthic and epiphytic cyanobacteria as well. High rates of N_2 fixation by attached benthic and epiphytic cyanobacteria often occur in shallow coastal marine ecosystems when sufficient light reaches the bottom to support photosynthesis (Howarth et al. 1988a). Grazing pressure on these cyanobacteria is perhaps low enough to compensate for slow growth from sulfate inhibition, although the proximity to reducing conditions in sediments likely increases the availability of both lower oxidation state Mo and Fe species (Howarth et al. 1988b). We have observed moderate to high rates of N_2 fixation by epiphytic cyanobacteria associated with seagrasses in a shallow estuary on Cape Cod, MA, with the highest rates per blade area in portions of the estuary where the abundance of animals that graze on epiphytes such as *Bittium* snails and amphipods is lowest (Reynolds et al. 2015; Marino et al., unpubl data). Results from a reciprocal transplant experiment across a nutrient-loading gradient in this estuary suggest that grazing in this system may be a more important control on N_2 fixation by epiphytes than bottom-up factors (Reynolds et al. 2015).

Our research summarized here has focused largely on the paradox of why planktonic, heterocystous N_2-fixing cyanobacteria that can bloom spectacularly in lakes and some brackish waters and add significant amounts of new N to those ecosystems are so unimportant in nutrient-enriched, N-limited estuaries at higher salinities. We have not explicitly studied the non-heterocystous cyanobacteria that fix N_2 in many areas of the oligotrophic, N-limited tropical and subtropical oceans, such as large, filamentous *Trichodesmium* or the unicellular diazotrophs. These species have high light requirements and low maximum growth rates, and their abundance is thought to be primarily controlled by low Fe and P bioavailability and a high energy requirement, in part due to necessarily rapid turnover of nitrogenase (Monteiro et al. 2010), although the inhibiting effect of sulfate on Mo uptake may also contribute (Karl et al. 2002). The availabilities of Fe and P are far higher in estuaries and coastal zones than in the open ocean, due to inputs from land, and water column light levels in general are lower due to shading by plankton blooms and more seasonality in temperate latitudes. Nonetheless, we believe our mechanistic model for N_2 fixation control by an interaction of grazing and sulfate as a factor slowing, but not preventing, the cells from meeting their Mo requirement is consistent with ecologically significant N_2 fixation by *Trichodesmium*. A few, very specific animals are known to graze on *Trichodesmium* (O'Neil and Roman 1994), but these are not generalist grazers as *Acartia* is in estuaries. Generalist grazers in subtropical gyres would be very small, adapted to feeding on the major primary producers in those waters, which are also very small, and would be unable to ingest the rather large *Trichodesmium* trichomes. The specialized grazers associated with *Trichodesmium* are not sufficient to completely suppress blooms even with slow maximum growth rates and as such can constrain, but not eliminate N_2 fixation. Of course, this argument would not seem to pertain to the very small unicellular fixers such as *Synechococcus* sp., which are likely to be readily consumed by generalist grazers such as protozoa. We suggest that a closer examination of the factors controlling growth and mortality on *Synechococcus* would be extremely helpful in further elucidating this part of the ocean nitrogen fixation story.

In summary, a variety of physical, biogeochemical, and ecological factors influence the growth and rate of N_2 fixation by planktonic cyanobacteria, yet few of these represent fundamental differences between freshwater and coastal marine ecosystems and the observed dichotomy of response to N limitation. Our research supports the concept that the relative absence of heterocystous N_2-fixing cyanobacteria from the plankton of nutrient-enriched coastal marine systems at salinities greater than 10 ppt is not the result of a single factor but rather the interaction of a biogeochemical constraint on growth (sulfate lowering Mo availability) coupled with an ecological control (grazing by generalist zooplankton). In our 1988 review, we wrote that "perhaps the greatest uncertainty at present is how various biogeochemical factors interact with each other and with other physical factors to regulate N fixation" (Howarth et al. 1988b). We are excited by the continued research progress on these interacting issues since that time, in terrestrial as well as aquatic ecosystems, and note an emerging confluence of thought that large-scale patterns in the rates and importance of N_2 fixation across ecosystems seem generally to be controlled by interactions of two or more factors—often including both ecological and biogeochemical ones—rather than by any single factor (Vitousek et al. 2002; Karl et al. 2002; Monteiro et al. 2010). Still, many questions remain unanswered, and continued incorporation of updated mechanisms into ecosystem models is essential for understanding N limitation of primary production, as well as the likely consequences of continued human acceleration of the N cycle, particularly in heavily impacted areas such as the coastal zone.

Acknowledgements We were fortunate during the development of the ideas and experiments described in the paper to have as colleagues several creative and broad-minded, cross-system comparative ecologists: Jon Cole, Gene Likens, Scott Nixon, and Mike Pace. Francis Chan's doctoral research was integral to our mesocosm experiments and our understanding of the role of grazing. We thank Norbert Jaworski and others at the EPA Narragansett Bay lab during the 1990s, and Scott Nixon and the members of his lab group at the University of Rhode Island GSO for use of facilities and technical assistance with the mesocosm experiments. Much of the research described here was supported by grants from the NSF Ecosystems Program and an endowment from David R. Atkinson to Cornell University for the support of our research group and salary for RWH.

Roxanne Marino and Robert W. Howarth

We met in 1980 when Roxanne, newly out of college with a BS in chemistry, interviewed for a job at the MBL Ecosystems Center in Woods Hole with Bob, who was supervising his first research project after earning his PhD in the MIT/WHOI Joint Program in Oceanography. Except for a brief period when Roxanne studied at the University of Washington, we have worked together ever since. We spent many hours bonding over radioactive, sulfurous mud before moving on in 1985 to our current biogeochemical cycling study element of choice, nitrogen, still priding ourselves on being two of only a handful of ecosystem researchers who were interested in, let alone could spell, molybdenum. We enjoyed the opportunity to work at the Cary Institute of Ecosystem Studies for several months before moving to Cornell in 1985, continuing all along to enrich our shared interests in comparative ecosystems ecology and biogeochemistry. We were married in 1987 on our farm outside Trumansburg, NY. Our daughter Marina was born in 1996 and is now studying environmental science and engineering at Smith College. In 2001, Roxanne earned her Ph.D. at Cornell in the employee degree program, with Gene Likens. We have worked on a variety of other research topics beyond those described in this paper and also share a passion for applying science to environmental policy and management, and for community service. Roxanne served on the Ulysses Town Board for several years, including as Town Supervisor. Bob was the founding Editor-in-Chief of *Biogeochemistry* for 21 years, the President of the Coastal & Estuarine Research Federation from 2007 to 2009, and currently serves as Editor-in-Chief of *Limnology and Oceanography*.

References

Bellenger JP, Xu Y, Zhang X, Morel FFM, Kraepiel AML (2014) Possible contribution of alternative nitrogenases to nitrogen fixation by asymbiotic N_2-fixing bacteria in soils. Soil Biol Biochem 69:413–420

Chan F (2001) Ecological controls on estuarine planktonic nitrogen fixation: the roles of grazing and cross-ecosystem patterns in phytoplankton mortality. PhD thesis, Cornell University, Ithaca, NY

Chan F, Pace ML, Howarth RW, Marino R (2004) Bloom formation in heterocystic nitrogen-fixing cyanobacteria: the dependence of colony size and zooplankton grazing. Limnol Oceanogr 49:2171–2178

Chan F, Marino R, Howarth RW, Pace ML (2006) Experimental tests of ecological constraints on planktonic nitrogen fixation in saline estuaries: 2. Mechanisms of trophic control. Mar Ecol Prog Ser 309:41–53

Cole JJ, Howarth RW, Nolan SS, Marino R (1986) Sulfate inhibition of molybdate assimilation by planktonic algae and bacteria: some implications for the aquatic nitrogen cycle. Biogeochemistry 2:179–196

Cole JJ, Lane JM, Marino R, Howarth RW (1993) Molybdenum assimilation by cyanobacteria and phytoplankton in freshwater and salt waters. Limnol Oceanogr 38:25–35

Eady RR (1996) Structure-function relationships of alternative nitrogenases. Chem Rev 96(7): 3013–3030

Goldman R (1964) Primary productivity and micro-nutrient limiting factors in some North American and New Zealand lakes. Int Ver Theor Angew Limnol Verh 15:365–374

Howarth RW (1988) Nutrient limitation of net primary production in marine ecosystems. Annu Rev Ecol Syst 19:89–110

Howarth RW, Cole JJ (1985) Molybdenum availability, nitrogen limitation, and phytoplankton growth in natural waters. Science 229:653–655

Howarth RW, Marino R (2006) Nitrogen as the limiting nutrient for eutrophication in coastal marine ecosystems: evolving views over 3 decades. Limnol Oceanogr 51:364–376

Howarth RW, Teal JM (1979) Sulfate reduction in a New England salt marsh. Limnol Oceanogr 24:999–1013

Howarth RW, Marino R, Lane JM, Cole JJ (1988a) Nitrogen fixation in freshwater, estuarine, and marine ecosystems. l. Rates and importance. Limnol Oceanogr 33:669–687

Howarth RW, Marino R, Cole JJ (1988b) Nitrogen fixation in freshwater, estuarine, and marine ecosystems. 2. Biogeochemical controls. Limnol Oceanogr 33:688–701

Howarth RW, Chan F, Marino R (1999) Do top-down and bottom-up controls interact to exclude nitrogen-fixing cyanobacteria from the plankton of estuaries: explorations with a simulation model. Biogeochemistry 46:203–231

Howarth RW, Billen G, Chan F, Conley D, Doney SC, Garnier J, Marino R (2011) Coupled biogeochemical cycles: eutrophication and hypoxia in coastal marine ecosystems. Front Ecol Environ 9:18–26

Jones GJ, Blackburn SI, Parker NS (1994) A toxic bloom of *Nodularia spumigena* Mertens in Orielton Lagoon. Tasmania Aust J Mar Freshwat Res 45:787–800

Karl D, Michaels B, Bergman D, Capone E, Carpenter R, Letelier F, Lipschultz H, Paerl DS, Stal L (2002) Dinitrogen fixation in the world's oceans. Biogeochemistry 57(58):47–98

Lukatelich RJ, Mccomb AJ (1986) Nutrient levels and the development of diatom and blue-green algal blooms in a shallow Australian estuary. J Plankton Res 8:597–618

Marino R (2001) An experimental study of the role of phosphorus, molybdenum, and grazing as interacting controls on planktonic nitrogen fixation in estuaries. PhD thesis, Cornell University, Ithaca, NY

Marino R, Howarth RW, Shamess J, Prepas EE (1990) Molybdenum and sulfate as controls on the abundance of nitrogen-fixing cyanobacteria in saline lakes in Alberta. Limnol Oceanogr 35:245–259

Marino R, Chan F, Howarth RW, Pace ML, Likens GE (2002) Ecological and biogeochemical interactions constrain planktonic nitrogen fixation in estuaries. Ecosystems 5:719–725

Marino R, Howarth RW, Chan F, Cole JJ, Likens GE (2003) Sulfate inhibition of molybdenum-dependent nitrogen fixation by planktonic cyanobacteria under seawater conditions: a non-reversible effect. Hydrobiologia 500:277–293

Marino R, Chan F, Howarth RW, Pace ML, Likens GE (2006) Experimental tests of ecological constraints on planktonic nitrogen fixation in saline estuaries: 1. Nutrient and trophic controls. Mar Ecol Progr Ser 309:25–39

Monteiro FM, Follows MJ, Dutkiewicz S (2010) Distribution of diverse nitrogen fixers in the global ocean. Glob Biogeochem Cycles 24, GB3017. doi:10.1029/2009GB003731

O'Neil JM, Roman MR (1994) Ingestion of the cyanobacterium *Trichodesmium* spp. by pelagic harpacticoid copepods *Macrosetella*, *Miracia* and *Oculosetella*. Hydrobiologia 292(293): 235–240

Paerl HW (1996) A comparison of cyanobacterial bloom dynamics is freshwater, estuarine and marine environments. Phycologia 35:25–35

Paerl HW, Zehr JP (2000) Nitrogen fixation. In: Kirchman DL (ed) Microbial ecology of the oceans. Wiley-Liss, Inc, Chichester, pp 387–426

Reynolds LK, Marino R, Muth MF, McLenaghan N, Hayn M, Tyler AC, McGlathery KJ, Howarth RW (2015) Evidence of grazer control on nitrogen fixation by eelgrass epiphytes in a temperate coastal bay. Mar Ecol Prog Ser 526:11–19. doi:10.3354/meps11234

Schaffner WR, Hairston NG Jr, Howarth RW (1994) Feeding rates and filament clipping by crustacean zooplankton consuming cyanobacteria. Verh Internat Verein Limnol 25:2375–2381

Schindler DW (1977) Evolution of phosphorus limitation in lakes. Science 195:260–262

Vitousek PM, Howarth RW (1991) Nitrogen limitation on land and in the sea. How can it occur? Biogeochemistry 13:87–115

Vitousek PM, Cassman K, Cleveland C, Crews T, Field CB, Grimm NB, Howarth RW, Marino R, Martinelli L, Rastetter EB, Sprent JI (2002) Towards an ecological understanding of biological nitrogen fixation. Biogeochemistry 57(58):1–45

Wolk CP, Ernst A, Elhai J (1994) Heterocyst metabolism and development. In: Bryant DA (ed) The molecular biology of cyanobacteria. Kluwer Academic Publishers, Berlin, pp 769–823

Where Light and Nutrients Collide: The Global Distribution and Activity of Subsurface Chlorophyll Maximum Layers

Greg M. Silsbe and Sairah Y. Malkin

At the Confluence of Light and Nutrients

Subsurface chlorophyll maximum layers (SCMLs) are contiguous regions of elevated phytoplankton chlorophyll *a* (Chl*a*) found beneath the surface mixed layer. Their distribution and persistence are important for understanding global carbon cycling, particularly with regard to their potential role in supporting secondary production and export of carbon to the deep ocean (i.e., the biological pump). It has long been established that community composition within SCMLs can be distinct from those in the overlying surface layer, suggesting that a unique suite of drivers structure communities at these depths (Fairbanks and Weibe 1980; Venrick 1988). The specific mechanisms that regulate the community composition, distribution, and productivity of SCMLs can be variable through space and time (reviewed in Cullen 2014), but in general SCMLs are controlled by the opposing gradients of light supplied from above and nutrients supplied from below (Klausmeier and Litchman 2001; Cullen 2014). Switches between heterotrophic and autotrophic metabolisms, variable grazing pressure, and hydrodynamic processes can also influence SCML structure and phenology, to a secondary degree (Tittel et al. 2003; Cullen 2014; Villareal et al. 2014).

G.M. Silsbe (✉)
Horn Point Laboratory, University of Maryland Center for Environmental Science, Cambridge, MD, USA
e-mail: gsilsbe@umces.edu

Department of Botany and Plant Pathology, Oregon State University (OSU), Corvallis, OR, USA

S.Y. Malkin
Horn Point Laboratory, University of Maryland Center for Environmental Science, Cambridge, MD, USA

Department of Marine Sciences, University of Georgia, Athens, GA, USA

P.M. Glibert, T.M. Kana (eds.), *Aquatic Microbial Ecology and Biogeochemistry: A Dual Perspective*, DOI 10.1007/978-3-319-30259-1_12

In the progressively darker waters beneath the surface mixed layer, light limitation elicits genotypic (photo adaptation) and phenotypic (photo acclimation) responses in phytoplankton communities. For simplicity, we have chosen photo acclimation as an umbrella term to refer to both responses, although we recognize that each are important. Photo acclimation has been well characterized across light gradients in culture (Geider et al. 1998) and in situ (Cullen 2014). The most prominent physiological response to diminished light is an upregulation of cellular pigmentation and this response can be substantial. Phytoplankton Chl*a* to carbon biomass ratios (Chl*a*:C) can increase by up to a factor of 5 in the SCML, relative to surface waters, and becomes maximal at the bottom of the euphotic zone (Cullen 2014). In addition to changes in the quantity of light, the spectral quality of light also changes through depth. Phytoplankton residing in SCMLs are often chromatically adapted to their optical environment with accessory pigments that match the in situ light spectrum (e.g., phycocyanin and Chl*b*; Hickman et al. 2009; Malmstrom et al. 2010).

The relative importance of photo acclimation on observed changes in Chl*a* through depth varies across a trophic gradient. In the most oligotrophic and clearest waters (Uitz et al. 2006), elevated chlorophyll at the SCML almost entirely reflects a photo acclimation response, with no matching increases in phytoplankton carbon at depth (Fennel and Boss 2003; Cullen 2014; Mignot et al. 2014). In stratified waters that support a higher flux of nutrients, SCMLs are promoted at shallower optical depths (i.e., at depths with higher irradiance; Uitz et al. 2006), and the increase in Chl*a* observed at the SCML is progressively more associated with an increase in phytoplankton biomass (Cullen 2014). In eutrophic waters, variability in Chl*a* through depth can nearly entirely reflect variation in phytoplankton biomass, with nearly constant Chl*a*:C (Uitz et al. 2006, Cullen 2014).

The various mechanisms that deliver nutrients to the SCML have implications for the distribution, persistence, and primary productivity of these layers. As discussed in later sections, the frequency and magnitude of nutrient flux to the SCML is important for fuelling subsurface phytoplankton blooms with implications for rates of primary production. In stably stratified waters, nutrients diffuse from below steep nutrient gradients (i.e., nutriclines) which develop in the pycnocline. Mesoscale cyclonic eddies in the oligotrophic gyres can also entrain nutrients to subsurface waters, promoting development of SCMLs (Seki et al. 2001). In coastal seas, a variety of local hydrographic and topographic features further influences local nutrient injections (Estrada 1996). Internal tides have been proposed to fuel nutrient flux to subsurface phytoplankton (Richardson et al. 2000), and additionally, episodic events driven by wind and nonlinear internal waves also appear to be important in fuelling nutrients to subsurface phytoplankton layers and may temporarily displace phytoplankton to higher light environments (Pannard et al. 2011; Williams et al. 2013).

Distribution of Marine SCMLs

Satellite measurements of ocean color have enabled estimates of ocean Chl*a* concentrations with previously unprecedented coverage and resolution. As observed from space, the mean annual distribution of Chl*a* concentration spans more than two orders of magnitude, from the desert-like subtropical gyres to the fertile coastal oceans (Fig. 1a). However, this view is limited because the optical properties detectable by satellites are restricted to the surface mixed layer. In most of the global ocean, sufficient light passes through the surface mixed layer, systematically enabling phytoplankton growth in subsurface waters that are hidden from satellite view. Recognizing this deficiency, empirical models were developed to predict the vertical distribution of Chl*a* through the euphotic zone from surface Chl*a* concentrations (Uitz et al. 2006). Using remotely sensed Chl*a* data and these predictive relationships, we can extend satellite-based measurements through depth. To provide a more complete picture of its global distribution, we used surface Chl*a* from Fig. 1a to calculate the depth distribution globally. In Fig. 1b, we illustrate the maximum Chl*a* concentrations in the euphotic zone. By contrasting Fig. 1a against Fig. 1b, it becomes immediately evident that the vast majority of Chl*a* in the oceans is not represented in satellite-based measurements.

Satellite data can also enable a better understanding of the phenology of SCMLs. To the best of our knowledge, the potential distribution of SCMLs across the global ocean and through seasons has not been systematically presented in the literature. To predict the presence of an SCML, we used monthly climatological satellite and hydrographic data to quantify the flux of light that reaches mixed layer depths. Following the methods of Westberry et al. (2008), we use MODIS satellite data and mixed layer depths from the Fleet Numerical Meteorology and Oceanography Center (FNMOC) to map the monthly averaged light reaching the mixed layer depth. We defined regions as being capable of supporting SCMLs if the light at the mixed layer depth exceeds a minimum threshold required to sustain photoautotrophic growth. Previous findings from intensive sampling have shown that subsurface Chl*a* maxima can be maintained in light as low as 0.1 mol photons m^{-2} $days^{-1}$ (Mignot et al. 2014); we have taken a threshold of 0.5 mol photons m^{-2} $days^{-1}$ (~6 μmol m^{-2} s^{-1} over the full day) to be conservative. *Prochloroccocus* strains isolated from SCMLs and grown at 0.5 mol photons m^{-2} $days^{-1}$ can have division rates up to 0.3 $days^{-1}$ (Moore and Chisholm 1999). The predicted occurrence of an SCML at each satellite pixel was calculated for each month, and the seasonal persistence was then expressed as a proportion (where 0 = never present and 1 = permanently present; Fig. 2a). To show the seasonality of predicted SCMLs, in Fig. 2b the daily flux of light reaching the mixed layer depth is shown for seven selected oceanic regions (white boxes in Fig. 2a). The dashed line in Fig. 2b (0.5 mol photons m^{-2} $days^{-1}$) represents our threshold to infer the occurrence of an SMCL.

We found that over the course of a year, between 59 and 73 % of the ocean is predicted to support an SCML. With the exception of shallow coastal regions, and a narrow region of the equatorial divergence, SCMLs are predicted to be permanent features of tropical and subtropical regions. It is worth noting that even in these central regions, our analysis demonstrates a seasonal oscillation in light reaching the mixed layer depth. These predictions are well supported by long-term monitoring records at the fixed stations BATS and HOT and autonomous profiling data of Chl*a* fluorescence from oligotrophic gyres (Letelier et al. 2004; Mignot et al. 2014). Progressing poleward, the light climate has stronger seasonality and SCMLs are predicted to occur for shorter durations. Additionally, less light is predicted to reach

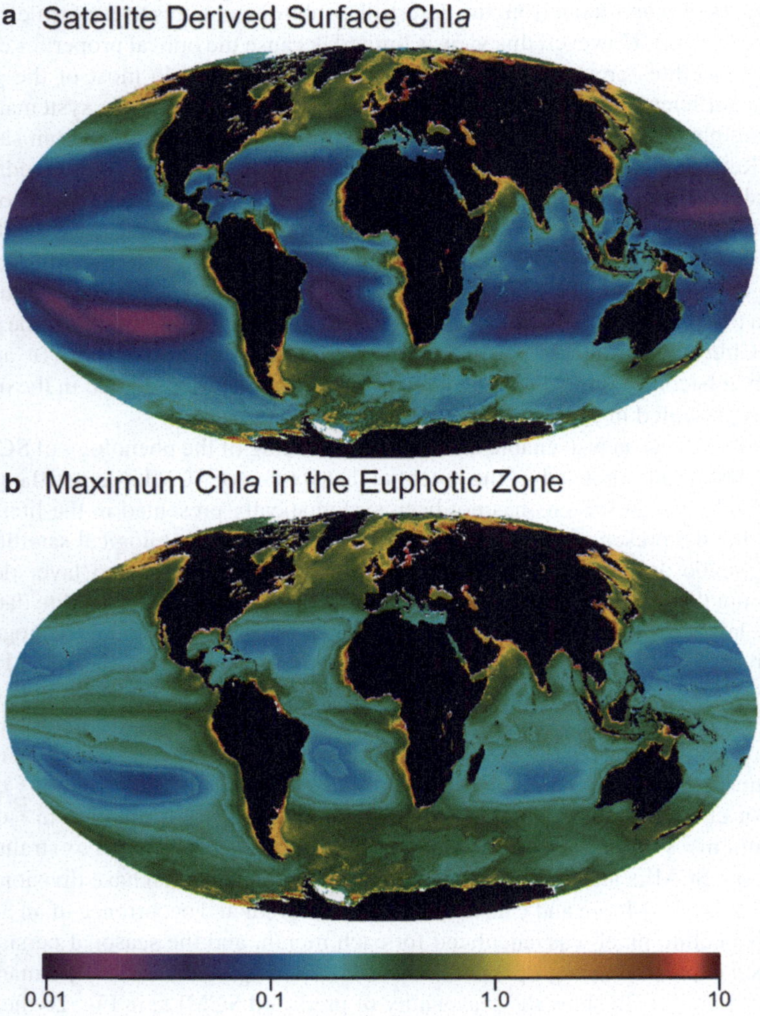

Fig. 1 (**a**) The annual average surface Chl*a* as measured from space (MODIS Aqua monthly climatology), and (**b**) the corresponding maximum Chl*a* in the euphotic zone predicted by Uitz et al. (2006)

SCMLs relative to the tropics. In arctic waters (shown here for 66–80°), our model predicts SCMLs persist for approximately 2.5 months each year. We caution that our estimates for polar waters may be less accurate than elsewhere because the FNMOC model is less reliable in the polar oceans. Nevertheless, recent measurements from Canadian arctic waters consistently observed the presence of SCMLs between 14 and 32 m which are believed to be nearly ubiquitous and persistent throughout the ice-free period (Martin et al. 2010).

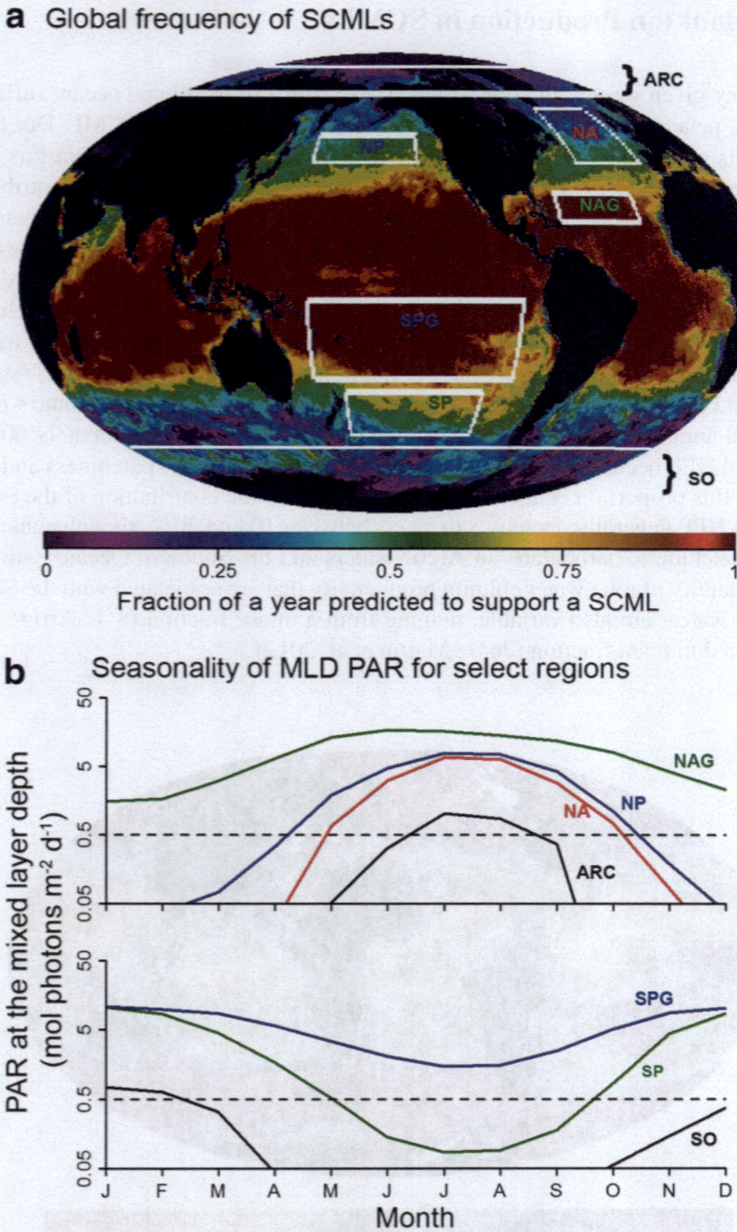

Fig. 2 (**a**) The annual duration of subsurface chlorophyll maximum layers (SCMLs) calculated from the monthly climatology of light reaching the mixed layer depth (MLD). *White boxes* correspond to the plots in (**b**) that show the mean monthly variations in light at the mixed layer depth for boreal (*top*) and austral (*bottom*) regions of the ocean. The dashed line demarcates the estimated minimum light level at the bottom of the mixed layer that is required to support an SCML. ARC, Arctic Ocean; NA, North Atlantic; NP, North Pacific; NAG, North Atlantic Gyre; SPG, South Pacific Gyre; SP, South Pacific; SO, Southern Ocean

Phytoplankton Production in SCMLs

During any given season, in approximately two-thirds of the global ocean, sufficient irradiance penetrates beneath the surface mixed layer to support an SCML. Due to the constraints of lower light levels in the SMCL, primary production within this layer could conceivably be extremely low. We therefore sought to quantify the contribution of SCMLs to global net primary production (NPP). A variety of vertically resolved models have been constructed to predict global NPP (Saba et al. 2011). We chose the model of Westberry et al. (2008) because it takes into account physiological differences in phytoplankton between the surface mixed layer and the SCML, including photo acclimation (i.e., changes in Chl*a*:C) and the consequences of the nutricline depth. We used this model to predict the fraction of annual NPP that occurs beneath the mixed layer (Fig. 3). Overall, according to this model, SCMLs contribute 47 % of the global annual NPP. In the permanently stratified regions, approximately 60 % of the annual NPP occurs within the SCML, although the degree of patchiness and variability in this proportion is notable. Moving poleward, the contribution of the SCML to annual NPP generally decreases to levels between 10 and 30 %, though again with notable patchiness, particularly in Arctic waters and the Southern Ocean. Estimates for the quantity of total water column productivity that are associated with the SCML in Arctic waters are also variable, ranging from a minor fraction (8 %; Arrigo et al. 2011) to a dominant fraction (76 %; Martin et al. 2012).

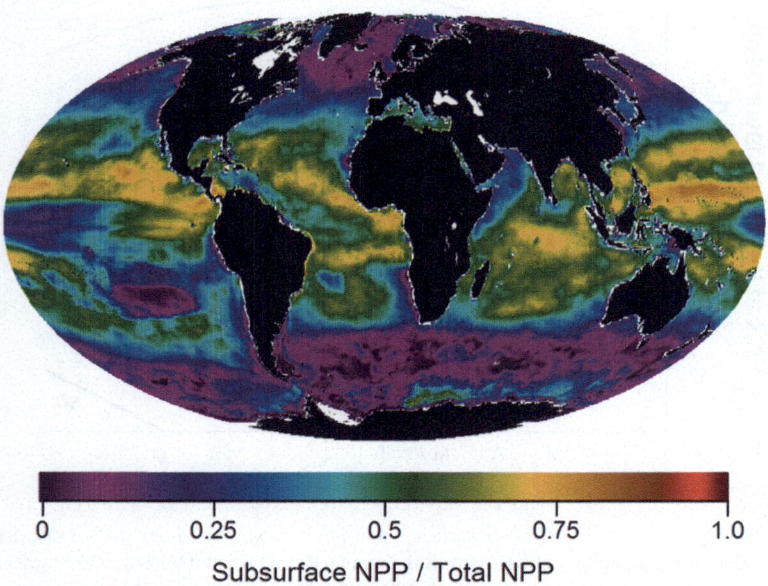

Fig. 3 The fraction of annual net primary production that occurs beneath the mixed layer depth calculated using the model of Westberry et al. (2008)

Though NPP within SCMLs can be substantial, this metric alone undervalues the role SCMLs may play in carbon cycling. Global NPP models do not distinguish between new production (i.e., fueled by the injection of nutrients from below the euphotic zone) and recycled production (i.e., fueled by nutrients regenerated within the euphotic zone; Dugdale and Goering 1967). This distinction is important because new production provides an energy base for higher trophic levels and is proportional to carbon export to the deep ocean (i.e., the biological pump). During stratification, SCMLs can intercept nearly all ascending nitrate arising from below the nutricline (Johnson et al. 2010; Martin et al. 2012). SCMLs may thereby act as filters, limiting the supply of nutrients to the surface mixed layer. Consequently, new production may be largely restricted to the SCML, while production in the nutrient-limited surface waters is mostly sustained by recycled production (Martin et al. 2012). At the same time, accelerated nutrient supply to the SCML enables an accumulation of carbon within this zone (Mignot et al. 2014) and so the mechanisms and magnitude of nutrient delivery to SCMLs may be key drivers of new productivity across systems. Here, we explore marine literature that has examined primary productivity within SCMLs and highlight the mechanisms of nutrient delivery in the context of new production in three diverse systems.

In the permanently stratified waters of the central oligotrophic gyres, deepening of the mixed layer even during the winter is insufficient to erode the nutricline, and surface phytoplankton blooms are absent. Below the mixed layer, SCMLs are a permanent feature of these gyres, and here they almost entirely reflect elevated chlorophyll content (Mignot et al. 2014). Subsurface chlorophyll maxima have traditionally been modeled by tracking a fixed percentage of surface insolation. However, it has more recently emerged that the depth of maximal Chl*a* in SCMLs seasonally tracks fixed isolumes (0.1–1.2 mol photons m^{-2} $days^{-1}$), where an isolume is a constant level of daily integrated light (Letelier et al. 2004; Cullen 2014; Mignot et al. 2014). These fixed isolumes deepen from spring to summer, migrating downward into the nutricline (Letelier et al. 2004; Mignot et al. 2014). When SCMLs intercept the nutricline, particulate carbon accumulates, suggesting a subsurface phytoplankton bloom during this time. Although phototrophic primary productivity is constrained by the low light at these depths, the carbon fixed here likely represents a significant portion of the new production (Johnson et al. 2010; Mignot et al. 2014). From summer to autumn, the isolumes shoal due to higher light attenuation in the surface mixed layer, and particulate carbon within the SCML recedes once again and remains low until the following spring, suggesting that primary productivity for the rest of the year is largely sustained by nutrient regeneration (Mignot et al. 2014).

In temperate oceans, spring blooms in the surface waters are an important feature of phytoplankton phenology and can supply a dominant portion of the annual new production. In some regions, however, annual surface primary production, including the spring bloom, appears to be insufficient to account for estimated rates of secondary production. In such regions, SCMLs, which appear seasonally (Fig. 2), have been proposed as an additional source of new production. For example, the Dogger Bank in the North Sea supports high densities of mesozooplankton and fish larvae, and is known for high fisheries production. During stratification, SCMLs are present at around 30 m depth, but unlike in the oligotrophic gyres, direct measurements identify that these SCMLs are unevenly distributed horizontally and through time. This heterogenous distribution likely reflects local variations in vertical mixing and variable bottom topography (Richardson et al. 2000). Internal tidal pumping, sweeping from shallow to deep waters over a 2-week period, has been suggested as the mechanism that injects a periodic supply of deepwater nutrients to the SCML (Richardson et al. 2000). Although primary production rates within the SCML are much lower than the surface spring bloom, these lower rates are compensated by their longer duration over the course of the stratified season. In this area, subsurface blooms were estimated to fix 24–48 gC m^{-2} $year^{-1}$ while the spring bloom was estimated to fix ~20 gC m^{-2} $year^{-1}$ (Richardson et al. 2000). As another example, in the Celtic Sea, seasonal SCMLs are reportedly responsible for 40–50 % of the annual water column primary productivity, where they fix an estimated 10–19 gC m^{-2} $year^{-1}$ (Hickman et al. 2012). Here, phytoplankton within the SCML are also demonstrably elevated in carbon relative to the surface waters (Moore et al. 2006), consistent with the idea that higher nutrient fluxes to the SCML (as seen in these coastal temperate seas) enhances carbon accumulation in this layer.

In Canadian coastal arctic seas, freshwater inputs create a shallow mixing depth (14–32 m) and a strong pycnocline (Martin et al. 2010). Extensive surveys in coastal waters revealed nearly ubiquitous SCMLs during the stratified open-water season. In these regions, the SCML is associated with a biomass (C) maximum (Martin et al. 2010). Within days after ice-out, a bloom of fast-growing phytoplankton exploits initially high nutrient concentrations, which become rapidly depleted from the mixed layer. Once the nutrients in the mixed layer are exhausted, phytoplankton biomass then proliferates within the SCML (Martin et al. 2010). Seasonally, the SCML proceeds downward, progressively consuming limiting nutrients beneath the surface mixed layer, until the pycnocline is destabilized by extensive deep mixing in the fall (Martin et al. 2010, 2012). The role of the SCML in filtering nitrate from deep waters in the Arctic may be immense, with potentially 98 % of the nitrate trapped at the SCML (Martin et al. 2012). During the development of the SCML, primary productivity is initially dominated by new production, but this is eventually replaced by a dominance of regenerated production which persists for the remainder of the ice-free stratified season (Martin et al. 2012).

The observation that nutrient additions to the SCML stimulate phytoplankton blooms at these depths appears to be widespread. We conceptualize the SCML as a gradient environment (i.e., rather than a well-mixed reactor), where nutrients arising from below are progressively intercepted by phytoplankton, and light provided from above is progressively attenuated with depth. Within this framework it is possible that the phytoplankton within the SCML are co-limited by nutrients and light, such that an influx in nutrients stimulates phytoplankton growth rates. Mignot et al. (2014) further postulate that when SCMLs progressively intercept the nutricline (i.e., when nutrient fluxes to the SCML are increasing), phytoplankton division rates increase faster than grazing rates, and it is this escape from predation control that enables the SCML phytoplankton biomass to bloom. Recently, it has also been suggested that nitrate-storing phytoplankton within the SCML, such as the mat-forming diatom *Rhizosolenia*, may migrate upward in the water column following a pulse of nutrients (e.g., from a mesoscale eddy; Johnson et al. 2010; Villareal et al. 2014). Nutrient translocation to the surface waters by phytoplankton may be quantitatively important for water column primary production budgets, but it does not explain enhanced primary production within the SCML. A robust conceptual model that resolves the mechanisms underlying changes in phytoplankton productivity within SCMLs, and the potential role of grazers, is still unresolved.

Subsurface Chlorophyll Maximum Layers in Lakes

The structure and maintenance of SCMLs in lakes largely parallels marine systems. As in marine systems, the contrasting gradients of light supply from above and nutrients from below are primary drivers of the distribution of SCMLs (Klausmeier and Litchman 2001). Similarly, the degree to which the elevated Chl*a* within the SCML reflects photo acclimation versus elevated phytoplankton biomass likely also varies predictably along a gradient of trophic status (Cullen 2014). Photo acclimation can be entirely responsible for the elevated Chl*a* at depth in oligotrophic waters (Barbeiro and Tuchman 2001; Hodges and Rudnick 2004), while elevated phytoplankton biomass may be responsible for elevated Chl*a* at depth in more productive waters (Abbott et al. 1984). As in the marine realm, phytoplankton assemblages within the SCML are chromatically adapted to the spectral qualities of their light environment (Silsbe et al. 2012). Finally, as in marine systems, the delivery of limiting nutrients to SCMLs in lakes during stratification affects the metabolic activity of the phytoplankton residing there, potentially suggesting nutrient and light co-limitation within the SCML (Abbott et al. 1984). Lakes exhibit a large range in optical and nutritional properties across systems, and consequently, lakes may be particularly well suited for testing hypotheses related to the mechanisms of how SCML productivity is regulated.

At times SCMLs in lakes have been demonstrably associated with elevated primary production (e.g., Abbott et al. 1984; Moll et al. 1984; Fahnenstiel and Scavia 1987). Yet generalities about the occurrence and magnitude of phytoplankton blooms below the epilimnion are lacking. The majority of published datasets are limited to ex situ incubations of surface waters, and measurements from deeper waters, when examined, are often susceptible to spectral bias. The contribution of SCML phytoplankton productivity to whole lake productivity may be substantial in clear-water lakes during stratified seasons, but our understanding of their role in geochemical cycling is still evolving.

Because primary productivity within SCMLs in lakes can be elevated relative to the nutrient-starved epilimnion, phytoplankton production within an SCML may be important for lacustrine consumer production. The food quality of phytoplankton for zooplankton may be elevated within the SCML, and this can even be independent of whether carbon is elevated within the SCML (Williamson et al. 1996). In a survey of 25 small temperate lakes, primary production derived from SCML phytoplankton was found to account for up to 80 % of total zooplankton production, with the greatest contribution by SCML primary production associated with the clearest lakes (Francis et al. 2011).

The intersection of SCMLs with the sessile benthic suspension feeders may also be an important pathway linking primary production to secondary consumption. In Lake Ontario, for example, seasonal growth rates of the invasive Quagga Mussels (*Dreissena bugensis*) were maximal when the SCML intersected the benthos (Malkin et al. 2012). Net positive growth rates of the Quagga Mussels were observed only in the spring, when an SCML was detectable in proximity to the littoral zone benthos, suggesting that the nutritional subsidy provided by subsurface phytoplankton may be important on an annual basis (Malkin et al. 2012). Such benthic–pelagic coupling may in part explain the depth distribution of benthic filter feeders in other Laurentian Great Lakes. In Lake Michigan, for example, that *Dreissena* abundance is maximal within the 30–50 m depth interval (Nalepa et al. 2009) may reflect a dependence on elevated sub-epilimnetic phytoplankton commonly observed during stratification. A better understanding of the seasonal distribution, persistence, rates of productivity, and nutrient delivery to SCML phytoplankton are needed to improve our general understanding of the role of subsurface phytoplankton in lake ecosystem functioning.

Acknowledgements We wish to thank our past and present mentors, particularly Bob Hecky and Stephanie Guildford, for encouraging us through our careers. We also extend our appreciation to Sébastien Gardoll for a brief residency in Paris, to formulate the ideas that became this chapter. Data and the model code of Westberry et al. (2008) were obtained from the NASA-funded Oregon State University Ocean Productivity project (http://www.science.oregonstate.edu/ocean.productivity/).

Greg M. Silsbe and Sairah Y. Malkin

We met the day Sairah was interviewing for an M.Sc. position with Bill Taylor at the University of Waterloo. In the hopes of impressing her, I (Greg) informed her I had just returned from a CIDA-funded internship program on the tropical shores of Lake Victoria, East Africa, where I was soon to be returning as an M.Sc. student. And then, I proceeded to show her data tables of CTD profiles. Nevertheless, we would be dating a year later, and we remained at U. Waterloo, to pursue Ph.D.s in limnology with Bob Hecky and Stephanie Guildford, supervisors whose wisdom and humanity has informed much of our careers. Bob had once suggested that if we wanted to better understand lakes, we should take a look at coastal oceans. We took this to heart, and so from U. Waterloo, we went on to pursue postdocs at The Netherlands Institute of Sea Research (NIOZ; with Jacco Kromkamp and Filip Meysman), living variously in The Netherlands, and then across the Schelde in Antwerp, Belgium (where the roads were worse, but the food was better). We were becoming increasingly specialized in the benthos (Sairah) and pelagia (Greg). Logically, our first manuscript together would examine benthic-pelagic coupling (Malkin et al. 2012). After brief stops at the University of Georgia (Sairah; with Mandy Joye), and the University of Oregon (Greg; with Toby Westberry), we are now about to embark on faculty positions with the University of Maryland Center for Environmental Sciences (UMCES), Horn Point Laboratory. We are grateful to institutes that recognize the synergistic benefits of dual spousal hires, and we are looking forward to continuing our research careers together at UMCES.

References

Abbott MR, Denman KL, Powell TM, Richerson PJ, Richards RC, Goldman CR (1984) Mixing and the dynamics of the deep chlorophyll maximum in Lake Tahoe. Limnol Oceanogr 29:862–878. doi:10.4319/lo.1984.29.4.0862

Arrigo KR, Matrai PA, van Dijken GL (2011) Primary productivity in the Arctic Ocean: impacts of complex optical properties and subsurface chlorophyll maxima on large-scale estimates. J Geophys Res: Oceans 116, C11022. doi:10.1029/2011JC007273

Barbeiro RP, Tuchman ML (2001) Results from the U.S. EPA's biological open water surveillance program of the Laurentian great lakes: II. Deep chlorophyll maxima *J*. Great Lakes Res 27:155–166. doi:10.1016/S0380-1330(01)70629-6

Cullen JJ (2014) Subsurface chlorophyll maximum layers: enduring enigma or mystery solved? Ann Rev Mar Sci 7:19.1–19.33. doi:10.1146/annurev-marine-010213-135111

Dugdale RC, Goering JJ (1967) Uptake of new and regenerated forms of nitrogen in primary productivity. Limnol Oceanogr 12:196–206. doi:10.4319/lo.1967.12.2.0196

Estrada M (1996) Primary production in the northwestern Mediterranean. Sci Mar 60(Suppl 2): 55–64

Fahnenstiel GL, Scavia D (1987) Dynamics of lake Michigan phytoplankton: the deep chlorophyll layer. J Great Lakes Res 13:285–295. doi:10.1016/S0380-1330(87)71652-9

Fairbanks RG, Weibe PH (1980) Foraminifera and chlorophyll maximum: vertical distribution, seasonal succession, and paleoceanographic significance. Science 209:1524–1526. doi:10.1126/science.209.4464.1524

Fennel K, Boss E (2003) Subsurface maxima of phytoplankton and chlorophyll: steady-state solutions from a simple model. Limnol Oceanogr 48:1521–1534. doi:10.4319/lo.2003.48.4.1521

Francis TB, Schindler DE, Holtgrieve GW, Larson ER, Scheuerell MD, Semmens BX, Ward EJ (2011) Habitat structure determines resource use by zooplankton in temperate lakes. Ecol Lett 14:364–372. doi:10.1111/j.1461-0248.2011.01597.x

Geider RJ, MacIntyre HL, Kana TM (1998) A dynamic regulatory model of phytoplankton acclimation to light, nutrients and temperature. Limnol Oceanogr 43:679–694. doi:10.4319/lo.1998.43.4.0679

Hickman AE, Holligan PM, Moore CM, Sharples J, Krivtsov V, Palmer MR (2009) Distribution and chromatic adaptation of phytoplankton within a shelf sea thermocline. Limnol Oceanogr 54:525–536. doi:10.4319/lo.2009.54.2.0525

Hickman AE, Moore CM, Sharples J, Lucas MI, Tilstone GH, Krivtsov V, Holligan PM (2012) Primary production and nitrate uptake within the seasonal thermocline of a stratified shelf sea. Mar Ecol Prog Ser 463:39–57. doi:10.3354/meps09836

Hodges BA, Rudnick DL (2004) Simple models of steady deep maxima in chlorophyll and biomass. Deep-Sea Res Part I 51:999–1015. doi:10.1016/j.dsr.2004.02.009

Johnson KS, Riser SC, Karl DM (2010) Nitrate supply from deep to near-surface waters of the North Pacific subtropical gyre. Nature 465:1062–1065. doi:10.1038/nature09170

Klausmeier CA, Litchman E (2001) Algal games: the vertical distribution of phytoplankton in poorly mixed water columns. Limnol Oceanogr 46:1998–2007. doi:10.4319/lo.2001.46.8.1998

Letelier RM, Karl DM, Abbot MR, Bidagare RR (2004) Light driven seasonal patterns of chlorophyll and nitrate in the lower euphotic zone of the North Pacific Subtropical Gyre. Limnol Oceanogr 49:508–519. doi:10.4319/lo.2004.49.2.0508

Malkin SY, Silsbe GM, Smith REH, Howell ET (2012) A deep chlorophyll maximum nourishes benthic filter feeders in the coastal zone of a large clear lake. Limnol Oceanogr 57:735–748. doi:10.4319/lo.2012.57.3.0735

Malmstrom RR, Coe A, Kettler GC, Martiny AC, Frias-Lopez J, Zinser ER, Chisholm SW (2010) Temporal dynamics of Prochlorococcus ecotypes in the Atlantic and Pacific oceans. ISME J 4:1252–1264. doi:10.1038/ismej.2010.60

Martin J, Tremblay J-É, Gagnon J, Tremblay G, Lapoussière A, Jose C, Poulin M, Gosselin M, Gratton Y, Michel C (2010) Prevalence, structure and properties of subsurface chlorophyll maxima in Canadian Arctic waters. Mar Ecol Prog Ser 412:69–84. doi:10.3354/meps08666

Martin J, Tremblay J-É, Price NM (2012) Nutritive and photosynthetic ecology of subsurface chlorophyll maxima in Canadian Arctic waters. Biogeosciences 9:5353–5371. doi:10.5194/bg-9-5353-2012

Mignot A, Claustre H, Uitz J, Poteau A, D'Ortenzio F, Xing X (2014) Understanding the seasonal dynamics of phytoplankton biomass and the deep chlorophyll maximum in oligotrophic environments: A Bio-Argo float investigation. Global Biogeochem Cycles 28:856–876. doi:10.1002/2013GB004781

Moll RA, Brahce MZ, Peterson TP (1984) Phytoplankton dynamics within the subsurface chlorophyll maximum of Lake Michigan. J Plankton Res 6:751–766. doi:10.1093/plankt/6.5.751

Moore LM, Chisholm SW (1999) Photophysiology of the marine cyanobacterium *Prochlorococcus*: Ecotypic differences among cultured isolates. Limnol Oceanogr 44:628–638. doi:10.4319/lo.1999.44.3.0628

Moore CM, Suggett DJ, Hickman AE, Kim Y-N, Tweddle JF, Sharples J, Geider RJ, Holligan PM (2006) Phytoplankton photoacclimation and photoadaptation in response to environmental gradients in a shelf sea. Limnol Oceanogr 51:936–949. doi:10.4319/lo.2006.51.2.0936

Nalepa TF, Fanslow DL, Lang GA (2009) Transformation of the offshore benthic community in Lake Michigan: recent shift from the native amphipod *Diporeia* spp. to the invasive mussel *Dreissena rostriformis* bugensis. Freshwater Biol 54:466–479. doi:10.1111/j.1365-2427.2008.02123.x

Pannard A, Beisner BE, Bird DF, Braun J, Planas D, Bormans M (2011) Recurrent internal waves in a small lake: Potential ecological consequences for metalimnetic phytoplankton populations. Limnol Oceanogr: Fluids Environ 1:91–109. doi:10.1215/21573698-1303296

Richardson K, Visser AW, Pedersen FB (2000) Subsurface phytoplankton blooms fuel pelagic production in the North Sea. J Plankton Res 22:1663–1671. doi:10.1093/plankt/22.9.1663

Saba VS, Friedrichs MA, Carr M-E, Antoine D, Armstrong RA et al (2011) Challenges of modeling depth-integrated marine primary productivity over multiple decades: A case study at BATS and HOT. Glob Biogeochem Cycles 24, GB3020. doi:10.1029/2009GB003655

Seki MP, Polovina JJ, Brainard RE, Bidigare RR, Leonard CL, Foley DG (2001) Biological enhancement at cyclonic eddies tracked with GOES thermal imagery in Hawaiian waters. Geophys Res Lett 28:1583–1586. doi:10.1029/2000GL012439

Silsbe GM, Smith REH, Hecky RE (2012) Improved estimation of carbon fixation rates from active fluorometry using spectral fluorescence in light-limited environments. Limnol Oceanogr Methods 10:736–751. doi:10.4319/lom.2012.10.736

Tittel J, Bissinger V, Zippel B, Gaedke U, Bell E, Lorke A, Kamjunke N (2003) Mixotrophs combine resource use to outcompete specialists: Implications for aquatic food webs. Proc Natl Acad Sci U S A 100:12776–12781. doi:10.1073/pnas.2130696100

Uitz J, Claustre H, Morel A, Hooker SB (2006) Vertical distribution of phytoplankton communities in open ocean: an assessment based on surface chlorophyll. J Geophys Res 111:C08005. doi:10.1029/2005JC003207

Venrick EL (1988) The vertical distributions of chlorophyll and phytoplankton species in the North Pacific central environment. J Plankton Res 10:987–998. doi:10.1093/plankt/10.5.987

Villareal TA, Pilskaln CH, Montoya JP, Dennet M (2014) Upward nitrate transport by phytoplankton in oceanic waters: balancing nutrient budgets in oligotrophic seas. Peer J 2, e302. doi:10.7717/peerj.302

Westberry T, Behrenfeld MJ, Siegel DA, Boss E (2008) Carbon-based primary productivity modelling with vertically resolved photoacclimation. Global Biogeochem Cycles 22:1–18. doi:10.1029/2007GB003078

Williams C, Sharples J, Green M, Mahaffey C, Rippeth T (2013) The maintenance of the subsurface chlorophyll maximum in the stratified western Irish Sea. *Limnol*. Oceanogr: Fluids Environ 3:61–73. doi:10.1215/21573689-2285100

Williamson CE, Sanders RW, Moeller RE, Stutzman PL (1996) Utilization of subsurface food resources for zooplankton reproduction: Implications for diel vertical migration theory. Limnol Oceanogr 41:224–233. doi:10.4319/lo.1996.41.2.0224

Part IV
Looking in the Rear View Mirror: The Long View on Changing Ecosystems

An Ecosystem in Transition: The Emergence of Mixotrophy in the Arabian Sea

Joaquim I. Goes and Helga do R. Gomes

Introduction

The multidisciplinary JGOFS program undertaken from 1994 to 1996 (Smith et al. 1998) was able to establish that biological productivity in the Arabian Sea (AS) was largely driven by monsoonal winds, which caused upwelling and fertilization of its normally nutrient-depleted euphotic zone. This multinational study established that the huge summer-time increase in biological productivity predominantly off the coast of Somalia, Yemen, and Oman resulted from Ekman forced upwelling of nutrient-rich, subsurface waters, whereas in winter, the large increase in biological production was the result of nutrient enhancement from convective mixing. During these two periods of high primary production, phytoplankton communities were dominated by large diatom blooms, which also were responsible for the high amounts of C export (Banse and McClain 1986; Garrison et al. 2000). In contrast, prior to and following the winter monsoon, both periods of low nutrient concentrations, surface waters were dominated by autotrophic picoplankton *Prochlorococcus* and *Synechococcus* and measured C export efficiencies were significantly lower.

Post-JGOFS, because of the rise in piracy and the difficulty of mounting shipboard cruises to the region, most recent estimates of phytoplankton biomass (estimated as Chl *a*), phytoplankton productivity, and C export have come from satellite-based observations (Goes et al. 2005; Gomes et al. 2008; Lévy et al. 2007), small targeted area-specific shipboard cruises (Al-Azri et al. 2007; Gomes et al. 2008, 2009; Gomes et al. 2014; Dwivedi et al. 2015), or coupled biophysical models (Wiggert et al. 2005). One such study (Goes et al. 2005) based solely on satellite data was the first to show that the AS was becoming more productive during summer due to intensification of the southwest monsoon winds. The authors were able

J.I. Goes (✉) • H.d.R. Gomes
Marine Biology and Paleoenvironment, Lamont Doherty Earth Observatory at Columbia University, Palisades, New York, NY 10964, USA
e-mail: jig@ldeo.columbia.edu

P.M. Glibert, T.M. Kana (eds.), *Aquatic Microbial Ecology and Biogeochemistry: A Dual Perspective*, DOI 10.1007/978-3-319-30259-1_13

to establish that the loss of snow cover in the Himalayan-Tibetan period during winter was causing the southwest summer monsoon winds to intensify and escalate coastal upwelling, which in turn was causing a large increase in phytoplankton biomass during the summer monsoon.

More recently using shipboard and satellite data, Gomes et al. (2014) were able to demonstrate that the northern AS was experiencing a weakening of winter convective mixing, implying a potential decline in the input of subsurface nutrients to the euphotic column. However, contrary to this observation, satellite-derived Chl *a* concentrations revealed that the AS was becoming more productive in winter, raising questions about this apparent anomaly.

Materials and Methods

Arabian Sea Cruises and Sample Collection

Samples for this study were obtained from four cruises on board the Indian Research vessels *RV Sagar Kanya and FORV Sagar Sampada* to the AS conducted from between the months of February and March in 2009, 2010, and 2011. On board seawater samples were collected with a 5-L Niskin® sampler mounted on a Sea Bird Electronics® CTD Rosette using prescribed clean techniques. Calibrated sensors on the CTD provided vertical profiles of salinity, temperature, and dissolved O_2. The salinity sensor was calibrated against Autosal® Salinometer measurements, while the dissolved O_2 sensor was calibrated against measurements using the Winkler and colorimetric end point automated titration procedures in seawater samples from discrete depths. From 2009 onwards, dissolved O_2 was measured exclusively by the Winkler method with high-precision amperometric end point detection (Langdon 1984).

Phytoplankton Cell Counts

Microscopic identification and enumeration of phytoplankton were undertaken in samples fixed with 1 % Lugol's iodine in 3 % buffered formaldehyde solution and stored in cool and dark conditions for cells to settle. Prior to microscopy, samples were concentrated to 5–10 ml by siphoning out the top layer through a 10 μm Nitex® filter wrapped around the tip of a thin tube. Replicates of the concentrated sample were transferred on to a Sedgwick-Rafter® slide and cell numbers were enumerated with an Olympus Inverted microscope (Model IX 50) at 200× magnification. Identification of phytoplankton was based on standard taxonomic keys (Tomas 1997), and was in most cases undertaken up to the species level. Historic phytoplankton data was downloaded from the Open Access library PANGAEA.

Photosynthetic Rate Measurements

The photosynthetic competency of natural *Noctiluca*-dominated populations of phytoplankton in this study was measured using a Kimoto® Fast Repetition Rate Fluorometer (FRRF). The FRRF has two optical chambers for measurements of actinic light-induced fluorescence under dark and ambient light conditions. The light chamber is equipped with a dichroic cyan filter that prevents the red wavelengths of ambient light from penetrating the chamber. This design allows for accurate measurements of red fluorescence signals, even under strong ambient light in near-surface waters. Chl *a* fluorescence transients were obtained using a single-turnover protocol as in Fujiki et al. (2008) which allows for variable fluorescence normalized to maximum fluorescence (F_v/F_m; unitless), the functional absorption cross section of photosystem 2 (σ_{PSII}; units of (A^2 quanta^{-1})), and the turnover time of the primary PSII electron acceptor Q_a (τ_{Qa}), conducted in the light and dark chambers. Fluorescence response data obtained were fitted to the model of Kolber et al. (1998) from which the photosynthetic parameters F_o, F_m, F_v/F_m, σ_{PSII}, and τ_{Qa} were calculated. For each parameter, values were derived for both dark and ambient light conditions, and the latter corresponds to F', F_m' F_q'/F_m', and σ_{PSII} which can be used to estimate the chlorophyll *a* (Chl *a*)-specific gross primary productivity (P^B) by phytoplankton using the equation

$$P^B = \mathrm{PAR} \times \sigma_{\mathrm{PSII}} \times qP \times f \times n_{\mathrm{PSII}} \times \varphi e \times S,$$

of Kolber and Falkowski (1993), where PAR is the ambient irradiance; n_{PSII} the ratio of PSII reaction centers (RCII) to Chl *a*; *f* the fraction of RCII that are capable of evolving oxygen (O_2), also known as the "functional" state and assumed to be $(F_v/F_m)/0.65$; q_P the photochemical quenching coefficient, which is a measure of the fraction of open RCII; φ_e the quantum yield for oxygen evolution; and *S* a scaling factor.

Quantum yields and photosynthetic efficiencies of bulk *Noctiluca*-dominated phytoplankton populations and of two other size classes of phytoplankton (i.e., nanoplankton (<20 μm) and picoplankton (<3 μm)) were measured using a "photosynthethron." This temperature-controlled incubator allows measurements of ^{14}C-$NaHCO_3$ uptake-based photosynthesis against a gradient of light (Lewis and Smith 1983) that can be used for estimating photosynthetic rate parameters (α—a measure of photosynthetic efficiency indicated by the initial slope of the P vs. E curve, P_{max}—the maximum photosynthetic rate of a population, and β—the photoinhibition parameter).

Autotrophy Versus Heterotrophy in Noctiluca

Measurements of the growth rates of *Noctiluca* were made separately in deck incubators for 96 h to assess the relative importance of internal nutrition from endosymbionts and external phagotrophy for growth of *N. scintillans*. Groups of ten cells of

Noctiluca were incubated in 70 ml tissue culture flasks under reduced sunlight (see above) or in darkness (flask wrapped in aluminum foil) in either 0.45 μm Nybolt membrane-filtered seawater (no food) or whole seawater with natural plankton assemblage increased to five times the normal concentration (food) by gentle reverse filtration through a 20 μm mesh. Ambient nutrient concentrations in these experiments ranged from 0.37 to 0.66 μM for inorganic nitrate, 0.20 to 0.38 μM for inorganic phosphate, and 1.99 to 5.09 μM for silicate. Parallel experiments were performed in which nutrient concentrations within the tissue culture flasks were enhanced with 17.6 μM inorganic nitrate and 0.7 μM inorganic phosphate. At the end of the incubation period, cells were counted using an Olympus SZX10 stereomicroscope. Specific growth rates (μ day^{-1}) were calculated as the natural logarithm of the number of cells present at the end of the experiment minus the natural logarithm of the number of cells at the beginning of the experiment divided by the duration of the experiment in days. Although *Noctiluca* is capable of cannibalism which could compromise growth rate estimates, it is mainly a problem when *Noctiluca* densities are high and food is scarce. Our experiments had only ten cells per culture flask. In addition we observed the light treatments with and without food daily and found no increase in *Noctiluca* numbers followed by a decline as would be expected if initial growth followed by cannibalism after starvation or food depletion occurred.

Salp Grazing Experiments

Clean 10 L buckets were used to collect surface *N. scintillans* bloom populations. Triplicate samples of 50 ml were filtered onto GF/F glass fiber filters and immediately transferred into a liquid nitrogen Dewar for HPLC measurements of pigment concentrations at time zero. Live salps were collected separately from which single adult salps were carefully picked, and transferred into filtered seawater for about 2 h to empty their guts and then into the buckets containing the *Noctiluca* bloom. After a 4-h incubation period when the buckets were incubated in running seawater in the dark, 50 ml samples were collected in triplicate for post-incubation pigment analysis by HPLC (Bidigare et al. 2002). In addition the fecal pellets that were formed were carefully picked and placed in petri dishes.

Lipid Accumulation in Noctiluca

Lipid accumulation within *Noctiluca* was studied by staining the lipids with the red lysochrome azo dye, Sudan Red G. Staining of lipid droplets requires the cells to be fixed immediately with 2.5 % glutaraldehyde (15 min), followed by rinsing the cells with deionized water to rid the sample of excess glutaraldehyde. The cells were then transferred to 60 % isopropanol. When stained with Sudan Red G, lipid droplets can be viewed as red globules under a light microscope.

Statistical Analysis

The hydrological and chemical conditions under which *Noctiluca* thrived were established with the help of principal component analysis using the statistical package PRIMER® ver. 6 (Clarke et al. 2005). PRIMER® is ideal for ecological data sets that have a combination of biotic, chemical, and physical parameters. In particular, it was designed for use with biotic data in which there are a very large number of species, and most species are present infrequently across samples. The significance of differences among treatments and within treatments for the grazing experiments was assessed using Kruskal-Wallis one-way or Friedman's two-way ANOVA. Both these methods assume non-normal distribution of the data.

Results and Discussion

Emergence of Noctiluca and Shift in Phytoplankton Biodiversity in the Arabian Sea

Starting from 2003 onwards, data from cruises in the AS by our colleagues in India (which we were able to join from 2009 to 2011) established that the composition of winter blooms in the AS had changed drastically from those sampled during the JGOFS era. The data also provided an answer to the contradictory situation of increasing blooms in spite of a weakening of convective mixing which implies a decrease in nutrient inputs. These datasets, in particular those on phytoplankton taxonomy (Parab et al. 2006; Gomes et al. 2008, 2009; Gomes et al. 2014), collected as part of the validation program of Indian Space Research Organization's ocean color satellites, OCEANSAT-I and -II (Dwivedi et al. 2015), showed that the northern AS was experiencing a startling shift in the composition of winter-time phytoplankton blooms, from diatoms to those of green *Noctiluca* (Gomes et al. 2014). These taxonomic datasets revealed that this shift in bloom populations occurred around 2000. Earlier satellite ocean color and altimetry based studies helped show that *Noctiluca* blooms appear first off the coast of Oman in the month of November and with the aid of mesoscale eddies start shifting eastward as the season progresses, leading to their dispersal over the AS by early February (Fig. 1a–f). By mid-February, *Noctiluca* blooms engulf the entire northern AS (Gomes et al. 2008, 2009). Since they were first detected, *Noctiluca* blooms have become increasingly pervasive and widespread, throughout the AS occurring with remarkable consistency every year from December to mid-March (Parab et al. 2006; Gomes et al. 2008, 2009, Gomes et al. 2014; Thibodeau et al. 2014; Werdell 2014; Werdell et al. 2014). The *Noctiluca* bloom of 2015 was the thickest and largest on record, covering an area ~1.8×10^6 km^2 almost thrice the size of Texas (Fig. 1f).

Detailed historic taxonomic records (Krey et al. 1971; Piontkovski 2002), available for the winter monsoons of 1965 and 1972, show no indication of *Noctiluca*

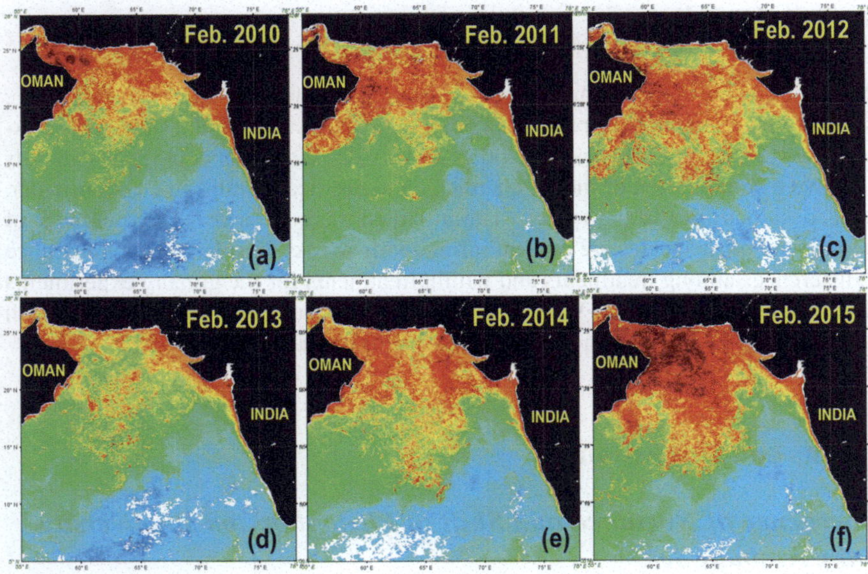

Fig. 1 **(a–f)** NASA's MODIS-Aqua monthly composite images of Chl *a* in the Arabian Sea showing the spatial expanse of *Noctiluca* blooms in February of 2010, 2011, 2012, 2013, 2014, and 2015

even as a minor component of the winter phytoplankton community. Similarly, taxonomic data from all earlier winter monsoonal expeditions such as the British ARABESQUE program in 1994 (Burkill 1999), the US JGOFS Expeditions from 1994 to 1996 (Garrison et al. 2000; Smith 2005), and the Indian JGOFS effort in 1995 (Sawant and Madhupratap 1996) do not cite *Noctiluca* as a component of the winter phytoplankton community, but instead show diatom dominance. Instead during all these cruises, diatoms were the dominant bloom-forming group fuelled by nutrient inputs from winter convective mixing (Banse and McClain 1986). Research undertaken during the JGOFS revealed that diatom blooms significantly influenced grazing (Landry et al. 1998; Smith et al. 1998; Caron and Dennett 1999) and C export (Garrison et al. 2000).

Environmental Factors Associated with Outbreaks of Noctiluca Blooms

Gomes et al. (2014) observed that *Noctiluca* thrives in waters that are undersaturated with respect to oxygen; and in shipboard experiments were able to show that endosymbionts of *Noctiluca* fix carbon more efficiently under suboxic conditions as compared to normal oxygen conditions. A synthesis of all available cruise

Fig. 2 Principal component analysis of environmental data collected during cruises in the AS showing the association of *Noctiluca* (**NOC**) with low-oxygen waters (**O2**), inorganic nutrients, nitrate (**NO3**), phosphate (**PO4**), and silicate (**SiO3**), diatoms (**DIA**), and dinoflagellates (**DIN**)

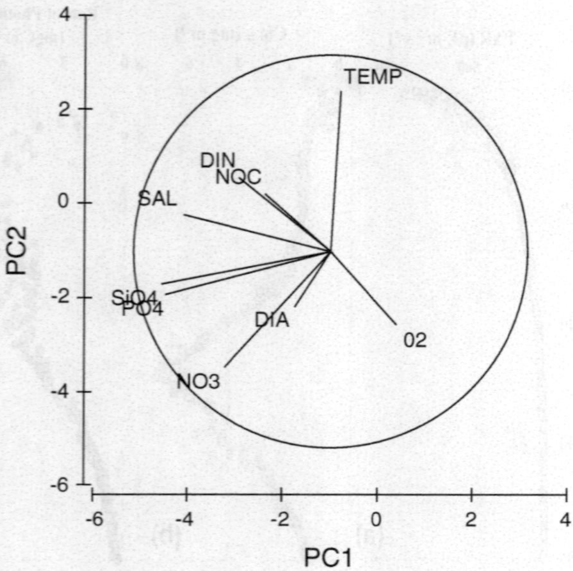

data does indeed confirm that *Noctiluca* has a propensity low-oxygen waters (Fig. 2a–c).

In October and November, prior to their appearance as surface blooms, *Noctiluca* cells are generally found at depth, close to the oxycline, oxygen levels are low (<3.5 ml L^{-1}) and nutrient levels high (Gomes et al. 2014). Photosynthesis-irradiance (P-E) experiments (Lewis and Smith 1983) revealed that the endosymbionts within both subsurface and surface populations of *Noctiluca* are indeed photosynthetically most efficient at incident irradiance levels of ~250 μE m^{-2} s^{-1}, lower than surface irradiance and closer to subsurface light intensities (Gomes et al. 2014) (Fig. 3a, b). At higher light intensities, the photosynthetic performance of *Noctiluca* cells declined sharply, seen in the form of a reduction in maximum quantum yield of photochemistry in PSII (F_v/F_m) and ^{14}C-based carbon fixation rates (Figs. 3c, d and 4a–c). This is paradoxical given that for at least 2 months later during the winter monsoon, *Noctiluca* exists as thick blooms at the surface where ambient light levels are sometimes five to six times higher than their optimal conditions, and oxygen levels are close to saturation.

Microscopic analysis invariably revealed large numbers of diatoms in the food vacuoles of surface-dwelling populations of *Noctiluca,* in contrast to subsurface populations, which were generally devoid of ingested diatoms (Fig. 5a). These observations lead us to hypothesize that *Noctiluca* can thrive under high light and saturated O_2 at the sea surface in the AS because of their switch to a substantial dependence on heterotrophy.

Preliminary feeding experiments conducted on the 2011 cruise allowed us to observe a highly complex feeding behavior in *Noctiluca* because of its mixotrophic adaptation (Gomes et al. 2014). In a mixed bloom of diatoms and *Noctiluca,* diatoms

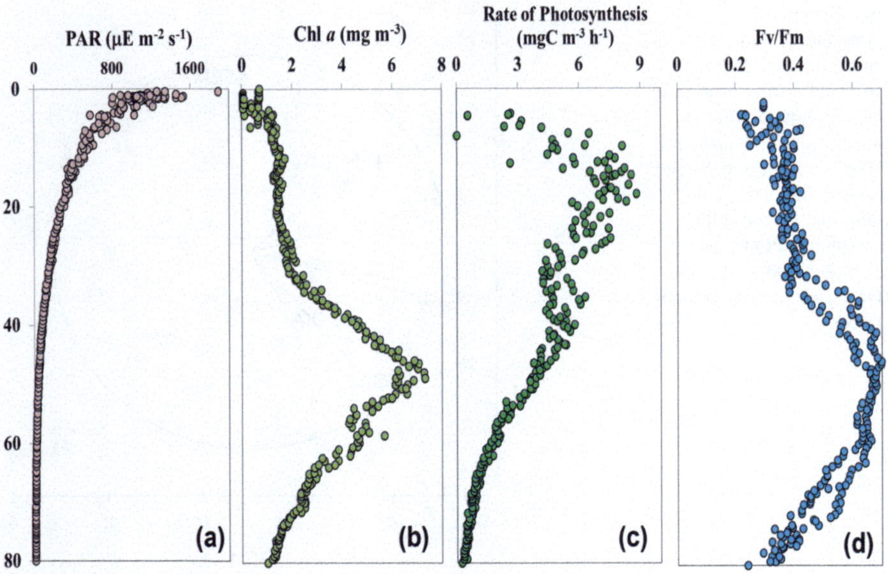

Fig. 3 Profiles of (**a**) PAR, (**b**) fluorescence-derived Chl *a*, (**c**) rates of photosynthesis, and (**d**) F_v/F_m obtained with a profiling FRRF, at a typical *Noctiluca*-dominated station in the Arabian Sea

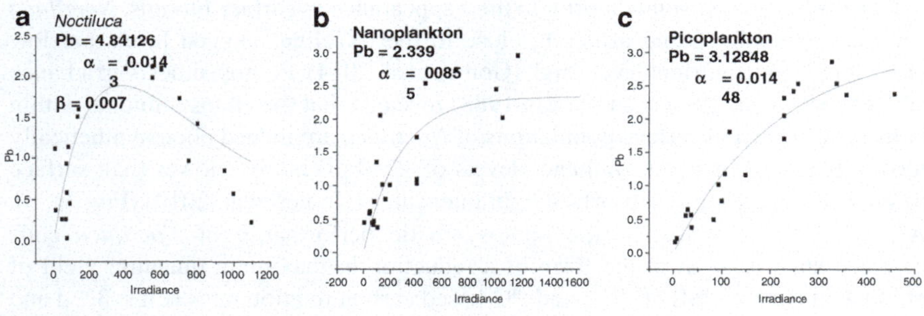

Fig. 4 Typical photosynthesis versus light curves for surface-dwelling populations of *Noctiluca*, nanoplankton (<20 µm), and picoplankton (<3 µm) populations in the Arabian Sea

were rapidly grazed by *Noctiluca*. When grown in the dark with or without food, *Noctiluca* rates were much lower than in the light, indicating the importance of the autotrophic endosymbionts for their growth. When nutrients were added to the "No Food" set of bottles, growth rates were significantly enhanced over controls reinforcing the view that *Noctiluca* can meet its metabolic needs via its autotrophic endosymbionts. If however extraneous food and light were available, growth rates of *Noctiluca* were comparable to treatments with only nutrients and light indicating that *Noctiluca* can also sustain its growth via phagotrophy. The lack of any further increase in growth rates, when nutrients were added to flasks containing food and exposed to light, suggests that when food is available, *Noctiluca* is capable of meet-

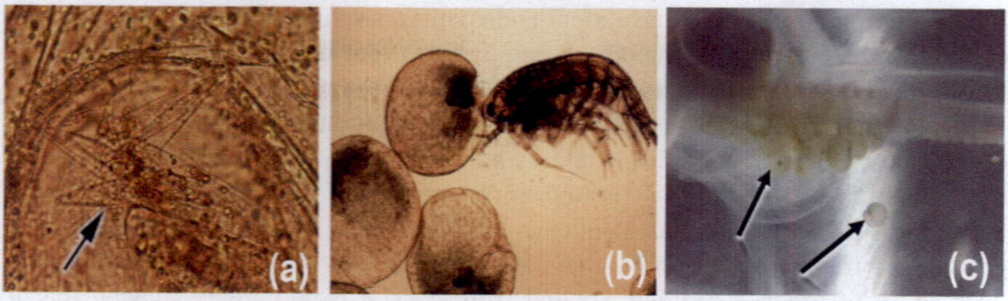

Fig. 5 (**a**) Microscopic image of surface-dwelling populations of *Noctiluca* showing the presence of ingested diatoms (*arrow*) in the food vacuoles. (**b**) *Noctiluca* cell depicting its large size (up to 1 mm in diameter) shown side by side of an amphipod. (**c**) Close-up of a salp after feeding on *Noctiluca*. *Arrows* showing *Noctiluca* cells ingested by salps

ing a large fraction of its metabolic requirements via phagotrophy. Conversely, when food becomes scarce but nutrients are available, *Noctiluca* meets its metabolic needs via carbon fixation by its autotrophic endosymbionts.

Furthermore, *Noctiluca* did not appear to be a preferred food source for both micro- and mesozooplankton possibly because of its large size (Fig. 5b). Instead, its major consumers were salps and jellyfish (Gomes et al. 2014). Clearance rates of 200–800 ml per hour were estimated for the salp *Pegea confoederata,* suggesting that high concentrations of salps (up to hundreds of zooids per m^2) (Fig. 5c) could rapidly consume dense aggregations of *Noctiluca.* Using pigments as a measure of food consumed in the first 3 h, a single salp could remove on average 71 % of Chl *a* (average decrease from 11.23 to 3.3 μg/L) and 78 % of Chl *b* (average decrease from 6.5 to 1.4 μg/L) from seawater dominated by *Noctiluca* cells with the deposition of large pellets. This short-circuiting of the food web brought about by the emergence of *Noctiluca* is consistent with the idea that when a system is dominated by mixotrophs, its food chain is much shorter and its trophic structure fundamentally different from the traditional planktonic food web in its flow of energy and cycling of nutrients (Mitra et al. 2014).

Noctiluca and Mixotrophy

There are many mixotrophic phytoplankton, particularly dinoflagellates, some that have permanent chloroplasts, others that can sequester (steal) chloroplasts, and a third group that forms symbioses with algae (Hansen 2011; Stoecker et al. 2009). However few have a permanent and independent, free-swimming endosymbionts as observed in green *Noctiluca,* which makes it particularly unique. Most mixotrophs have permanent chloroplasts and the host retains full control of the chloroplast repair and division (Hansen 2011). There are cases of permanent symbiosis of dinoflagellates with diatoms, but there is no evidence that these species are truly mixotrophic (Hansen 2011). Other open-ocean mixotrophs such as planktonic foraminifera and radiolaria have algal endosymbionts that are coccoid and each is enclosed in a cytoplasmic vacuole (symbiosome) that is under the control of the host (Lee and Anderson 1991; Anderson 2014).

Although the permanence of *Noctiluca's* endosymbionts (Sweeney 1976; Furuya et al. 2006a; Harrison et al. 2011) and its mixotrophic nature have been established (Gomes et al. 2014), there is limited understanding of the complex relationship between *Noctiluca* and its green endosymbionts or the role of exogenous food in its growth and transfer of C and N. One of the first reports by Sweeney (1971) based on a strain of green *Noctiluca* isolated from coastal New Guinea showed that it could survive and divide in the light without the addition of food for a month leading the author to posit that *Noctiluca* either ingested its symbionts or utilized their photosynthetic products as nutrition. However long-term growth of *Noctiluca* required the presence of prey since the symbionts were gradually lost. In a more recent study, Hansen et al. (2004) were not able to culture *Noctiluca* isolated from the Manila Bay for more than 3 weeks with or without food. However Saito et al. (2006) and Furuya et al. (2006a, b) reported two types of *Noctiluca* from the Gulf of Thailand and the Manila Bay, one that could grow without food for more than 3 years, while the other required feeding to survive beyond 2 weeks. In all these studies the symbionts did not survive outside of the host cell (Okaichi and Nishio 1976).

The exact details of the symbiotic relationship between the host *Noctiluca* and its green endosymbionts are not well understood, neither are the conditions under which this organism is able to shift its dependence from autotrophy towards a greater dependence on heterotrophy. What we know is that light conditions at the surface where *Noctiluca* occurs as large surface blooms and survives for almost 3 months are not ideal for C fixation by the endosymbionts of *Noctiluca*. The presence of large amounts of diatoms in the digestive vacuole observations suggests that green *Noctiluca* is able to switch to heterotrophy when conditions for growth of its endosymbiont are not ideal. What is unclear however is how *Noctiluca* with its lack of significant flagella (Lucas 1982) and its limited ability to swim is able to make its way to the surface. In the field, when bloom samples were transferred to a beaker, the cells invariably accumulated at the surface within a few minutes (Fig. 6a) suggesting that surface-dwelling populations of *Noctiluca* are highly buoyant. From the more extensively studied red *Noctiluca* which also lacks well-developed flagella, it has been reported that individual cells are capable of modulating their buoyancy via accumulation of ammonia content (Okaichi and Nishio 1976; Elbrachter and Qi 1998) as well as ionic regulation of the specific gravity by decreasing the concentration of heavier ions (Tiselius and Kiorboe 1998). Buoyancy in red *Noctiluca* allows it to ascend the water column enhancing its encounters with prey. Additionally, on account of its poor swimming capacity, red *Noctiluca* tends to feed only on immobile prey, explaining its preference for diatoms over motile dinoflagellates that possess whiplike flagella, capable of swimming, and more capable of escaping predation (Buskey 1995; Kiorboe and Titelman 1998). We have not yet explored ammonia accumulation in green *Noctiluca*, but what we have observed however is that surface-dwelling populations contain large amounts of lipid droplets within its cytoplasm

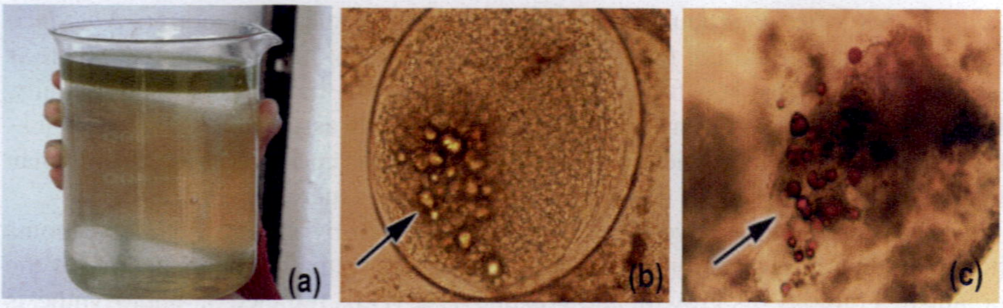

Fig. 6 (**a**) Seawater samples exhibiting the buoyant nature of *Noctiluca* and their tendency to float and accumulate near the surface. (**b**) Microscopic image of *Noctiluca* picked from sea surface in the AS showing lipid droplets (*arrows*). (**c**) Microscopic image of Sudan IV-stained cell of Noctiluca showing lipid droplets stained red (*arrow*)

when growing with an exogenous source of food (Fig. 6b) and red when stained with Sudan Red G (Fig. 6c).

Further advancements in our knowledge about the influence of *Noctiluca* on the food chain, especially its role in C and N cycling in the AS, will require a better understanding of the ecophysiology of this organism and clarity on questions that underpin its role in the ecosystem. For instance we need to know the following: (1) What are the optimal conditions for growth of *Noctiluca* and how does this impact its cellular C and N content? (2) Under what conditions of the environment does *Noctiluca* shift its dependence from autotrophy towards heterotrophy? (3) Is there a substantial change in the growth rates and C and N content of *Noctiluca* when it transitions from autotrophy to heterotrophy? Answers to these and other questions related to *Noctiluca* will be important for future ecosystem model constructs of the food web of the AS and other regions experiencing *Noctiluca* blooms.

Socioeconomic and Global Significance of Noctiluca Blooms

Recent observations of large green *Noctiluca* blooms in other regions of Indian Ocean such as off the coasts of Tanzania (Lugomela 2007) and Seychelles (David Rowat, Marine Conservation Society, Seychelles, pers. comm.) suggest potential range expansion of green *Noctiluca* in the Indian Ocean. *Noctiluca* has recently been reported during the southwest monsoon (Al-Hashmi et al. 2015) raising the spectre that this organism may also be expanding its temporal range to include the highly productive summer period when diatoms dominate and form blooms that support large coastal fisheries. Elsewhere in temperate waters of the Indian Ocean, there is evidence that red *Noctiluca* has expanded its range from Sydney, Australia, to the Southern Ocean, a direct consequence of the increased poleward penetration of the East Australian Current and the subsequent increase in frequency of warm core eddies traveling to Tasmania and towards the Southern Ocean (Mcleod et al. 2012).

In the AS, what is particularly worrisome is that if viable populations of *Noctiluca* become established during the summer monsoon, when diatoms dominate, there is every likelihood that the food web and fisheries will be impacted substantially. Reports of a decline in fisheries along the coast of Oman associated with phytoplankton blooms and oxygen-deficient waters suggest that the coastal ecosystem off Oman may be witnessing changes that mirror a trend towards increasing mixotrophy.

The portended disruptive impact on the food web by green *Noctiluca* blooms however remains a question that requires a more systematic understanding of the organism's biology, its growth, and grazing *vis-à-vis* its reliance on photosynthesis of its endosymbionts, especially as the blooms transition from colder nutrient-rich waters in late December to warmer, stratified, high insolation, and warmer temperature conditions of late March. This kind of ecophysiological information is required to build a holistic perspective of how phytoplankton communities respond to environmental change.

What we have established so far is that *Noctiluca* has become a major player in the Arabian Sea planktonic food web and this realization warrants a revision of earlier understanding of the web dynamics and allied biogeochemistry of the Arabian Sea gained from the JGOFS era of the 1990s (Smith 2005) which involved significant US participation and resources (~$50 M) (Sharon Smith, pers. comm.). The potential negative impacts on the food web by *Noctiluca* foreshadow critical impacts on the fisheries economy of the many countries that border the AS, many of which are underdeveloped and in a state of unrest from poverty and deprivation. If fisheries, the primary source of income for millions of people, is debilitated, then the current turmoil will only exacerbate and create even more problems for developed countries.

Acknowledgments This work was supported by grants NNX09AE48G and NNX07AK82G from National Aeronautical and Space Agency (NASA) and grants OCE-0824632 from the National Science Foundation (NSF), USA, to JIG and HRG. This collaborative study was catalyzed by a Virtual Centre grant to HRG and JIG from the Indo-US Science and Technology Foundation (IUSSTF), New Delhi. We are grateful to Dr. S. Nayak, Secretary, Ministry of Earth Sciences, Govt. of India, for providing the Indian research vessels and Dr. V. N. Sanjeevan, Director, Centre for Marine Living Resources and Environment (CMLRE), India, Dr. R.M. Dwivedi, former Head, Marine Living Resources Division, SAC and Dr. Prabhu Matondkar, Senior Scientist, National Institute of Oceanography, India, for enabling participation in cruises. We are thankful to Dr. Sean Solomon, Director, Lamont Doherty Earth Observatory, USA, Dr. A. Mitra, former Director IUSSTF Dr. Navalgund, former Director, SAC, India Space Research Organization

Joaquim I. Goes and Helga do R. Gomes

Helga and Joaquim were born in Nairobi, Kenya, to parents who formed part of a large diasporic community from a Portuguese colony in India named Goa. Political turmoil forced them to return to Goa where they completed their education. They met at the National Institute of Oceanography (NIO) in Goa where Helga was pursuing her doctorate and Joaquim was a researcher. Oddly, they did not notice each other at the undergraduate college where they both studied. Joaquim believes that it was the free rides on his motorbike, and the many days spent on cruises in the balmy Indian Ocean that brought them together. After her Ph.D. on the characterization of phytoplankton extracellular products, Helga stayed on as a postdoctoral researcher at the NIO because she firmly believed that there was no better or more beautiful place to live in India than Goa. She still believes so! In 1992, Joaquim was offered a Doctoral fellowship by the Japanese Ministry of Education and they moved to Nagoya University, Japan. Clueless in a pre-internet age, they dived headlong into a new culture and language, and loved every bit of it from the sashimi and onsen to kanji. After Japan lost its ocean color satellite, Joaquim changed the course of his research. Molina and Rowland had just won the Nobel Prize for their work on the formation of the Ozone Hole. Little was then known on how ocean biology would respond to excess solar UV radiation. Working with the late and well-known geochemist, Prof. Nobuhiko Handa, Joaquim showed how enhanced exposure to UV radiation would profoundly impact phytoplankton photosynthesis. Later, Joaquim pursued his Postdoctoral studies under the late Prof. Toshiro Saino developing an algorithm to estimate nitrate distribution from remotely sensed products. A larger than life character, Prof. Saino loved science, good food, and wine and Joaquim and Helga had found the perfect mentor and friend. Fortuitously, Prof. Saino invited Joaquim to Hawaii for the Japan-USA workshop on ocean color where Joaquim met Barney Balch, who invited him to Bigelow Laboratory, Maine. Once again they embarked on a new journey, this time in a land of lobsters and vast expanses of untouched land often covered in a foot of snow. A year into his Postdoc, Joaquim was appointed as a Senior Research Scientist while Helga continued as a Research Associate. In 2010, they both felt the need for another adventure so they moved to the big city of New York where they now work at Lamont Doherty, Columbia University in the Palisades, New York.

References

Al-Azri A, Al-Hashmi K, Goes J, Gomes HDR, Rushdi AI, Al-Habsi H, Al-Khusaibi S, Al-Kindi R, Al-Azri N (2007) Seasonality of the bloom-forming heterotrophic dinoflagellate *Noctiluca scintillans* in the Gulf of Oman in relation to environmental conditions. Int J Oceans Oceanogr 2:51–60

Al-Hashmi K, Claereboudt M, Piontkovski S, Al Azri A, Amin SMN (2015) Seasonal variability of size-classes of phytoplankton biomass in a sub-tropical embayment, Muscat. Sea Oman Iranian J Fisheries Sci 14:136–149

Anderson OR (2014) Living together in the plankton: a survey of marine protist symbioses. Acta Protozool 53:29–38

Banse K, Mcclain CR (1986) Winter blooms of phytoplankton in the Arabian Sea as observed by the coastal zone color scanner. Mar Ecol Prog Ser 34:201–211

Bidigare RR, Heukelem LV, Trees CC (2002) HPLC phytoplankton pigments: sampling, laboratory methods and quality assurance procedures. In: Mueller JL, Fargion GS (eds) Ocean optics protocols for satellite ocean color, sensor validation. NASA Goddard Space Flight Center, Greenbelt, MD, pp 154–160

Burkill PH (1999) ARABESQUE: an overview. Deep Sea Res II 46:529–547

Buskey EJ (1995) Growth and bioluminescence of *Noctiluca scintillans* on varying algal diets. J Plankton Res 17:29–40

Caron DA, Dennett MR (1999) Phytoplankton growth and mortality during the 1995 northeast monsoon and spring intermonsoon in the Arabian Sea. Deep Sea Res II 46:1665–1690

Clarke KR, Warwick RM, Somerfield PJ, Gorley RN (2005) Methods manual: Changes in marine communities. PRIMER v6

Dwivedi R, Rafeeq M, Smitha BR, Padmakumar KB, Thomas L, Sanjeevan VN, Prakash P, Raman M (2015) Species identification of mixed algal bloom in the Northern Arabian Sea using remote sensing techniques. Environ Monit Assess 187:1–11

Elbrachter M, Qi Y-Z (1998) Aspects of *Noctiluca* (Dinophyceae) population dynamics. In: Andersen DM, Cembella A, Hallegraeff GM (eds) Physiological ecology of harmful algal blooms. NATO ASI Series, Berlin, pp 315–335

Fujiki T, Hosaka T, Kimoto H, Ishimaru T, Saino T (2008) *In situ* observation of phytoplankton productivity by an underwater profiling buoy system: use of fast repetition rate fluorometry. Mar Ecol Prog Ser 353:81–88

Furuya K, Saito H, Rujinard S, Vijayan AK, Omura T, Furio EE, Borja VM, Sopana B, Thaithaworn L (2006a) Persistent whole-bay red tide of *Noctiluca scintillans* in Manila Bay. Philipp Coast Mar Sci 30:74–79

Furuya K, Saito H, Sriwoon R, Omura T, Furio E, Borja V, Lirdwitayaprasit T (2006b) Vegetative growth of *Noctiluca scintillans* containing the endosymbiont *Pedinomonas noctilucae*. Afr J Mar Sci 28:305–308

Garrison DL, Gowing MM, Hughes MP, Campbell L, Caron DA, Dennett MR, Shalapyonok A, Olson RJ, Landry MR, Brown SL, Liu H-B, Azam F, Steward GF, Ducklow HW, Smith DC (2000) Microbial food web structure in the Arabian Sea: a US JGOFS study. Deep Sea Res II 47:1387–1422

Goes JI, Prasad TG, Gomes HDR, Fasullo JT (2005) Warming of the Eurasian landmass is making the Arabian Sea more productive. Science 308:545–547

Gomes HDR, Goes JI, Matondkar SGP, Parab SG, Al-Azri ARN, Thoppil PG (2008) Blooms of Noctiluca miliaris in the Arabian Sea—An in situ and satellite study. Deep Sea Res Part I 55:751–765

Gomes H, DR, Goes JI, Matondkar SGP, Parab SG, Al-Azri A, Thoppil PG (2009) Unusual blooms of the green *noctiluca miliaris* (dinophyceae) in the Arabian Sea during the winter monsoon. In: Wiggert JD, Hood RR, Naqvi SWA, Smith SL, Brink KH (eds) Indian Ocean: biogeochemical processes and ecological variability, AGU Book Series. American Geophysical Union, Washington, DC, pp 347–363

Gomes H, Goes JI, Matondkar SGP, Buskey EJ, Basu S, Parab S, Thoppil P (2014) Massive outbreaks of *Noctiluca scintillans* blooms in the Arabian Sea due to spread of hypoxia. Nat Commun 5:4862. doi: 10.1038/ncomms5862

Kolber ZS, Prášil O, Falkowski, PG (1998) Measurements of variable chlorophyll fluorescence using fast repetition rate techniques: defining methodology and experimental protocols, Biochimica et Biophysica Acta (BBA) - Bioenergetics,1367, 1–3, 88–106

Hansen PJ (2011) The role of photosynthesis and food uptake for the growth of marine mixotrophic Dinoflagellates1. J Eukaryot Microbiol 58:203–214

Hansen PJ, Miranda L, Azanza R (2004) Green *Noctiluca scintillans*: a dinoflagellate with its own greenhouse. Mar Ecol Prog Ser 275:79–87

Harrison PJ, Furuya K, Glibert PM, Xu J, Liu HB, Yin K, Lee JHW, Liu H, Gowan R, Al-Azri AR, Ho AYT, Anderson DM (2011) Geographical distribution of red and green *Noctiluca scintillans*. Chin J Oceanol Limnol 29:807–831

Kiorboe T, Titelman J (1998) Feeding, prey selection and prey encounter mechanisms in the heterotrophic dinoflagellate *Noctiluca scintillans*. J Plankton Res 20:1615–1636

Kolber Z, Falkowski PG (1993) Use of active fluorescence to estimate phytoplankton photosynthesis in situ. Limnol Oceanogr 38:1646–1665

Krey J, Bole R, Gillbricht M, Lenz J (1971) Planktological-chemical observation during the International Indian Ocean Expedition (IIOE) with RV "Meteor" in 1964/65, http://doi.pangaea.de/10.1594/PANGAEA.611057

Landry MR, Brown SL, Campbell L, Constantinou J, Liu H (1998) Spatial patterns in phytoplankton growth and microzooplankton grazing in the Arabian Sea during monsoon forcing. Deep Sea Res II 45:2353–2368

Langdon C (1984) Dissolved oxygen monitoring system using a pulsed electrode: design, performance and evaluation. Deep Sea Res I 31:1357–1367

Lee JJ, Anderson OR (1991) Symbiosis in foraminifera. Biol Foraminifera 157–220

Lévy M, Shankar D, André JM, Shenoi SSC, Durand F, De Boyer Montégut C (2007) Basin-wide seasonal evolution of the Indian Ocean's phytoplankton blooms. J Geophys Res 112, C12014

Lewis MR, Smith JC (1983) A small volume, short-incubation-time method for measurement of photosynthesis as a function of incident irradiance. Mar Ecol Prog Ser 13:99–102

Lucas IAN (1982) Observations on *Noctiluca scintillans* Macartney (Ehrenb.) (Dinophyceae) with notes on an intracellular bacterium. J Plankton Res 4:401–409

Lugomela C (2007) *Noctiluca scintillans* (Dinophyceae) in central coastal waters of Tanzania: A new phytoplankton record for the area. West Indian Ocean J Mar Sci 6:117–124

Mcleod DJ, Hallegraeff GM, Hosie GW, Richardson AJ (2012) Climate-driven range expansion of the red-tide dinoflagellate *Noctiluca scintillans* into the Southern Ocean. J Plankton Res 34:332–337

Mitra A, Flynn KJ, Burkholder JM, Berge T, Calbet A, Raven JA, Granéli E, Glibert PM, Hansen PJ, Stoecker DK, Thingstad F, Tillmann U, Våge S, Wilken S, Zubkov MV (2014) The role of mixotrophic protists in the biological carbon pump. Biogeosciences 11:995–1005

Okaichi T, Nishio S (1976) Identification of ammonia as the toxic principle of red tide *Noctiluca miliaris*. Bull Plankton Soc Japan 23:75–80

Parab SG, Matondkar SGP, Gomes HDR, Goes JI (2006) Monsoon driven changes in phytoplankton populations in the eastern Arabian Sea as revealed by microscopy and HPLC pigment analysis. Cont Shelf Res 26:2538–2558

Piontkovski S (2002) The Global Plankton Database: an inventory and data from the former Soviet Union expeditions, Darwin Initiative for the Survival of Species (Intas INFO 00-059), NATO Science for Peace project (1999-2002), ed. (Plymouth Marine Laboratory, United Kingdom, Institute of Biology of the Southern Seas & Marine Hydrophysical Institute, Sevastopol, Ukraine; Institute of Zoology of Azerbaijan Academy of Sciences, Baku, Azerbaijan; Institute of Zoology, Almaty, Kazakhstan, CD-ROM)

Saito H, Furuya K, Lirdwitayaprasir T (2006) Photoautotrophic growth of *Noctiluca scintillans* with the endosymbiont *Pedinomonas noctilucae*. Plankton Benthos Res 1:97–101

Sawant SS, Madhupratap M (1996) Seasonality and composition of phytoplankton in the Arabian Sea. Curr Sci 71:869–873

Smith SL (2005) The Arabian Sea of the 1990s: New biogeochemical understanding. Prog Oceanogr 65:113–115

Smith S, Roman M, Prusova I, Wishner K, Gowing M, Codispoti LA, Barber R, Marra J, Flagg C (1998) Seasonal response of zooplankton to monsoonal reversals in the Arabian Sea. Deep Sea Res II 45:2369–2403

Stoecker DK, Johnson MD, Devargas C, Not F (2009) Acquired phototrophy in aquatic protists. Aquat Microb Ecol 57:279–310

Sweeney BM (1971) Laboratory studies of a green *Noctiluca* from New guinea. J Phycol 7:53–58

Sweeney BM (1976) *Pedinomonas noctilucae* (Prasinophyceae) the flagellate symbiotic in *Noctiluca* (Dinophyceae) in Southeast Asia. J Phycol 12:460–464

Thibodeau PS, Roesler CS, Drapeau SL, Prabhu Matondkar SG, Goes JI, Werdell PJ (2014) Locating *Noctiluca miliaris* in the Arabian Sea: An optical proxy approach. Limnol Oceanogr 59:2042–2056

Tiselius P, Kiorboe T (1998) Colonization of diatom aggregates by the dinoflagellate *Noctiluca scintillans*. Limnol Oceanogr 43:154–159
Tomas CR (1997) Identifying marine phytoplankton. Academic, San Diego
Wang L, Lin X, Goes, JI, Lin S (2016) Phylogenetic analyses of three genes of *Pedinomonas noctilucae*, the green endosymbiont of the marine dinoflagellate *Noctiluca scintillans*, reveal its affiliation to the Order Marsupiomonadales (Chlorophyta, Pedinophyceae) under the reinstated name *Protoeuglena noctilucae*. Protist 167:205–216
Werdell PJ (2014) Ocean color satellites observe a decade-long transition of dominant phytoplankton species types in the Arabian Sea. Ph.D. Thesis, University of Maine, Paper 2128, http://digitalcommons.library.umaine.edu/etd/2128
Werdell PJ, Roesler CS, Goes JI (2014) Discrimination of phytoplankton functional groups using an ocean reflectance inversion model. Appl Opt 53:4833–4849
Wiggert JD, Hood RR, Banse K, Kindle JC (2005) Monsoon-driven biogeochemical processes in the Arabian Sea. Prog Oceanogr 65:176–213

The Saint Lawrence Island Polynya: A 25-Year Evaluation of an Analogue for Climate Change in Polar Regions

Jacqueline M. Grebmeier and Lee W. Cooper

Introduction

Polynya is a Russian-origin word that is used to describe ice-free areas in otherwise ice-covered seas. Polynyas are a common feature in both Arctic and Antarctic waters, even at temperatures well below freezing, when prevailing winds blow sea ice away from islands or shorelines (latent heat polynyas), or where warmer waters are upwelled to the surface (sensible heat polynyas). Where these features are predictable from year to year, polynyas often attract higher trophic organisms such as diving ducks and marine mammals that require open water refuges for feeding and rest in frozen sea habitats (Stirling 1980; Gilchrist and Robertson 2000).

The year 2015 marks more than a quarter century since the first internationally coordinated effort to study polynyas, the International Arctic Polynya Program (IAPP; Mikkelsen 2015), undertaken by the Arctic Ocean Sciences Board, a non-governmental international science cooperation entity that has now been merged into the International Arctic Science Committee. Interest in the systematic study of polynyas was inspired in part by the potential for understanding the impacts of future climate change in the polar regions by studying persistent areas where sea ice was not present seasonally (Smith and Barber 2007).

The polynya that persists in the winter to the south of Saint Lawrence Island (SLIP) in the Bering Sea was one of the first candidate polynyas that was targeted for study in the IAPP (Alexander and Muench 1988), and was supported by the US National Science Foundation funding initiated in 1990 to the two authors of this chapter (Fig. 1a, b; Grebmeier and Cooper 1995). Although other studies of polynyas in succeeding years, in the Northeast Water (Greenland), in the North Water

J.M. Grebmeier (✉) • L.W. Cooper
Chesapeake Biological Laboratory, University of Maryland Center for Environmental Science, Solomons, MD 20688, USA
e-mail: jgrebmei@umces.edu

P.M. Glibert, T.M. Kana (eds.), *Aquatic Microbial Ecology and Biogeochemistry: A Dual Perspective*, DOI 10.1007/978-3-319-30259-1_14

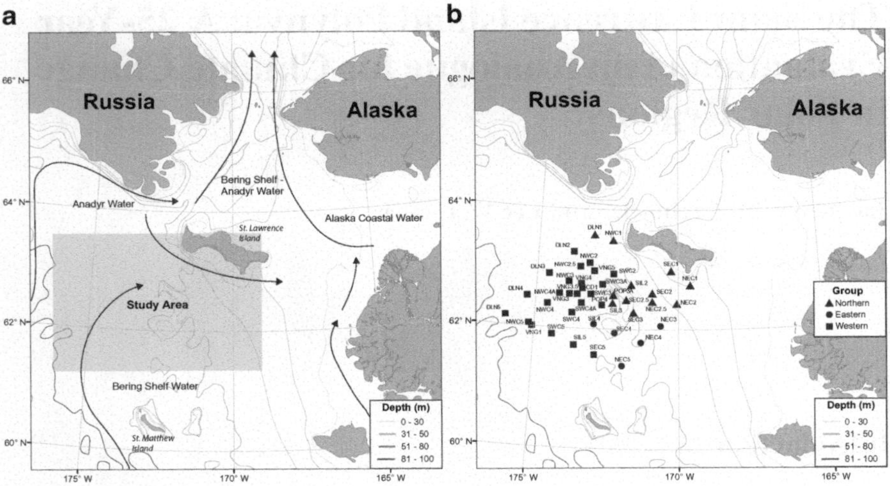

Fig. 1 (**a**) Study region in the northern Bering Sea south of St. Lawrence Island, USA, with general current patterns by water type. (**b**) Example of the standard macrofaunal distribution of three benthic community station cluster groups sampled during process cruises in the study region: northern (*triangles*), eastern (*circles*), western (*squares*)

(Baffin Bay), the Laptev Sea flaw lead system, and the Cape Bathurst region of the Canadian Arctic have been more intensive and more widely supported on an international scale, the initial study of the SLIP, which was summarized in a special issue of the Journal of Geophysical Research on leads and polynyas (Grebmeier and Cooper 1995), provided an initial understanding of the biological functioning of this shallow continental shelf winter polynya and its influence on benthic biology of the northern Bering Sea. The physics of the SLIP have also been broadly studied, particularly its importance as an area of sea ice production that influences seasonal ice cover in the Bering Sea (e.g., Pease 1987; Drucker et al. 2003; Danielson et al. 2006).

Our initial biologically oriented study (Grebmeier and Cooper 1995) was followed up by a number of other efforts that provide a fuller context of the importance of this polynya within the northern Bering Sea ecosystem (e.g., Cooper et al. 2002, 2009, 2012, 2013; Clement et al. 2004; Lovvorn et al. 2005, 2009, 2014; Grebmeier 2012; Grebmeier et al. 2006a, b; 2015; Grebmeier and Barry 2007; Jay et al. 2014). Data collection has continued over this past quarter century on a number of follow-on projects in the northern Bering Sea, and we use this opportunity to synthesize these findings from over the years with new insights that are developing from the internationally coordinated Distributed Biological Observatory (DBO) project that is annually sampling at biologically productive regions of the northern Bering Sea (including within the SLIP region), and adjoining Chukchi Sea.

The Northern Bering Sea, roughly from St. Matthew Island north, is a distinct ecosystem and functions differently than the pelagic system to the south. This difference is particularly striking in the winter, when walruses, ice seals, and spectacled eiders congregate in large numbers to take advantage of abundant food supplies on the seafloor, and also in some cases from under ice biota such as arctic cod and euphausiids. Productivity in the water column is low due to light limitations, but west-to-east decreases in chlorophyll and nutrients are already present as they are later in the seasonal cycle when massive sea ice edge blooms occur (Cooper et al 2012, 2013).

Our observations indicate the importance of the rich benthic communities south of St. Lawrence Island to top apex predators during winter foraging (Grebmeier 2012) and have contributed to understanding the seasonal cycle of this Arctic-subarctic system that has a winter polynya (Grebmeier and Barry 2007). The benthic macrofauna has historically high biomass levels on the western side of the system, dominated by tellinid, nuculid, and nuculanid bivalves (Grebmeier and Cooper 1995; Simpkins et al. 2003; Grebmeier et al. 2006a, 2015; Grebmeier 2012). By comparison, the eastern section has low infaunal biomass dominated by tunicates, amphipods, polychaetes, and various bivalves. We characterized both macrofaunal and epifaunal habitats and evaluated infaunal to epibenthic relationships, associated abiotic influences such as sediment types, and overlying water column hydrographic measurements. Particularly near Saint Lawrence and Saint Matthew Islands and along the northwest Alaska coast, distinct and diverse epifaunal communities are present that reflect system energy and geomorphology. For example, sand dollars are locally abundant inshore of the boundary between Bering Shelf Water (BSW) south of St. Lawrence Island and Alaska Coastal Water (ACW) (Fig. 1a). Large portions of the continental shelf of the Bering Sea are dominated by brittle stars (Ophiuroidea), including on the outer shelf southwest of Saint Lawrence Island where fine silt and clay sediments occur. By contrast, in sandier sediments to the southeast of Saint Lawrence Island, a wider range of mobile and sessile epifauna (e.g., crabs, gastropods, and tunicates) are present (Lovvorn et al. 2005). These patterns of macrofaunal and epibenthic community separation suggest a potential link between infaunal and epifaunal communities through trophic interactions or the influence of environmental parameters on both communities at similar scales. Patterns in hydrodynamics and/or sea ice and therefore carbon supply are potential driving factors.

Sediment oxygen demand (SOD) has also been determined at multiple sites during many of the cruises as an indicator of carbon supply to the benthos (Grebmeier and Cooper 1995; Cooper et al. 2012, 2013). Focused experiments on temperature controls on SOD in 2007 and 2009 indicated that SOD increased with both temperature and carbon supply (Bailey et al. 2009; Cooper et al. 2013). Nutrient exchanges between the sediments and the water column varied and suggested that interactions with benthic fauna affect rates of exchange, as is the case in other areas in the Pacific Arctic shelf region (Mathis et al. 2014). Sediment cores for these temperature-controlled experiments were collected solely from time series stations maintained since 1990 that had the best time series record for sediment respiration in this region of the Bering Sea. The object of the experiments was to test for changing environmental impacts that may be driving the long-term declines in benthic communities and populations that have now been documented over our full sampling record (Grebmeier et al. 2006b; Grebmeier 2012).

The implications of these declines are that it may be affecting apex predators such as walruses, bearded seals, and several species of diving ducks, including spectacled eiders, who dive to the shallow seafloor (40–60 m) to forage on the abundant clams and other benthic organisms. Moreover, our observations indicate that spectacled eiders can be excluded from optimal foraging areas by the position and extent of seasonal sea ice (Lovvorn et al. 2009, 2014). The benthic biomass declined over the past several decades (Grebmeier et al. 2006b; Grebmeier 2012), and also may bring additional changes to the ecosystem, including the movement northward of industrial fishing operations as water temperature warms, which will have further negative consequences for the ecosystem.

Another key higher trophic organism to consider are the walruses that use the region south of St. Lawrence Island in late winter for rest and feeding (Jay et al. 2012, 2014). Feeding patterns reflect benthic biomass and seafloor community structure. Pacific walruses forage on the seafloor, primarily for benthic invertebrates, and the wide continental shelves of the Chukchi and Bering Seas provide extensive areas of high benthic biomass as prey for these upper trophic organisms. Resource selection, for food or habitat, has a direct impact on walrus survival capabilities and partitioning of populations. The heterogeneity or patchiness of prey and/or habitat resources can occur over multiple temporal and spatial scales. Individual species adapt to the changing environment and in turn this can influence population structure and demographic trends. Both spectacled eider and walrus satellite tracking data can be matched with benthic prey data and have been used in habitat selection analyses by the predators (Lovvorn et al. 2014; Jay et al. 2014; Sexson et al. 2016).

Synthesis Results and Discussion

Overview for Synthesis

Water column temperature, salinity, chlorophyll *a* content, nutrients, macro- and epifaunal collections, and sediment parameters constitute much of the data collected at stations in the SLIP region over the multiple cruises (see methods in Grebmeier and Cooper 1995; Grebmeier 2012; Grebmeier et al. 2015; Cooper et al. 2012, 2013). Specific to the seasonal summaries below, the benthic sampling of bottom sediments and associated fauna at stations from 1990 to present included four replicates of a 0.1 m^2 van Veen grab for macrofaunal collections at each station (Grebmeier and Cooper 1995). Organisms were sieved from each of these grabs through 1 mm screens, preserved, and returned to the laboratory for species identification and determinations of biomass. Additional sediment grabs were also undertaken at each station for surface sediment analysis, including determinations of sediment chlorophyll *a* content, total organic carbon and nitrogen, and other sediment chemical parameters, when practical.

Summer Sampling in SLIP (July–September 1990–2015)

The first series of samples in the SLIP occurred during the summer period (June–Sept 1990–1999, $n = 5$ cruises), with an additional 15 cruises using a Canadian icebreaker that sampled a subset of five time-series sites in the western high benthic biomass region offshore of St. Lawrence Island (SLI) (2000–2015; see Grebmeier et al. 2015 for specific details on cruises and methods). A number of prior publications highlighted the spatial variability in water masses south of SLI, with nutrient-rich Anadyr Water (AW; salinity >32.5) occurring in the western region of the study area as well as in the northern area just south of SLI, Alaska Coastal Water (ACW; salinity <31.8) in the eastern region, and Bering Shelf Water (BSW; salinity = 31.8–32.5) occurring between AW and ACW (Grebmeier and Cooper 1995; Cooper et al. 2012, 2013). Most of the water column chlorophyll *a* descends to the benthos after the spring bloom, so by midsummer, water column stratification has sealed off the cold bottom waters (<–1 °C) remaining near bottom from the winter polynya formation period. This is a function of sea ice formation; when cold, dense saline water is injected into the water as sea ice forms, and by summer these cold bottom waters are separated from surface waters that warm to 4–8 °C. The bottom sediments have higher silt and clay content and sediment chlorophyll *a* content in the western region compared to the eastern, less nutrient-rich area (Grebmeier and Cooper 1995; Grebmeier 2012). Benthic macrofaunal biomass is highest in the western region, dominated by bivalves and polychaetes, with lower biomass of similar fauna, along with amphipods, in the eastern region (Grebmeier and Cooper 1995; Simpkins et al. 2003; Grebmeier 2012). The northern sector, just south of SLI, is dominated by amphipods and bivalves.

Spring (April–June 1999–2007)

We have had four process cruises in the SLIP region in spring (April–June from 1999 to 2007). These spring cruises required us to sample in the ice-covered seas south of SLI using US Coast Guard (USCG) icebreakers that allowed us to observe the ice edge and spring bloom period as the ice receded (Clement et al. 2004). A large part of the area on the central shelf was occupied by Bering Shelf Water with intermediate salinity (32–32.4). Stations close to the south side of SLI showed salinities of ~33 which may be characteristic for waters advected with the southern branch of the Anadyr current that bifurcates on the west side of SLI (the other

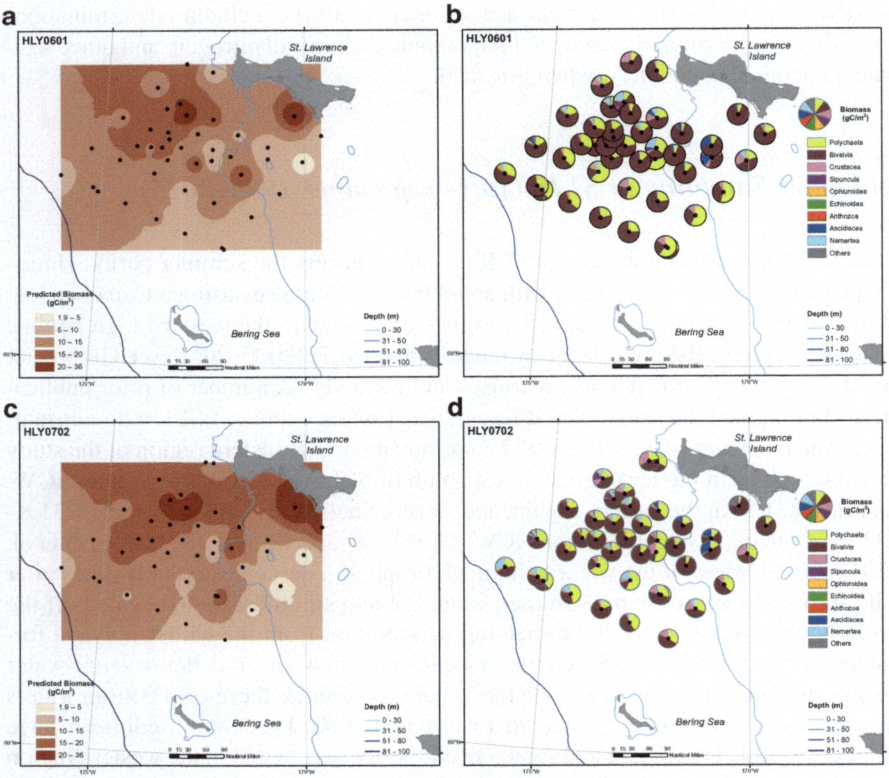

Fig. 2 Distribution of macrofaunal biomass (gC/m^2) and macrofaunal types, by biomass, in spring 2006 for USCGC Healy cruises (HLY0601; **a**, **b**, respectively) and 2007 (HLY0702; **c**, **d**, respectively)

branch flows northeast through Anadyr Strait separating the northwest cape of St. Lawrence Island from continental Asia; see Fig. 1a). Higher saline water was found to the southeast of SLI. The high salinity values (~33.5–34) of these waters are likely caused by brine rejection events during ice formation (Clement et al. 2004). A modeling study further elaborated into the processes driving physical and biological water column properties during this time (Clement et al. 2005).

This spring period as ice retreats is the most productive period for both ice algae and large diatoms that dominated the primary production and higher water column chlorophyll occurred in the western region (Cooper et al. 2002, 2012). Zooplankton populations are relatively low (Lovvorn et al. 2005), with high export of primary produced organic carbon directly to the benthos (Cooper et al. 2002; Grebmeier and Dunton 2000; Grebmeier 2012). This high export of labile carbon allows for the buildup of high bivalve and polychaete biomass in the western offshore area south of SLI (Fig. 2a–d). The eastern area, as mentioned previously for the summer period, has smaller benthic biomass of bivalves, amphipods, and polychaetes.

Winter (March 2008–2010)

Late winter observations in the northern Bering Sea have been rare, but three USCG icebreaker deployments during March in 2008–2010 have provided new insights on how the ecosystem functions and preconditions the productive spring bloom (Cooper et al. 2013) in comparison with spring studies in 2006 and 2007 (Cooper et al. 2012). Salinity is the dominant physical parameter that determines horizontal density variations in spring, with the warmest water offshore to the southwest, but the rest of the study area is influenced by cold <1 °C water produced as sea ice forms at the surface in a well-mixed water column. While water column chlorophyll concentrations were two orders of magnitude lower in late winter than observed during the spring bloom in May, sea ice algal inventories of chlorophyll were high and corresponded to the distribution of water masses with high nutrient content (Cooper et al. 2013). Brine injection while the late winter SLI polynya is active can mix inorganic nutrients from the bottom upward, and the specific vectors of the prevailing north to easterly winds influence the distribution of high nutrient waters that originate on the northwestern Bering Sea Shelf near and within the Gulf of Anadyr. Thus, conditions such as wind and air temperature in March have an influence on the intensity of the spring bloom, in addition to the actual pattern and timing of sea ice retreat. The underlying benthos continued to show the signature of highest benthic biomass of bivalves and polychaetes in the western region of the SLI region, with lower biomass to the east (Fig. 3a–d). After the spring bloom, water column stratification limits phytoplankton production, although the bottom benthic communities continue to build biomass over the summer based on organic carbon resources that were produced both in the SLI system and advected in from the southern Bering Sea.

The Northern Bering Sea: Interannual Variability and Change

In addition to the water column data discussed above, benthic infaunal abundance and biomass data for the area south of SLI were also compiled from the many cruises undertaken over the span of 25 years in the SLIP region (details summarized in Grebmeier et al. 2006b, 2015; Grebmeier 2012). Three distinct community groupings were identified on every cruise by cluster and statistical analyses of macrofaunal abundance and biomass (Fig. 1b; discussed previously). However, we have seen dramatic changes in benthic biomass for the dominant three species of bivalves in the western region that are key to the spectacled eider and walrus prey base (Grebmeier et al. 2006b; Grebmeier 2012; Lovvorn et al. 2014). Specifically, nuculanid bivalves continue to decline in biomass, particularly under the influence of nutrient-rich Anadyr water (Grebmeier 2012). However, smaller nuculid bivalves have increased in abundance closer to SLI, but it is not clear if the spectacled eiders that feed there in the winter can take advantage of this shift in food supplies due to ice conditions (Grebmeier 2012; Lovvorn et al. 2003, 2014).

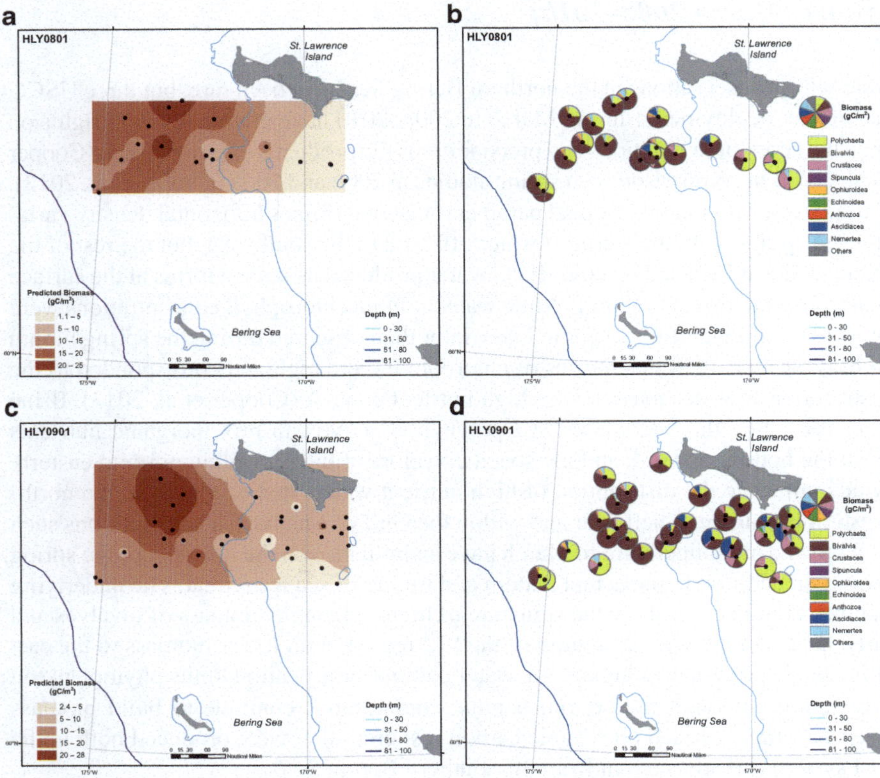

Fig. 3 Distribution of macrofaunal biomass (gC/m^2) and macrofaunal types, by biomass, in late winter 2008 for USCGC Healy cruises (HLY0801; **a**, **b**, respectively) and 2009 (HLY0901; **c**, **d**, respectively)

Time Series Stations Within the "Western" Cluster Group Under Anadyr Water

Time series analyses of five stations in the area of highest biomass show that these locations all cluster in a productive western group (Grebmeier et al. 2006b; Grebmeier 2012). Sampling at these sites indicates that the three most southern sites in this cluster group are changing both in infaunal composition and overall biomass in response to larger scale physical and biological forcing that is acting upon the system (Fig. 4a, b; Grebmeier 2012). Statistical analyses of macrofaunal populations and environmental parameters have identified water column depth and sediment grain size as the primary factors influencing benthic macrofaunal community structure, both for stations that are part of our formal time series ($r = 0.619$, $p \leq 0.05$) and for more spatially extensive cruises. Since depth is not changing over time, but

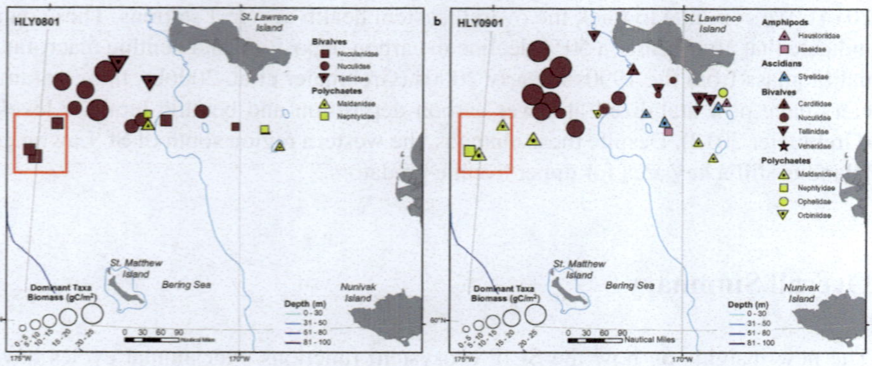

Fig. 4 Benthic macrofaunal biomass (gC/m^2) and dominant fauna (% of total station biomass) in (**a**) March 2008 (HLY0801) and (**b**) March 2009 (HLY0901) south of St. Lawrence Island in the northern Bering Sea. The red box surrounds the three time series sites where decadal biomass declines have been observed (Grebmeier 2012) and shows the change in dominance from bivalves to polychaetes that occurred in 2009

is highly correlated with greater silt and clay (Grebmeier 2012; Grebmeier et al. 2015), the primary factor leading to the changing benthic community composition is likely an increase in the percent composition of the silt and clay fraction. These sediment pattern changes suggest reduced currents that are increasing the silt and clay fraction of sediments and altering benthic community composition. Besides a decline in overall station biomass, we observe a shift in macrofaunal dominance from bivalves to polychaetes in 2009 (Fig. 4a, b-red box) where sediments fined. We are continuing tracking of these five time series sites annually as part of the DBO project (Grebmeier et al. 2010, 2015). These stations serve as the most southern line of the DBO network, which is being developed as an international cooperative approach to detecting change in arctic ecological systems.

Benthivores

Late winter is an important foraging period for benthic-feeding Pacific walrus and diving sea ducks, such as the king, spectacled, and common eider. As might be expected, feeding patterns reflect the underlying benthic biomass and seafloor community structure. Sea ice dynamics in addition to benthic food availability also plays a role in the scale and distributions in particular of spectacled eiders.

In our study we characterized benthic habitats and evaluated macrofaunal benthic community relationships to overlying water column hydrographic measurements, abiotic sediments, and sea ice cover. Sediment organic carbon content indicates a deposition zone to the southwest of SLI (Grebmeier 2012) that coincides with high benthic macrofaunal biomass and areas of walrus predation (Jay et al.

2014). We continue to track the overall system health at select stations. These data indicate that after nearly a 50 % decline in carbon deposition and benthic macrofaunal biomass from the 1990s to early 2000s (Grebmeier et al. 2006b), these parameters have now stabilized at lower carbon deposition and benthic biomass levels (Grebmeier 2012). Despite these changes, the western region south of St. Lawrence Island is still a key area for upper trophic predators.

Overall Summary

The new insights on how the SLIP ecosystem functions over annual cycles have required seasonal sampling to facilitate understanding the preconditioning required for development of the productive spring bloom. While water column chlorophyll concentrations are two orders of magnitude lower in late winter (March–early April) than observed during the spring bloom in May, sea ice algal inventories of chlorophyll are high and correspond to the distribution of water masses with high nutrient content. Nutrient availability clearly modulates open water and sea ice productivity and the transmission of organic material to the shallow benthos following sea ice retreat. In some areas, significant late winter re-suspension of ammonium also occurs from sediments. Active brine injection can mix these inorganic nutrients, and the specific vectors of the prevailing north to easterly winds influence the distribution of high nutrient waters that originate on the northwestern Bering Sea Shelf near and within the Gulf of Anadyr. Thus conditions such as wind and air temperature in March have an influence on the intensity of the spring bloom, in addition to the actual pattern and timing of sea ice retreat.

Feeding patterns of walruses and spectacled eiders reflect benthic biomass, seafloor community structure, and at least in the case of spectacled eiders that require sufficient open water to dive successfully sea ice dynamics. Kittlitz's murrelets, which feed on zooplankton, are another late-winter migrant that were observed to take advantage of open water in the polynya south of Saint Lawrence Island. In experimental incubations, high sediment uptake rates occurred with both higher seawater temperatures and increased organic carbon (food supply). Sediment organic carbon content indicates a deposition zone to the southwest of SLI, and matches up well with high sediment oxygen update and benthic infaunal biomass. We continue to track the overall system health at select stations. Times series data indicate a "fining" of sediment (increased ≥5phi size silt and clay particles), indicative of slowing currents and increasing settling of fine particles to the benthos. We are evaluating these sediment grain size trends with both macrofaunal biomass and declined sediment oxygen uptake rates.

The overall picture that emerges from these early seasonal observations is of a biologically active system with strong atmospheric and hydrographic controls. These controls predetermine nutrient distributions and water column mixing through brine rejection and ultimately influence the intensity of seasonal productivity and the subsequent seasonal activities of biological components from ice algae to zooplankton to the benthos, and to apex predators that are emblematic features of this subpolar ecosystem.

Acknowledgements We thank the technical support and graduate students over the years for environmental and biological measurements in the laboratory, most recently Linton Beaven and Stephanie Soques. Eva Bailey and Maria Ceballos kindly assisted in the time series sediment oxygen demand and nutrient exchange experiments in 2009. Alynne Bayard provided GIS mapping imagery for the benthic data. Funding for preparation of this chapter was provided by the US National Science Foundation (currently grant #ARC-1204082), National Oceanic and Atmospheric Administration, and North Pacific Research Board.

Jacqueline M. Grebmeier and Lee W. Cooper

Jackie Grebmeier and Lee Cooper met at the University of California (UC) Bodega Bay Marine Laboratory in spring 1976 where we studied marine worms and played in the mud. Lee graduated from UC Santa Cruz in 1978 and Jackie graduated from UC Davis in 1977, with a Masters from Stanford University in 1978. In the early 1980s we went to the University of Washington for Masters programs in Botany (Lee) and Marine Affairs/Oceanography (Jackie), then northward to Alaska where we undertook research on Arctic ecosystems (Lee on seagrasses and stable isotopes and Jackie on benthic communities and pelagic-benthic coupling). We both received our Ph.D. degrees in Oceanography from the University of Alaska Fairbanks (Lee in 1986, the year we got married, and Jackie in 1987). We then did postdoctoral work in California, Lee at the University of California, Los Angeles in stable isotope geochemistry and Jackie at the University of Southern California on Arctic benthic systems and sediment radioisotopes. We moved to Tennessee in 1988 where we worked at both Oak Ridge National Laboratory and the University of Tennessee in Knoxville. In 2008 we took up Research Professor positions at the Chesapeake Biological Laboratory at the University of Maryland Center for Environmental Sciences where we are currently located. Besides science, we have managed to raise one daughter, Ruth Cooper, who is a senior at the University of Notre Dame, and one very friendly and tolerant cat named Emily.

References

Alexander V, Muench R (1988) Arctic polynyas. Eos, Trans Am Geophys Union 69:572

Bailey EM, Ceballos MAC, Grebmeier JM, Boynton WR (2009) The effects of temperature and carbon addition on Arctic sediment oxygen and nutrient exchanges. Oral presentation at Estuaries and Coasts in a Changing World: Coastal and Estuarine Research Federation 20th Biennial Conference, November 1-5, 2009, Portland, Oregon

Clement JL, Cooper LW, Grebmeier JM (2004) Late-winter water column and sea ice conditions in the northern Bering Sea. J Geophys Res 109, C03022. doi:10.1029/2003JC002047

Clement JL, Maslowski W, Cooper LW, Grebmeier JM, Walczowski W (2005) Ocean circulation and exchanges through the northern Bering Sea: 1979-2001 model results. Deep-Sea Res II Top Stud Oceanogr 52:3509–3540

Cooper LW, Grebmeier JM, Larsen IL, Egorov VG, Theodorakis C, Kelly HP, Lovvorn JR (2002) Seasonal variation in water column processes and sedimentation of organic materials in the St. Lawrence Island polynya region, Bering Sea. Mar Ecol Prog Ser 226:13–26

Cooper LW, Lalande C, Pirtle-Levy RS, Larsen IL, Grebmeier JM (2009) Seasonal and decadal shifts in particulate organic matter processing and sedimentation in the Bering Strait Shelf Region. Deep-Sea Res II Top Stud Oceanogr 56:1316–1325

Cooper LW, Janout MA, Frey KE, Pirtle-Levy R, Guarinello ML, Grebmeier JM, Lovvorn JR (2012) The relationship between sea ice break-up, water mass variation, chlorophyll biomass, and sedimentation in the northern Bering Sea. Deep Sea Res Part II 65:141–162

Cooper LW, Sexson MG, Grebmeier JM, Gradinger R, Mordy CW, Lovvorn JR (2013) Linkages between sea ice coverage, pelagic-benthic coupling and the distribution of spectacled eiders: observations in March 2008, 2009 and 2010 from the northern Bering Sea. Deep-Sea Res II Top Stud Oceanogr 94:31–43

Danielson S, Aagaard K, Weingartner T, Martin S, Winsor P, Gawarkiewicz G, Quadfasel D (2006) The St. Lawrence polynya and the Bering shelf circulation: New observations and a model comparison. J Geophys Res-Oceans 111(C9): doi 10.1029/2005jc003268

Drucker, R., S. Martin, and R. Moritz. 2003. Observations of ice thickness and frazil ice in the St. Lawrence Island polynya from satellite imagery, upward looking sonar, and salinity/temperature moorings, J. Geophys. Res., Oceans 108(C5), 3149, doi:10.1029/2001jc001213

Gilchrist HG, Robertson GJ (2000) Observations of marine birds and mammals wintering at polynyas and ice edges in the Belcher Islands, Nunavut, Canada. Arctic: 61–68

Grebmeier JM (2012) Biological community shifts in Pacific Arctic and sub-Arctic seas. Ann Rev Mar Sci 4:63–78

Grebmeier JM, Barry JP (2007) Benthic processes in polynyas. In: Smith WO Jr, Barber DG (eds) Polynyas: Windows to the World, vol 74, Oceanography series. Elsevier, Amsterdam, pp 363–390

Grebmeier JM, Cooper LW (1995) Influence of the St. Lawrence Island polynya on the Bering Sea benthos. J Geophys Res 100:4439–4460

Grebmeier JM, Dunton KH (2000) Benthic processes in the northern Bering/Chukchi seas: status and global change. In: Huntington HP (ed) Impacts of Changes in Sea Ice and Other Environmental Parameters in the Arctic. Marine Mammal Commission Workshop, Girdwood, Alaska, 15–17 February 2000, pp 80–93

Grebmeier JM, Cooper LW, Feder HM, Sirenko BI (2006a) Ecosystem dynamics of the Pacific influenced Northern Bering and Chukchi Seas. Prog Oceanogr 71:331–361

Grebmeier JM, Overland JE, Moore SE, Farley EV, Carmack EC, Cooper LW, Frey KE, Helle JH, McLaughlin FA, McNutt L (2006b) A major ecosystem shift in the northern Bering Sea. Science 311:461–1464

Grebmeier JM, Moore SE, Overland JE, Frey KE, Gradinger R (2010) Biological response to recent Pacific Arctic sea ice retreats. Eos, Trans Am Geophys Union 91:161–162

Grebmeier JM, Bluhm BA, Cooper LW, Danielson S, Arrigo KR, Blanchard AL, Clark JT, Day RH, Frey KE, Gradinger RR, Kedra M, Konar B, Kuletz KJ, Lee SH, Lovvorn JR, Norcross BL, Okkonen SR (2015) Ecosystem characteristics and processes facilitating persistent macrobenthic biomass hotspots and associated benthivory in the Pacific Arctic. Prog Oceanogr 136:92–114

Jay CV, Udevitz MS, Kwok R, Fischbach AS, Douglas DC (2012) Divergent movements of walrus and sea ice in the northern Bering Sea. Mar Ecol Prog Ser 407:293–302

Jay CV, Grebmeier JM, Fischbach AS, McDonald TL, Cooper LW, Hornsby F (2014) Pacific Walrus (*Odobenus rosmarus divergens*) resource selection in the northern Bering Sea. PLoS One. doi:10.1371/journal.pone.0093035

Lovvorn JR, Richman SE, Grebmeier JM, Cooper LW (2003) Diet and body condition of spectacled eiders in pack ice of the Bering Sea. Polar Biol 26:259–267

Lovvorn JR, Cooper LW, Brooks ML, DeRuyck CC, Bump JK, Grebmeier JM (2005) Organic matter pathways to zooplankton and benthos under pack ice in late winter and open water in late summer in the north-central Bering Sea. Mar Ecol Prog Series 291:135–150

Lovvorn JR, Grebmeier JM, Cooper LW, Bump JK, Richman JG (2009) Modeling marine protected areas for threatened eiders in a climatically shifting Bering Sea. Ecol Appl 19: 1596–1613

Lovvorn JR, Anderson EM, Rocha AR, Larned WW, Grebmeier JM, Cooper LW, Kolts JM, North CA (2014) Variable wind, pack ice, and prey dispersion affect the long-term adequacy of protected areas for an Arctic sea duck. Ecol Appl 24:396–412

Mathis JT, Grebmeier JM, Hansell DA, Hopcroft RR, Kirchman D, Lee SH, Moran SB (2014) Carbon biogeochemistry of the western Arctic: production, export and ocean acidification. In: Grebmeier JM, Maslowski W (eds) The Pacific Arctic Region: Ecosystem Status and Trends in a Rapidly Changing Environment. Springer, Dordrecht, pp 223–268

Mikkelsen N (2015) History of the Arctic Ocean Sciences Board (AOSB) and Cooperation with other Organizations. In: Rogne O, Rachold V, Hacquebord L, Corell R. (eds) IASC after 25 years, Special Issue of the IASC Bulletin. ISBN 979-3-9813637-7-7. Potsdam, Germany

Pease CH (1987) The size of wind-driven coastal polynyas. J Geophys Res 92(C7):7049–7059

Sexson MG, Petersen MR, Breed GA, Powell AN (2016) Shifts in the distribution of molting Spectacled Eiders (Somateria fischeri) indicate ecosystem change in the Arctic. Condor 118:463–476

Simpkins MA, Hiruki-Raring LS, Sheffield G, Grebmeier JM, Bengston JL (2003) Habitat selection by ice-associated pinnipeds near St. Lawrence Island, Alaska in March 2001. Polar Biol 26:577–586

Smith WO Jr, Barber D (eds) (2007) Polynyas: Windows to the World, vol 74. Elsevier, Amsterdam

Stirling I (1980) The biological importance of polynyas in the Canadian Arctic. Arctic 33:303–315

Ecological Processes and Nutrient Transfers from Land to Sea: A 25-Year Perspective on Research and Management of the Seine River System

Josette Garnier and Gilles Billen

Introduction

Science is like a snowball, which grows by incorporating material extracted from the ground as it rolls. Research issues change over time, not only as a result of the internal dynamics of science itself, but also under the pressure of the changing expectations of an evolving society. This is what we wish to illustrate, following the thread of our nearly 30-year joint careers.

The starting point was in the early 1980s, the Golden Age of microbial ecology. At that time the progress of molecular biology made the direct measurement and observation of microbial life in aquatic environments possible. "And now, small is plentiful" was the title of a Nature Views and News article (Sherr 1989) highlighting the new position accorded to microorganisms (from protozoans to viruses) in our understanding of the ecological function of water environments. For the first time, these microorganisms were shown to be a quantitatively important compartment in the functioning of aquatic systems and the concept of the microbial loop was introduced, leading to an alternative to the linear trophic chain due to its role as a sink or a link for higher order consumers (Azam et al. 1983; Pace et al. 1984; Sherr and Sherr 1987, among the first). Microbial ecology of oceans and lakes thus developed rapidly, and together with the concept of bottom-up and top-down controls (Paine 1980) participated in the emergence of a comprehensive ecological theory including cascading effects and retroactive interactions (Carpenter et al. 2009).

River systems, however, long resisted such analysis, probably because of the complexity of these largely open systems: water quality and ecological function in a given river stretch are largely dependent on the upstream drainage network and

J. Garnier (✉) • G. Billen
Université Pierre et Marie Curie, CNRS UMR 7619 Metis,
BP 123, Tour 56-55, Etage 4, 4 Place Jussieu, 75005 Paris, France
e-mail: josette.garnier@upmc.fr

P.M. Glibert, T.M. Kana (eds.), *Aquatic Microbial Ecology and Biogeochemistry: A Dual Perspective*, DOI 10.1007/978-3-319-30259-1_15

watershed functioning. Dealing with these complex relationships remained a real scientific challenge, in spite of the development of the conceptual framework offered by the River Continuum Concept (RCC, Vannote et al. 1980), which proposed an interpretation of longitudinal functional changes along river systems from headwaters to river outlets.

This scientific challenge was also very much in line with a strong social demand, as in Europe this period was the beginning of water resource management at the watershed scale. This approach recognized that water quality in any stretch of river reflects the human activities in the upstream watershed. Agriculture, which feeds the watershed population, uses mineral fertilizers leaching to ground- and surface water. Drinking water comes from the river itself or from groundwater wells contaminated by agricultural pollution. Wastewater produced by domestic or industrial activities is returned to the river directly or after treatment. The resulting sludge is only partly used for fertilization of agricultural fields and the rest is landfilled. Nutrient pollution from point and diffuse sources accumulated in the river reaches the coastal sea where it triggers algal growth on which fish and shellfish production depends.

We have conducted environmental research on the Seine River system and adjacent marine areas since 1989 (Fig. 1). During this long period of time, the environmental issues and the corresponding management stakes have considerably changed,

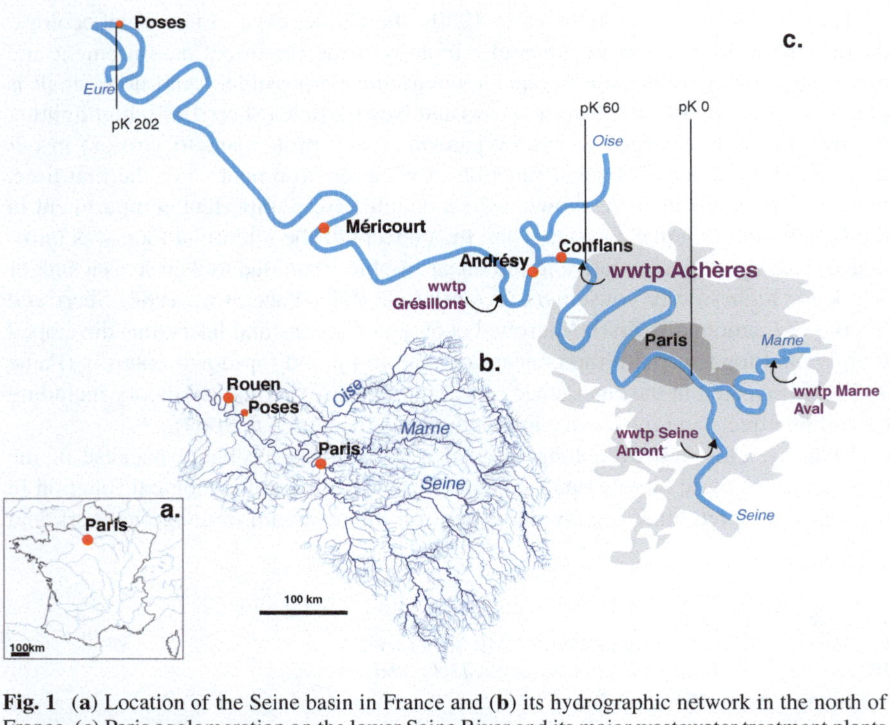

Fig. 1 (**a**) Location of the Seine basin in France and (**b**) its hydrographic network in the north of France. (**c**) Paris agglomeration on the lower Seine River and its major wastewater treatment plants

as did the basic and applied research studies required to help decision making by the authorities in charge of this system. This chapter aims at summarizing the long-term alteration of the Seine River by human activity, the actions taken by managers to reduce this alteration, the success or failure of these actions, and the research questions that arose from all this.

1850–1990: Organic Pollution and Oxygen

As early as the end of the nineteenth century, increasing urban populations and generalization of wastewater collection systems discharging directly into the rivers caused severe organic pollution, often resulting in complete anoxia of long stretches of rivers downstream from large urban centers. At one time, aquatic systems and especially rivers were considered to be able to evacuate all the pollution generated by industrial and domestic activities. Several urban rivers were covered to hide their black color and putrid smell (Billen et al 1999; Garnier et al. 2013). The pollution of the Seine River, which was still directly used as a drinking water source, led to dramatic cholera epidemics causing the death of 30,000 people from 1832 to 1866. From 1850 on, with the development of Paris and its agglomeration, a long race against time started between the water needs of the population and equipping the river for water supply and sanitation (Mouchel et al. 1998). In 1964, the Seine-Normandy Water Agency was created (together with five other agencies for each of the largest water districts in France). Their main concern at the time was to solve the problem of oxygen depletion related to organic matter and ammonium contamination by urban effluent directly discharged into the surface waters or incompletely treated. All efforts were devoted to the implementation of urban wastewater treatment plants (WWTPs).

Since its publication in 1925, the Streeter and Phelps model was used by sanitation engineers to connect river dissolved oxygen concentration and point discharge of organic matter (expressed in biochemical oxygen demand, BOD). The representation of the organic matter degradation process by a simple first-order kinetic equation could not, however, account for the (micro)-biological nature of the processes involved. Indeed, together with organic matter and ammonium, WWTPs also released microorganisms that play a direct role in the metabolism of these substances once released in surface waters (Garnier et al. 1991). Particularly striking is the dynamics of nitrifying organisms which, in the Seine River, develop only slowly after the release of ammonium by Paris WWTPs, so that their effect is only apparent 200 km downstream, in a second, delayed oxygen depletion area at the entrance of the estuarine sector of the river, whereas the river has completely recovered from the first zone of anoxia, immediately downstream of Paris (Fig. 2) (Garnier et al. 2007).

Such phenomena could only be accurately simulated by a second-generation model, explicitly taking into account the dynamics of microorganisms, such as the RIVE model that we developed for that purpose (Billen et al. 1994; Garnier et al. 2002). This model consisted of a detailed description of the processes related to substrate uptake, growth, and mortality of autotrophic and heterotrophic microorganisms

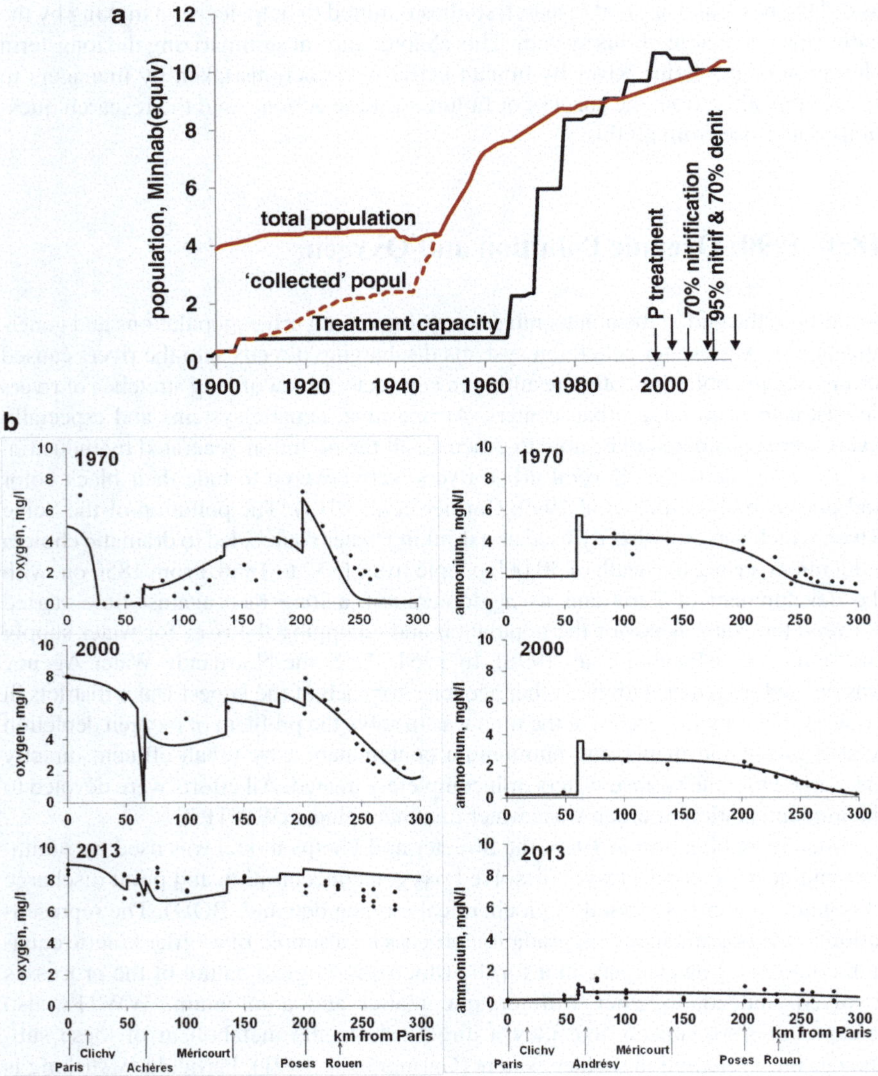

Fig. 2 The long-term evolution of urban organic pollution of the Seine River downstream from Paris. (**a**) Organic and ammonium loading from the Paris agglomeration from the mid-nineteenth century and existing treatment capacity. The wastewater treatment capacity could not be implemented with the increase of pollution loading until the very recent years. (**b**) Simulation, with the model of in-stream microbial processes, of oxygen and ammonium concentration profiles along the Seine River from Paris to the sea, from the mid-nineteenth century to recent years. The improvement of organic matter elimination from wastewater through conventional activated sludge processes finally solved the oxygen depletion problem immediately downstream from the Paris agglomeration in the early 1990s. However, until the recent implementation of nitrification and denitrification treatment of wastewater (2007), a second oxygen depletion zone occurred 200 km downstream, due to nitrification of ammonium

present in aquatic systems in a limited sector of the river. Most parameters involved in the corresponding kinetic equations were directly measured either in the field or in the laboratory using the methods developed in aquatic microbial ecology, so that the model offers a generic representation of microbial metabolism in aquatic systems and does not require any calibration steps.

1990–2000: Eutrophication and Algal Blooms

The problem of eutrophication, i.e., excessive development of algae due to excess nutrients, was noticed in stagnant aquatic systems as early as the 1960s (Vollenweider 1968), and the peak of disturbance was reached in the 1980s, before efficient programs of nutrient abatement measures were implemented. Awareness of eutrophication problems came only later for rivers (Descy 1992; Garnier et al. 1995) and coastal waters (Cugier et al. 2005; Lancelot et al. 2011; Passy et al. 2013; Turner and Rabalais 1994).

In the case of the Seine River system, heavy blooms of diatoms occurred regularly in spring, reaching a biomass above 100 μg/l chlorophyll *a*, severely hindering drinking water production, by clogging sand filters, increasing water pH above 8, which precluded the use of aluminum salts as flocculating agents, and increasing the level of dissolved organic matter in distributed treated water. These blooms generally collapsed after 2 or 3 weeks, resulting in oxygen depletion in the river.

Since these blooms are not generated in the main branch of the Seine River crossing Paris, but in the upstream drainage network (Fig. 3a), a new modeling approach had to be developed to understand their dynamics and predict their response to phosphorus (P) abatement programs. The Riverstrahler model, developed for that purpose, encapsulated the RIVE model of ecological processes, describing the dynamics of nutrients and microorganisms including several types of bacteria and organic matter, three taxonomic classes of phytoplankton (diatoms, Chlorophyceae, and Cyanobacteria), and two groups of zooplankton, into a description of the hydrology of the upstream part of the basin, where the complex network of tributaries is replaced by a regular river confluence scheme of increasing stream order (Strahler 1957) with mean morphological characteristics.

The model correctly simulates the timing of algal development and its geographical distribution in the river network (Garnier et al. 1995; Passy et al. 2013), allowing one to predict the distribution of autotrophic and heterotrophic metabolisms along the river continuum from the description of river network morphology and hydrology as well as the distribution of point and diffuse sources of nutrients (Billen et al. 1994; Garnier and Billen 2007) (Fig. 3a).

Banning polyphosphates from laundry powders in European countries (Billen et al. 1999; Van Drecht et al. 2009), followed by systematically implementing P treatment of urban wastewater, reduced point sources of P tenfold in the Seine watershed. Even though diffuse sources of P, originating from arable soil erosion,

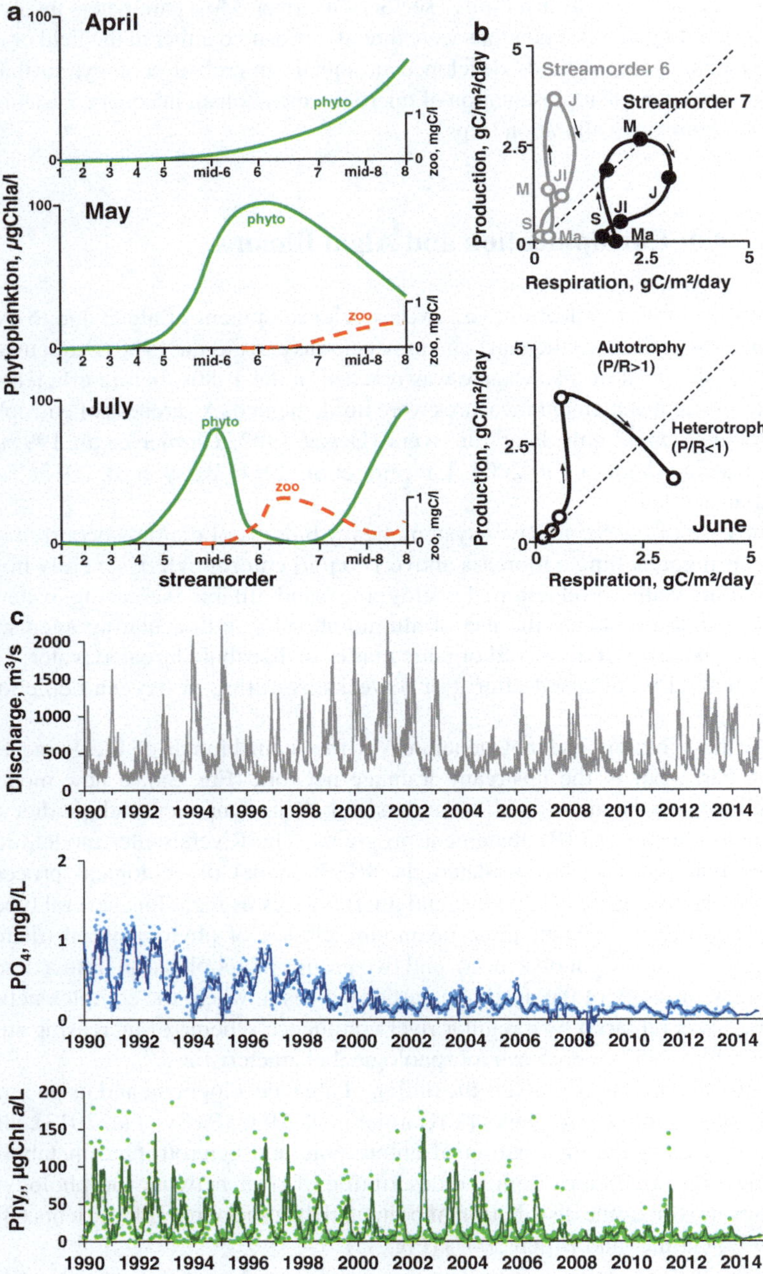

Fig. 3 (**a**) Modeling the development pattern of phytoplankton (chlorophyll concentration) along the river continuum from headwater to estuary in different seasons (redrawn from Garnier et al. 1995) and (**b**) the autotrophic/heterotrophic activities (trajectories in the classical P/R diagram (redrawn from Billen et al. 1994)). (**c**) Long-term reduction of P contamination of the drainage network (variations of total P concentration at the entrance of the Seine estuarine zone (Poses) over the last 25 years, and the resulting decrease in the frequency and intensity of algal blooms)

continue to provide P to sustain algal growth, the general reduction of P loading clearly decreased both the frequency of occurrence and the general level of algal blooms in the downstream sectors of the Seine River system (Fig. 3b).

In the coastal zone of the Seine Bight, algal development is also highly dependent on nutrient fluxes discharged by the Seine River. However, the situation is quite different. In rivers, nitrogen (N), mainly originating from diffuse agricultural sources, is present in large excess with respect to the requirements of algal growth, and P is the most limiting factor of algal growth. Silica (Si), stemming from rock weathering, is most often in excess but can become limiting for diatoms at a high P load (Billen et al. 2007). In marine coastal areas, after mixing with nutrient-poor seawater, all three nutrients, N, P, and Si, are able to limit algal development at some stage of the seasonal cycle and to determine the taxonomic composition of algal communities. We showed that the amount of N or P discharged by rivers in excess over Si with respect to the stoichiometry of diatoms is a good indicator of coastal eutrophication potential (ICEP, Billen and Garnier 2007), characterizing the risk of development of undesirable, often harmful, non-diatom blooms. Since the strong reduction of P fluxes by the Seine was not accompanied by a similar reduction of N, coastal eutrophication remained and is manifested by summer blooms of toxic dinoflagellates following the spring diatom bloom. The occasional occurrence of toxic *Pseudo-nitzschia* blooms, preventing the commercialization of shellfish, is also likely to be a consequence of unbalanced river inputs of nutrients; there is indeed evidence that their domoic acid toxin production is controlled by N (Trainer et al. 2012). Due to the excess in N over P and Si at the coast of most developed countries with intensive agriculture, a substantial reduction of nitrate concentration in river water could decrease eutrophication problems (Passy et al. 2015).

2000–2015: Agricultural Pollution and Nitrate Contamination

The concern about nitrate concentrations in freshwater is not motivated only by the need to reduce coastal eutrophication. It also arose from preoccupations about the drinking water supply. More than 300 dwellings were closed in the Seine basin during the last 10 years because of nitrate levels above the drinking water standard (11 mgN/l). More generally, the environmental losses of N along the whole N cascade from agricultural soils through the atmosphere and hydrosphere cause a large number of problems, such as atmospheric pollution (namely fine particles of NH_4NO_3), greenhouse gas emissions (N_2O, mainly emitted by agricultural soils and the third-ranking cause of atmospheric warming), and loss of terrestrial and aquatic biodiversity (Sutton et al. 2011).

Environmental N losses from agriculture can be estimated from the soil N balance of arable soils integrated over the entire crop rotation cycle. The N balance is calculated as the difference between total fertilization (N inputs to the soil as synthetic or organic fertilizer, manure, symbiotic N_2 fixation, and atmospheric deposition) and export of N with harvested crops (Anglade et al. 2015). For arable soils, in the absence of systematic winter cover by catch crops, more than 70 % of the N balance is leached during the winter drainage period, so that the average nitrate concentration of infiltrating water can be easily predicted from the values of the N balance based on experimental measurement of N leaching (Benoit et al. 2014, 2015). Organic cash crop farms in the Seine watershed, practicing long and diversified rotation where cereals alternate with legume feed crops such as alfalfa, are often thought to be an alternative. We have instrumented a number of the few existing commercial organic farms in the Seine watershed and demonstrated that they produce significantly lower N losses than conventional farms, with, however, very similar yields in terms of total protein content (Benoit et al. 2014; Anglade et al 2015).

The historical reconstruction of the N balance at the scale of the Seine watershed showed a period of rapid increase in the second half of the twentieth century, corresponding to the transition from traditional agriculture, based on a close connection between crop farming and animal husbandry, to industrial cash crop farming dependent on synthetic fertilizers. This increase was followed, with a few decades' delay due to the inertia of the vadose and aquifer reservoirs, by a considerable increase of nitrate contamination of ground- and river water (Fig. 4a). After the 1980s, improvement of farming practices, namely under the incentive of European environmental regulations and efforts to calculate the required N fertilization based on the needs of crop development, resulted in a decrease in the N balance, which stabilized nitrate contamination of ground- and surface waters. The N surplus remains too high, however, to meet the drinking water standard in infiltrating water, suggesting that good agricultural practices, based on equilibrated fertilization at the very high yield expected, now have reached their limits. Further improvement of water quality will therefore require more radical change not only in farming practices, but also in the general organization of the whole agro-food system at a global scale (Billen et al. 2015).

The current agro-food system of the Seine basin is today characterized by a strong disconnection of crop farming and livestock farming. While the basin exports 90 % of its crop production, it has to import most of its animal protein requirements for human consumption from the Brittany and Pays de la Loire regions, where most of the livestock is now concentrated and fed to a large extent with South America-imported soybeans and cakes. In both regions, the environmental N losses are considerable (Fig. 4b). A conversion of agricultural systems to organic farming would lead, however, to a large part of the production consisting of forage legumes instead of cereals. Consequently, the generalization of this type of crop farming would require a reconnection with livestock, as a local outlet for forage production as well as a way to recycle P as a finite resource (Garnier et al. 2015). This reconnection, which implies redistributing livestock at the scale of French regions, should also be accompanied by a reduction of the proportion of animal proteins in the human diet, both for ethical and public health reasons. Thus, a group of scientists published the Barsac declaration (http://www.nine-esf.org/

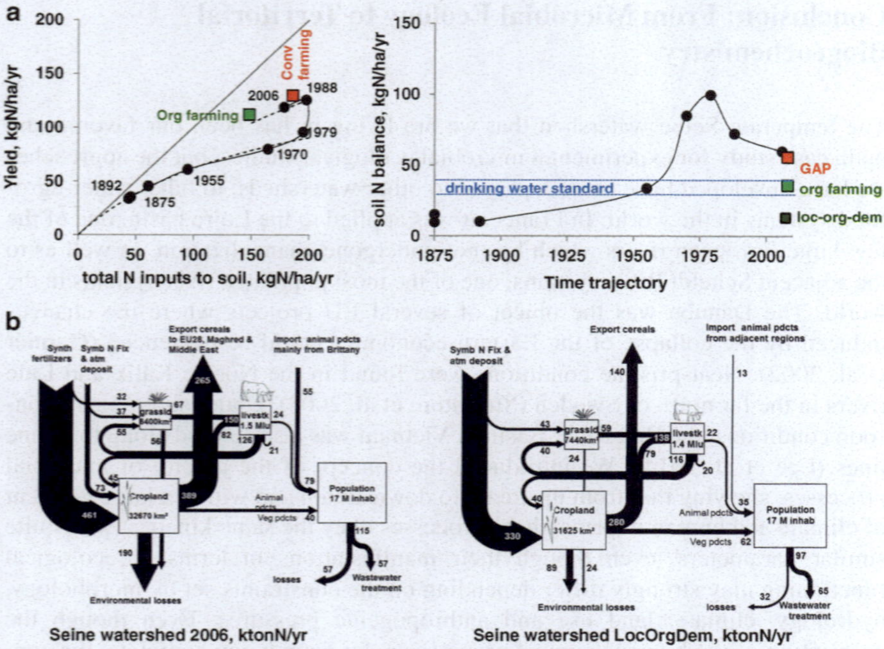

Fig. 4 (**a**) (*Left*) Long-term reconstruction of the N yield vs. N fertilizer inputs of cropland in the Seine watershed. (**a**) (*Right*) The current average N balance (i.e. N fertilizer inputs minus N yield) of conventional and organic crop rotations is also indicated, as well as the maximum value compatible with nitrate concentration standards for drinking water. (**b**) Schematic representation of nitrogen transfers through the agro-food chain of the Seine watershed in the current situation and in a hypothetical organic-local and Demitarian scenario

barsac-declaration), advocating a "Demitarian diet," i.e., a reduction by a factor two of the portion of meat and milk in the Western human diet. Based on these considerations, we constructed an "organic-local and Demitarian" scenario for the Seine watershed (Billen et al. 2012) and extended it to the whole of France. We thus demonstrated its ability to feed the population and still export part of its production while producing much less environmental contamination (Fig. 4b).

The scenario is constructed in such a way as to be easily translated into the input files required to run the Riverstrahler model. It was therefore possible to compare the water quality resulting from this radical scenario with the current scenario and that of a scenario of improved sustainable agriculture based on the generalization of catch crop implementation during winter. The results show that the margins of improvement by sustainable agriculture, although significant, are much more limited than what could result from the organic-local and Demitarian scenario, in terms of groundwater contamination, river water quality, and N fluxes to the coastal zone (Fig. 4c).

Conclusion: From Microbial Ecology to Territorial Biogeochemistry

The temperate Seine watershed that we are living in has been our favored and main case study for experimental microbial ecological studies, but the approaches we have developed have been deployed to other watersheds, to other water-agro-food systems in the world. In France, it was applied to the Loire basin, one of the few large European rivers which has not undergone channelization, as well as to the adjacent Scheldt River systems, one of the most populated river systems in the world. The Danube was the object of several EU projects where the changes induced by the collapse of the Eastern economies could be evidenced (Garnier et al. 2002). Near-pristine conditions were found in the Nordic Kalix and Lule rivers in the far north of Sweden (Sferratore et al. 2008). Under subtropical monsoon conditions, the Red River Basin in Vietnam was also studied along the same lines (Lee et al. 2014). We introduced the concept of the unicity of microbial processes, showing that from upstream to downstream and within a large gradient of climate and human impacts, these processes obey the same kinetics, with quite similar parameters, even though their manifestations in terms of ecological functioning may strongly differ depending on the constraints set by morphology, hydrology, climate, land use and anthropogenic pressures. Even though the description of the biogeochemical processes is far from being complete, the concept of their unicity in the large range of environmental conditions met in aquatic systems from headwater streams to the ocean is particularly fruitful, helping to generalize local field studies.

In addition, challenging the application of concepts and methods developed in marine and lacustrine environments to river systems, we finally developed an original modeling approach (Riverstrahler) for studying not only the river system, but also the terrestrial watershed it drains, with its agricultural and urban systems impacting the coastal marine zone into which it flows. We recently coined the term "water-agro-food systems" to designate this complex mosaic of aquatic and terrestrial ecosystems, deeply modified, exploited, and managed by a society. As such, water-agro-food systems can be viewed as territories, and we have named "territorial biogeochemistry," the branch of science that describes and tries to understand the functioning of such complex systems, their internal and external exchanges of material, and the (physical, chemical, biological, or social-economic) mechanisms controlling these exchanges. Here we open the way to a comprehensive and, why not, citizen-oriented way of practicing science, helping to clarify the societal choices to which we are confronted to address the threat of global change.

Acknowledgments We are deeply grateful to all our Ph.D. students and postdocs who contributed so much to our scientific trajectory. We also extend our thanks to Professor G. de Marsily who welcomed both of us in the laboratory he was directing (UMR 7619-Sisyphe which became UMR 7619-Metis in 2014).

Josette Garnier and Gilles Billen

Connecting lakes and coastal seas with rivers and their watersheds is one of our most fulfilling accomplishments. In the mid-1980s, Josette, with a position at the CNRS in Paris, was studying the ecological functioning of lakes (especially urban sand-pit lakes) while Gilles was involved in marine research at the University of Brussels (Belgium). In 1989, when a new interdisciplinary program (the PIREN-Seine Program) was launched by the CNRS on the Seine River, we both jumped on board: Josette, keen to escape her 42-ha lake, Gilles, prone to seasickness, delighted to step onto solid ground. In June 1989, finding themselves alone together, at midnight, in a small boat in the middle of the Seine River for a 24-h sampling cycle, they already hoped that the PIREN-Seine Program would be long lasting.

We married in 1992 in Brussels, then Gilles left Belgium for a position at the French CNRS. For more than 25 years, we have been co-constructing the now tried-and-true biogeochemical/ecological Riverstrahler model, which we continue to improve and have implemented beyond the Seine to a wide variety of watersheds, from Nordic to subtropical systems. Field work has remained a major occupation, to which we rapidly associated our young daughter, even before she could walk. Thanks to faithful and friendly collaborations, we linked Riverstrahler to coastal zone models, making it possible to assess the measures required in the terrestrial watershed and the river network to mitigate marine eutrophication problems: and thus the loop was closed. Today, most of our energy is devoted to collaborative research projects, focusing on nutrient losses from agriculture and promoting alternative and sustainable water-agro-food-systems management.

We have jointly supervised a large number of graduate students. Many of them have been awarded academic positions, and just as they feel a member of a family during their Ph.D., we still take an interest in their career and keep close contact with them all.

References

Anglade J, Billen G, Garnier J, Makridis T, Puech T, Tittel C (2015) Agro-environmental performance of organic compared to conventional cash crop farming in the Seine watershed. Agr Syst 139:82–92

Azam F, Fenchel T, Gray JG, Field JS, Meyer-Reil LA, Thingstad F (1983) The ecological role of water-column microbes in the sea. Mar Ecol Prog Ser 10:257–263

Benoit M, Garnier J, Anglade J, Billen G (2014) Nitrate leaching from organic and conventional arable crop farms in the Seine Basin (France). Nutr Cycl Agroecosyst 100:285–299. doi:10.1007/s10705-014-9650-9

Benoit M, Garnier J, Billen G, Tournebize J, Gréhan E, Mary B (2015) Nitrous oxide emissions and nitrate leaching in an organic and a conventional cropping system (Seine basin, France). Agr Ecosyst Environ 213:131–141. doi:10.1016/j.agee.2015.07.030

Billen G, Garnier J (2007) River basin nutrient delivery to the coastal sea: assessing its potential to sustain new production of non siliceous algae. Mar Chem 106:148–160. doi:10.1016/j.marchem.2006.12.017

Billen G, Garnier J, Hanset P (1994) Modelling phytoplankton development in whole drainage networks: the RIVERSTRAHLER model applied to the Seine river system. Hydrobiologia 289:119–137

Billen G, Garnier J, Deligne C, Billen C (1999) Estimates of early-industrial inputs of nutrients to river systems: implication for coastal eutrophication. Sci Total Environ 243(244):43–52

Billen G, Garnier J, Némery J, Sebilo M, Sferratore A, Barles S, Benoit P, Benoît M (2007) Nutrient transfers through the Seine river continuum: mechanisms and long term trends. Sci Total Environ 375:80–97

Billen G, Garnier J, Silvestre M, Thieu V, Barles S, Chatzimpiros P (2012) Localising the nitrogen imprint of Paris food supply: the potential of organic farming and changes in human diet. Biogeosciences 9:607–616

Billen G, Lassaletta L, Garnier J (2015) A vast range of opportunities for feeding the world in 2050: trade-off between diet, N contamination and international trade. Environ Res Lett 10(2015):025001. doi:10.1088/1748-9326/10/2/025001

Carpenter SR, Cole JJ, Kitchell JF, Pace ML (2009) Trophic cascades in Lakes: lessons and prospects. Chapter 4. In: Terborgh J, Estes JA (eds) Trophic cascades. Island Press, Washington, DC

Cugier P, Billen G, Guillaud JF, Garnier J, Ménesguen A (2005) Modelling the eutrophication of the Seine Bight (France) under historical, present and future riverine nutrient loading. J Hydrol 304:381–396

Descy J-P (1992) Eutrophication in the River Meuse. In: Sutcliff DW, Jones JG (eds) Eutrophication: research and application to water supply. Freshwater Biological Association, Ambleside, pp 132–142

Garnier J, Billen G (2007) Production vs. respiration in river systems: an indicator of a "good ecological status" evaluation. Sci Total Environ 375:110–124

Garnier J, Servais P, Billen G (1991) Bacterioplankton in the Seine River: impact of the Parisian urban effluents. Can J Microbiol 38:56–64

Garnier J, Billen G, Coste M (1995) Seasonal succession of diatoms and Chlorophyceae in the drainage network of the river Seine: observations and modelling. Limnol Oceanogr 40:750–765

Garnier J, Billen G, Hannon E, Fonbonne S, Videnina Y, Soulie M (2002) Modeling transfer and retention of nutrients in the drainage network of the Danube River. Estuar Coast Shelf Sci 54:285–308

Garnier J, Billen G, Cébron A (2007) Modelling nitrogen transformations in the lower Seine river and estuary (France): impact of wastewater release on oxygenation and N_2O emission. Hydrobiologia 588:291–302

Garnier J, Brion N, Callens J, Passy P, Deligne C, Billen G, Servais P, Billen C (2013) Modelling historical changes in nutrient delivery and water quality of the Zenne River (1790s–2010): the role of land use, waterscape and urban wastewater management. J Mar Syst 128:62–76

Garnier J, Lassaletta L, Billen G, Romero E, Grizzetti B, Némery J, Le QLP, Pistocchi C, Aissa-Grouz N, Luu MTN, Vilmin L, Dorioz J-M (2015) Phosphorus budget in the water-agro-food system at nested scales in two contrasted regions of the world (ASEAN-8 and EU-27). Global Biogeochem Cycl 28(9):1348–1368. doi:10.1002/2015GB005147

Lancelot C, Thieu V, Polard A, Garnier J, Billen G, Hecq W, Gypens N (2011) Ecological and economic effectiveness of nutrient reduction policies on coastal *Phaeocystis* colony blooms in the Southern North Sea: an integrated modeling approach. Sci Total Environ 409:2179–2191. doi:10.1016/j.scitotenv.2011.02.023

Lee TPQ, Billen G, Garnier J (2014) Long-term evolution of the biogeochemical functioning of the Red River (Vietnam): past and present situations. Reg. Environ Change. doi:10.1007/s10113-014-0646-4

Mouchel J-M, Boët P, Hubert G, Guerrini M-C (1998) Un Bassin et des Hommes. Une histoire tourmentée. Chapitre 16. In: De Marsily D, Fustec E, Meybeck M (eds) La Seine en son bassin Fonctionnement écologique d'un système fluvial anthropisé. Elsevier, Paris, pp 77–125

Pace ML, Glasser JE, Pomeroy LR (1984) A simulation analysis of continental shelf food webs. Mar Biol 82:47–63

Paine RT (1980) Food webs: linkage, interaction strength and community infrastructure. J Anim Ecol 49:666–685

Passy P, Gypens N, Billen G, Garnier J, Lancelot C, Thieu V, Rousseau V, Callens J (2013) A Model reconstruction of riverine nutrient fluxes and eutrophication in the Belgian Coastal Zone since 1984. J Mar Syst 128:106–122, http://dx.doi.org/10.1016/j.jmarsys.2013.05.005

Passy P, Le Gendre R, Garnier J, Cugier P, Callens J, Paris F, Billen G, Riou P, Romero E (2015) How can eutrophication be modelled and management strategies improved to prevent algal blooms in the Bay of the Seine? Mar Ecol Prog Ser

Sferratore A, Billen G, Garnier J, Humborg C, Rahm L (2008) Modelling nutrient fluxes from sub-arctic basins: comparison of pristine vs. dammed rivers. J Mar Syst 73:236–249

Sherr EB (1989) And now, small is plentiful. Nature 340:429

Sherr EB, Sherr BF (1987) High rates of consumption of bacteria by pelagic ciliates. Nature 325:710–771. doi:10.1038/325710a0

Strahler AN (1957) Quantitative analysis of watershed geomorphology. T Am Geophys Union 38(6):913–960

Streeter HW, Phelps EB (1925) A study of the pollution and natural purification of the Ohio river. III. Factors concerned in the phenomena of oxidation and reaeration. Public Health Bulletin No. 146. Reprinted by U.S. Department of Health, Education and Welfare, Public Health Service, 1958, ISBN B001BP4GZI http://dspace.udel.edu:8080/dspace/bitstream/handle/19716/1590/C%26EE148.pdf?sequence=2

Sutton MA, Howarth CM, Erisman JW, Billen G, Bleeker A, Grennfelt P, van Grinsven H, Grizzetti B (2011) The European nitrogen assessment: sources effect and policy perspectives. Cambridge University Press, Cambridge, p 612

Trainer VL, Bates SS, Lundholm N, Thessen AE, Cochlan WP, Adams NG, Trick CG (2012) *Pseudo-nitzschia* physiological ecology, phytlogeny, toxicity, monitoring and impact on ecosystem health. Harmful Algae 14:271–300

Turner RE, Rabalais NN (1994) Evidence for coastal eutrophication near the Mississippi River Delta. Nature 368:619–621

Van Drecht G, Bouwman AF, Harrison J, Knoop JM (2009) Global nitrogen and phosphate in urban wastewater for the period 1970 to 2050. Global Biogeochem Cycl 23:GB0A03. doi:10.1029/2009GB003458

Vannote RL, Minshall GW, Cummins KW, Sedell JR, Cushing CE (1980) The river continuum concept. Can J Fish Aquat Sci 37:130–137. doi:10.1139/f80-017

Vollenweider R (1968) Scientific fundamentals of the eutrophication of lakes and flowing waters, with particular reference to nitrogen and phosphorous as factors in eutrophication. OECD Tech Rep. DAS/CSI/68.27. 30 cm. p 159, 34 figs

A Historical Perspective on Eutrophication in the Pensacola Bay Estuary, FL, USA

Jane M. Caffrey and Michael C. Murrell

"... about 11 o'clock I saw a bay, the best I have ever seen in my life" from the diary of Juan Jordan de Reina on entering Pensacola Bay on February 6, 1686 (Dunn 1916)

Introduction

The Pensacola Bay (Florida, USA) estuary shares characteristics with many estuaries that have a long history of human colonization. These human activities have the potential for fundamentally altering the patterns and magnitudes in the delivery of sediments, nutrients, and contaminants to coastal waters. In estuaries, it is well known that biogeochemical cycling of key elements such as carbon, oxygen, nitrogen, phosphorus, and silica varies tremendously along the land-sea gradient; however, we still do not fully understand the consequences of human alterations on these cycles. Similar to well-studied estuaries (e.g., the Chesapeake Bay, the Baltic Sea, and the Neuse River), Pensacola Bay has experienced increased nutrient delivery from urbanization and agricultural development of the landscape. Classically, increased nutrient delivery leads to increased organic matter production or eutrophication of the estuary (Nixon 1995; Kemp et al. 2005; Conley et al 2009). One frequent consequence of eutrophication is the occurrence of hypoxia (commonly defined as dissolved oxygen <2 mg L^{-1}) in bottom waters (Diaz and Rosenberg 2008), which renders the demersal zone uninhabitable by most fish and invertebrate species.

In this chapter, we profile the Pensacola Bay estuary, including a brief physical description and history of human colonization of the region leading up to modern times. To evaluate status and trends of eutrophication within the estuary, we assembled

J.M. Caffrey (✉)
Center for Environmental Diagnostics and Bioremediation, University of West Florida, 11000 University Parkway, Pensacola, FL 32514, USA
e-mail: jcaffrey@uwf.edu

M.C. Murrell
Gulf Ecology Division, US Environmental Protection Agency, Office of Research and Development, National Health and Environmental Effects Research Laboratory, 1 Sabine Island Drive, Gulf Breeze, FL 32561, USA

P.M. Glibert, T.M. Kana (eds.), *Aquatic Microbial Ecology and Biogeochemistry: A Dual Perspective*, DOI 10.1007/978-3-319-30259-1_16

available historical water quality data, asking whether changes in nutrient loading over the past 40 years are evident, and if so whether there are observable responses in the estuary, focusing on the distribution of phytoplankton, nutrients, and dissolved oxygen. We end with a discussion of our current understanding of the magnitude and factors influencing primary productivity, the likely factors affecting seasonal hypoxia, and the potential for nutrient recycling under hypoxic conditions.

Pensacola Bay Physical Setting

Pensacola Bay is a moderately sized (480 km^2) estuary located in northwestern Florida on the Gulf of Mexico (Fig. 1). Freshwater inputs to the Bay come predominately from three rivers: Escambia (71 %), Yellow (26 %), and Blackwater (4 %) (Thorpe et al. 1997). The watershed comprises ~18,000 km^2 and extends northward into Alabama; the Alabama portion represents 65 % of the total watershed area (Thorpe et al. 1997; Bricker et al. 2007). Seasonal patterns in freshwater flow are typical of many subtropical and temperate regions, with high winter and spring flows declining throughout the summer and fall. Monthly average flows vary approximately fourfold from 327 $m^3\ s^{-1}$ in March to 85 $m^3\ s^{-1}$ in November based on long-term records.

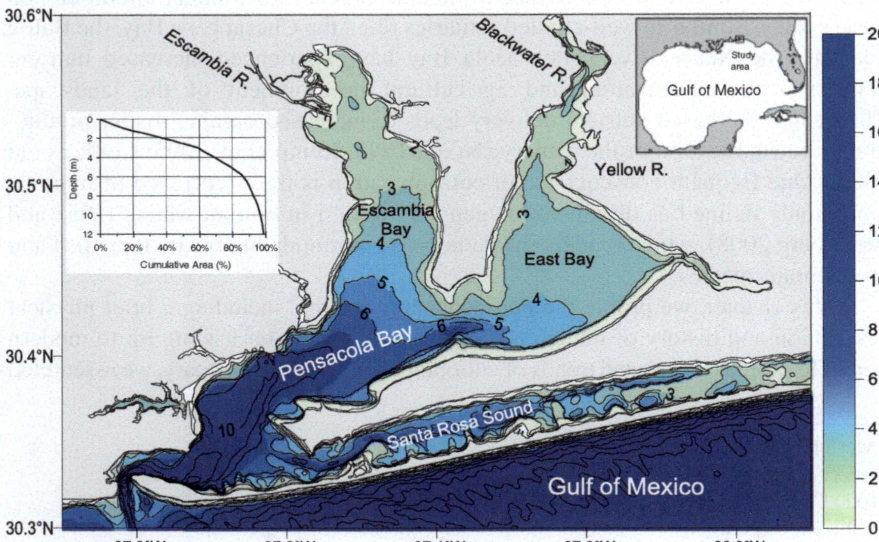

Fig. 1 Map of the Pensacola Bay estuary located in the northeastern Gulf of Mexico (*inset*). *Contours* and *color shading* depict bathymetry (m) of the system. A hypsographic curve (*inset*) depicts the areal distribution of depth within Pensacola Bay and Santa Rosa Sound. Adapted from Murrell et al. (2009)

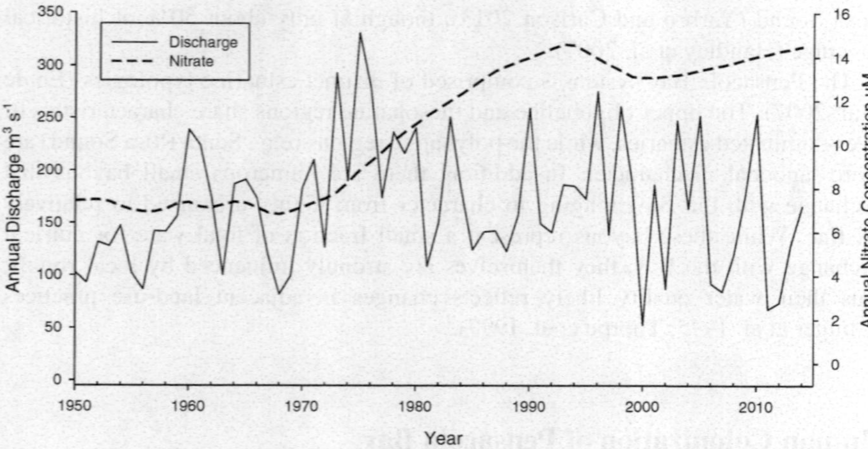

Fig. 2 Annual mean discharge (m^3 s^{-1}) and nitrate concentration (μM) from Escambia River at Century, FL between 1950–2013. Flow data from US Geological Survey, http://waterdata.usgs.gov/nwis, site number 02375500. Annual average nitrate concentration derived from weighted regression of monthly samples on time, discharge, and season using R (EGRET package). Original data obtained from WQP

Annual mean discharge has similarly high variability, ranging from 50 to 325 m^3 s^{-1} (Fig. 2). The average water residence time is 27 days (Bricker et al. 2007).

The upper estuary is comprised of two nearly symmetrical lobes, Escambia Bay on the west and Blackwater and East Bays on the east, which merge into Pensacola Bay proper. Santa Rosa Island separates the Bay from the Gulf of Mexico such that marine exchange is restricted to a narrow pass at the west end of the island. Pensacola Bay is shallow, with a 3.0 m mean depth, and tides are diurnal, having a small mean amplitude of ~0.4 m, common to many estuaries in the Gulf of Mexico (Bianchi et al. 1999). Despite its shallow depth, the high freshwater flow and low tidal mixing energy result in frequent and widespread stratification in the mesohaline reaches of both Escambia and East Bays, especially during spring and summer (Hagy and Murrell 2007).

Subtidal habitats within the Pensacola Bay system are dominated by sandy shoals graduating to finer grained sediments in the deeper central portions of the bay (George et al. 1988). In contrast to peninsular Florida where carbonate sediments dominate, the sands here are silica based (Isphording et al. 1989). Patchy subtidal oyster reefs occur along portions of Escambia and East Bays shoals, which have supported a commercial fishery during different periods of Pensacola's history (Olinger et al. 1975). Currently, seagrass and submerged aquatic vegetation cover 14.3 km^2 or 3 % of the bottom (Yarbro and Carlson 2013); however, this coverage is only 37 % of the 1960 estimate of 38.6 km^2 (8 % of the bay bottom), suggesting a significant decline in seagrass areal coverage over the past 60 years (Handley et al. 2007). Relatively seagrass healthy beds of *Thalassia testudinum* and *Halodule beaudettei* (formerly *Halodule wrightii*) currently occur in Santa

Rosa Sound (Yarbro and Carlson 2013), though at only about 50 % of historical coverage (Handley et al. 2007).

The Pensacola Bay system is comprised of distinct estuarine typologies (Engle et al. 2007). The upper oligohaline and mesohaline regions share characteristics of river-dominated estuaries, while the polyhaline regions (e.g., Santa Rosa Sound) are more lagoonal in character. In addition, there are numerous small bayous that exchange with the Bay, ranging in character from highly urbanized to relatively pristine. While these bayous represent a small fraction of total water or nutrient exchange with the Bay, they themselves are strongly influenced by local runoff; thus their water quality likely reflects changes in adjacent land-use practices (Olinger et al. 1975; Thorpe et al. 1997).

Human Colonization of Pensacola Bay

Human settlement of the Pensacola Bay region dates back to 4200 BP by Native Americans (Thompson and Worth 2011). Colonization by the Spanish began with the Tristan de Luna expedition in 1559. However, his fleet was largely destroyed within weeks of arrival by a hurricane, and the colony was ultimately abandoned in 1561 (Worth and Priestley 2010). Over a century later, in 1698, Spain successfully established a colony, marking the beginning of continuous occupation of the region by European cultures. The attraction of Pensacola Bay to early colonial settlements was strategic in that it provided a deep, safe, and defendable harbor. The region's abundant forests provided materials for ship building, and the presence of artesian springs provided high-quality freshwater. Despite these natural resources, early records of the Spanish colonists indicate that they were dependent on Native Americans for food.

During the colonial period, population remained low (<10,000), only seeing significant growth during the 1870s. Population growth in the late nineteenth and early twentieth centuries was supported by expanded exploitation of the forest resources including timber and naval stores (pitch and turpentine), and increased harvesting of shrimp, oyster, snapper, and mullet for food. In the mid-twentieth century, agricultural production in the watershed increased along with increased use of nitrogen fertilizers, which peaked in the late 1970s (Alexander and Smith 1990). Currently, slash pine remains an important natural resource, as are agricultural row crops including cotton, soybeans, and peanuts.

Following World War II, the population grew rapidly reaching approximately 90,000 by 1970. Fueled by industrialization, the local economy was transformed into a chemical manufacturing center in the 1950s. Heavy industrial installations included nylon and other chemical product manufacturing, paper and pulp production, and a coal-fired electrical power plant. Point source discharges from these industrial activities and municipal wastewater treatment plants dramatically affected water quality and habitats during this period including losses of seagrass beds, bay scallops, shrimp, and other fisheries, along with fecal contamination (Olinger et al. 1975).

Among the most dramatic effects of habitat degradation were the frequent and extensive fish kills that occurred during the 1950s and 1960s, observed predominantly in the upper estuarine bayous and inlets near the industrial outfalls (Olinger et al. 1975). Studies conducted during that era examined the problems, recommended solutions, and provided us the earliest known records of water quality for the area (FSBH 1969; Olinger et al. 1975; Thorpe et al. 1997; Gallagher et al. 1999; Livingston 2001). Following public outcry and enforcement actions under Clean Water Act provisions, point sources of nutrients and other contaminants into Escambia Bay declined rapidly after 1969 (Olinger et al. 1975). The frequency and magnitude of fish kills in Escambia Bay declined precipitously such that they have been infrequent since 1974 (Olinger et al. 1975). Also, the industries gradually transitioned their waste disposal practices toward deep well injection (Table 1). Despite these demonstrable improvements in water quality, the Florida Department of Environmental Protection currently identifies upper Escambia Bay as impaired by excess chlorophyll under the Clean Water Act reporting requirements.

The population in the watershed has increased about fivefold since the 1970s to approximately 460,000 based on the 2010 census. Currently, the major land uses in the watershed include forestry, agriculture, military installations including expansive conservation areas, and urban development; the latter is primarily concentrated

Table 1 Nitrogen sources to Escambia and lower Pensacola Bay in kg day^{-1} during three periods: 1950–1970, 1980–1990, and 2000–2010

Source	Form	1950–1970	1980–1990	2000–2010
River	NO_3[a]	1075–2362	2423–3316	2369–2378
River	TN	6200–10,400[b]	9600–16,100[c]	8000[d]
Non-point source to Escambia Bay	TN	92[b]		
WWTP	TN	1865[b]		635[e], ~0[f]
Monsanto	TN	1452[b]	71[c]	~0[f]
American Cyanamid	TN	2200[b]	102[c]	~0[f]
Air products	TN	2560[b]	19[c]	~0[f]

[a][NO3] × flow (our Fig. 2)
[b]Olinger et al. (1975)
[c]Gallagher et al. (1999)
[d]SPARROW model from Hoos and McMahon (2009)
[e]Input from Pensacola Main Street WWTP (Hagy, pers. comm.)
[f]Converted to deep well injection

to lands along the shores of Pensacola Bay. The local economy is strongly driven by tourism, due to the natural beauty of the clear emerald waters and "sugar-white" sandy beaches on Santa Rosa Island and along the shores of Pensacola Bay, which has earned it the moniker of "The Emerald Coast".

River and Estuarine Water Quality

Historical data for the Pensacola Bay are sparse compared to better studied estuaries. Therefore, our summary relies on assembling water quality data from a variety of monitoring efforts conducted over the years. The earliest water quality data come from the US Geological Survey (USGS) stream gage on the upper Escambia River at Century (near the Florida-Alabama border) that was established in 1935. From 1952 to 1994, USGS also collected monthly water grab samples from this site. For the lower Escambia River, the earliest available water quality data are from the early 1950s, and comprised a total of ten nutrient samples (five sites sampled on two dates) (ANSP 1953). The next available water quality data were collected from 1969 to 1974 as part of a large-scale effort to document compliance with regulations and recovery from point source discharges (Olinger et al. 1975). These data provide perhaps the most valuable and comprehensive picture of the historical water and sediment quality in Pensacola Bay, with an emphasis on Escambia Bay. The Florida State Board of Health (FSBH 1969) also sampled the Escambia River and Bay during this period, primarily motivated by concerns over fecal coliform pollution.

Since the 1980s, a variety of state and federal agencies have monitored water quality in the region. We assembled data from the following sources: Florida Department of Environmental Protection (FDEP), Florida Fish and Wildlife Research Institute, Florida Department of Agriculture and Consumer Services, Florida Department of Health, US Environmental Protection Agency, and a consulting company study of hydrodynamics, water quality, phytoplankton, benthic macrofauna, and fish communities (Gallagher et al. 1999; Livingston 2001). Most of these data were acquired using the Water Quality Portal (WQP, http://www.waterqualitydata.us/), a Web service that itself assembled data from a variety of federal agencies such as EPA (STORET) and USGS (NWIS). In addition to Escambia River data, we compiled the available chlorophyll and dissolved oxygen data from the mesohaline portion of Escambia Bay. We restricted our summary to this region for several reasons: (1) it receives most of the freshwater, (2) it is vulnerable to hypoxia, (3) historical data exist for the past 40 years, and (4) upper Escambia Bay is considered impaired by the Florida Department of Environmental Protection.

Based on the available data, nutrient concentrations in the Escambia River (Fig. 2) and Escambia Bay (Fig. 3) appear low relative to well-studied eutrophic estuaries. This result is consistent with the assessment of Bricker et al. (2007) who noted that nitrogen loading is among the lowest described for estuaries in the Gulf of Mexico. While the nitrate time series at Century, FL, revealed an upward trend,

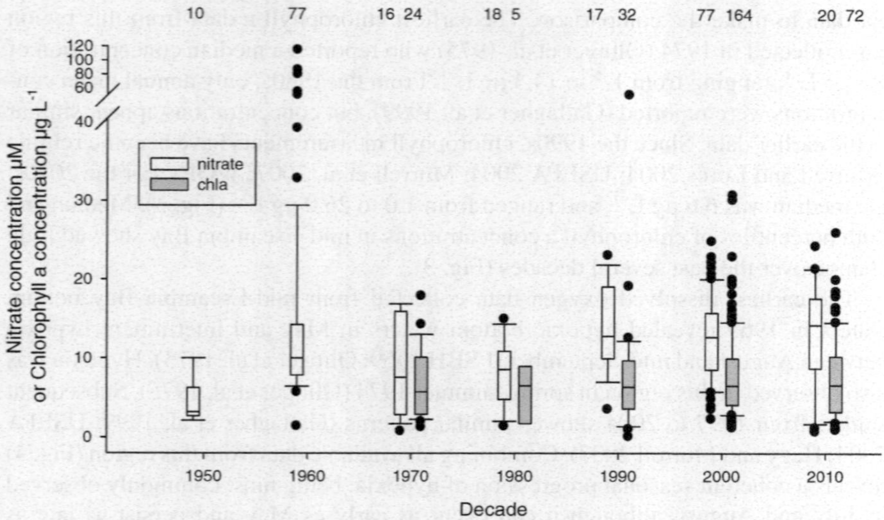

Fig. 3 Box plot by decade of nitrate (μM) (*open bars*) in the lower Escambia River and chlorophyll a (μg L^{-1}) (*shaded bars*) from mid-Escambia Bay. Shown are the interquartile ranges (*shaded*) bisected by the median (*line*). The *whiskers* represent the 10th and 90th percentiles and *symbols* are outliers. The number of observations comprising each bin is labeled. The extremely high nitrate values in 1960s were collected immediately downstream of Monsanto outfall. Data sources include FSBH (1969), Hopkins (1969), Hagy and Murrell (2007), Murrell et al. (2007, 2009), Smith and Caffrey (2009), Caffrey (unpublished), EPA (unpublished), and WQP

doubling from 7.5 to 14.3 μM between the 1970s and 1990s (Fig. 2), these concentrations are still much lower than the 100 μM or greater levels that are typical of rivers with large agricultural inputs (Bianchi et al. 1999; Kemp et al. 2005; Conley et al. 2009). Over this same period, median ammonium and Kjeldahl nitrogen concentrations increased from 2 to 8 μM and 20 to 34 μM, respectively (data not shown, WQP). Compared to the upper Escambia River, nitrate concentrations near the mouth were about twofold higher; yet median concentrations were still relatively low, rarely exceeding 25 μM (Fig. 3). The highest reported nitrate concentration in the lower Escambia River was ~120 μM during the 1960s; however these samples likely reflect highly localized conditions near industrial point source outfalls (Fig. 3). In mid-Escambia Bay, surface water nitrate and ammonium concentrations rarely exceeded 10 μM (Murrell et al. 2007). While not shown, the nitrate concentrations in the Yellow and Blackwater Rivers were similar to the Escambia River (Lewis 2010) and surface water nutrient concentrations in the mesohaline East Bay were similar to Escambia Bay (EPA, unpublished).

To evaluate changes in the phytoplankton response to river nutrient loads, we examined historical chlorophyll data in Escambia Bay surface waters (Fig. 3). We focused on the mesohaline region (salinity ~12–24) because this was the region where chlorophyll typically peaked, where salinity mixing diagrams indicated strong nutrient uptake (Murrell et al. 2007), and where there were sufficient historical data to make the comparison. The earliest chlorophyll a data from this region was collected in 1974 (Olinger et al. 1975) who reported a median concentration of 5.8 μg L^{-1}, ranging from 1.3 to 14.3 μg L^{-1}. From the 1980s, only annual mean concentrations were reported (Gallagher et al. 1999), but concentrations appear similar to the earlier data. Since the 1990s, chlorophyll measurements have become routine (Murrell and Lores 2004; USEPA 2004; Murrell et al. 2007; WQP). For the 2010s, the median was 6.6 μg L^{-1}, and ranged from 1.0 to 26.0 μg L^{-1} (Fig. 3). Median and 90th percentiles of chlorophyll a concentrations in mid-Escambia Bay showed little change over the past several decades (Fig. 3).

The earliest dissolved oxygen data collected from mid-Escambia Bay bottom waters in 1969 revealed hypoxic bottom waters in May and intermittent hypoxia between August and mid-September (FSBH 1969; Olinger et al. 1975). Hypoxia was also observed in this region in spring-summer 1974 (Olinger et al. 1975). Subsequent studies from 1997 to 2004 showed similar patterns (Gallagher et al. 1999; USEPA 2004; Hagy and Murrell 2007). Combining all available data from this region (Fig. 4) reveals a coherent seasonal progression of hypoxia, being most commonly observed in July and August, although it can occur as early as May and persist as late as October. In this regard, the seasonality of hypoxia is very similar to well-studied,

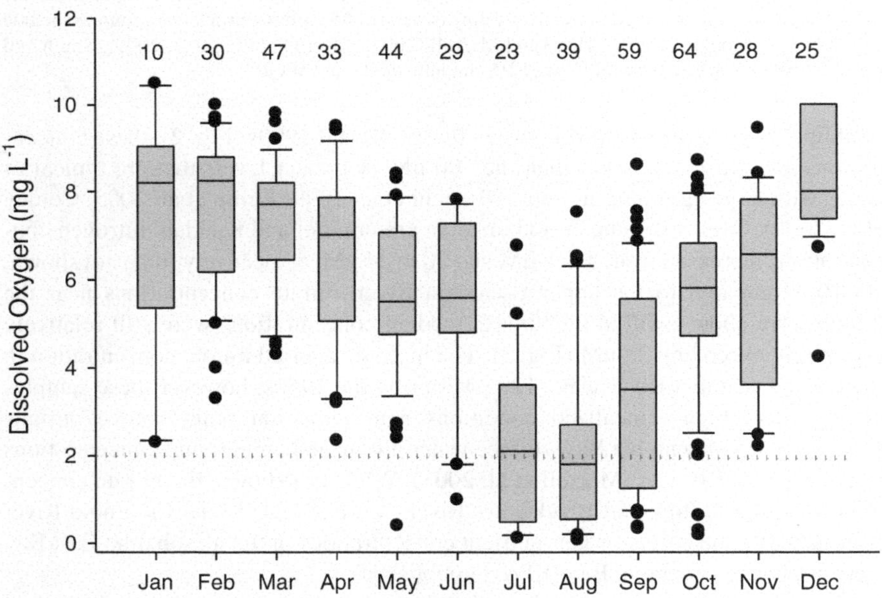

Fig. 4 Box plot by month of dissolved oxygen (mg L^{-1}) from mid-Escambia Bay bottom waters. Shown are the interquartile ranges (*shaded*) bisected by the median (*line*). The *whiskers* represent the 10th and 90th percentiles and *symbols* are outliers. The number of observations comprising each bin is labeled. The *horizontal dashed line* depicts the hypoxic threshold of 2 mg L^{-1}. Data include years 1969–2014: FSBH (1969), Hopkins (1969), Hagy and Murrell (2007), Murrell et al. (2009), Smith and Caffrey (2009), Caffrey (unpublished), WQP

eutrophic estuaries. Hagy and Murrell (2007) attributed the prevalence of hypoxia in Pensacola Bay to the combination of strong vertical stratification and sluggish horizontal transport; these physical characteristics have likely remained similar over the past 40 years and perhaps longer. In fact, there are notable examples of warm microtidal estuaries that exhibited seasonal hypoxia prior to large-scale human occupation (May 1973; Pennock et al. 1994; Stanley and Nixon 1992), suggesting that the incidence of hypoxia is not solely attributable to cultural eutrophication.

Controls on Primary Production, Organic Matter, and Nutrient Cycling

A series of process-oriented studies conducted in Pensacola Bay from 1999 to 2005 provided some insights into spatial patterns in planktonic and benthic production. Collectively, the studies suggest that Pensacola Bay is a moderately productive estuary, with important phytoplankton and benthic components. While baywide measurements of phytoplankton productivity have not been made, Murrell et al. (2007) reported annual productivity of 290 g C m^{-2} $year^{-1}$ from Escambia Bay, placing it at the 70th percentile of a recent global compilation (Cloern et al. 2014). As with most estuaries, we found that phytoplankton productivity in Escambia Bay strongly varied with seasonal and interannual variation in freshwater flow. During summer, nitrate concentrations rapidly declined along the estuarine gradient, such that nitrate in lower Escambia Bay surface waters was typically ~2 μM. The maximum nitrate depletion zone coincided with peak phytoplankton biomass and productivity (Murrell et al. 2007). Additionally, patterns of nutrient limitation appear similar to other estuaries with incidence of phosphorus limitation in winter and spring, particularly in the upper estuary switching to nitrogen limitation during summer and fall (Murrell et al. 2002; Juhl and Murrell 2008).

High water transparency throughout the system means that much of the water column and the benthic shoal regions are euphotic. Thus, sufficient light penetrates to support primary production in the lower water column and the benthos (Murrell et al. 2009). Thus oxygen production in sub-pycnocline waters and the benthos can potentially offset hypoxia (Hagy and Murrell 2007). Microphytobenthic productivity was significant in the extensive sandy shoals of Pensacola Bay accounting for 16–32 % of total system productivity (Murrell et al. 2009). This estimate did not include seagrass productivity which likely contributes additional significant productivity, particularly in Santa Rosa Sound, though likely a less significant component of baywide averages compared to seagrass-dominated systems like Tampa and Florida Bay.

One additional observation from this series of process studies was the regular summertime blooms in pico-phytoplankton (i.e., *Synechococcus*), which can comprise upwards of 90 % of total phytoplankton biomass in Escambia Bay (Murrell and Lores 2004). Similar blooms were also observed in nearby Weeks Bay, AL, and Apalachicola Bay, FL (Murrell and Caffrey 2005), and in other warm shallow systems such as Florida Bay (Phlips et al. 1999) and the Neuse River Estuary (Pinckney et al. 1998). The reason for this dominance is not altogether clear, but likely reflect a combination of factors that favor picoplankton over larger phytoplankton. First, the relatively low nitrate concentrations observed during summer combined with higher residual ammonium may favor picoplankton over diatoms, owing to the well-known inhibitory effect of ammonium on nitrate uptake by diatoms (Glibert et al. 2016). Additionally, copepod zooplankton can exert a strong grazing control on diatoms and other large phytoplankton, but picoplankton generally escape predation due to their small size. Third, growth rates of *Synechococcus* appear to be strongly temperature dependent, similar to other bacterioplankton; thus warm summer conditions may favor maximal growth rates (Li 1998).

The effects of hypoxia on nutrient cycling in this system have some similarities and differences to other estuaries. In the Escambia Bay channel, hypoxia coincided with an accumulation of ammonium in bottom waters, but not dissolved inorganic phosphorus (DIP) (Fig. 5a). Similar conditions existed in 1974 where maximum bottom water ammonium concentrations occurred at the stations with the lowest bottom water dissolved oxygen concentrations (Olinger et al. 1975). Similarly, we observed that ammonium fluxes were highest with low dissolved oxygen concentrations, while DIP fluxes were not (Fig. 5b). This implies active recycling of nitrogen, but not DIP in hypoxic waters of the Pensacola Bay system. It is perhaps surprising that hypoxia does not stimulate DIP release from sediments or leads to a buildup of DIP in bottom water, since this has been observed in other systems such as Chesapeake Bay (Kemp et al. 1990; Testa and Kemp 2012) and the Baltic Sea (Conley et al. 2002). In these systems, DIP fluxes out of sediments can triple when dissolved oxygen is less than 1 mg L^{-1} (Koop et al. 1990; Testa and Kemp 2012). Unfortunately, the lowest overlying oxygen concentration in flux experiments from the Pensacola Bay system was only 0.7 mg L^{-1} (Murrell et al. 2009; Smith and Caffrey 2009), so the response of sediments to lower oxygen concentrations is unknown. In addition, iron and iron-bound phosphorus concentrations also influence mobilization of phosphorus under hypoxia and anoxia, but the role of iron cycling has not been examined in the Pensacola Bay system.

Summary

Key aspects of Pensacola Bay biogeochemistry are similar to other eutrophic estuaries. These include an increase in phytoplankton productivity fueled by river nutrients, the occurrence of summer hypoxia in the bottom waters, the loss of seagrass habitat, and the increased recycling of ammonium but not phosphate in hypoxic bottom waters. Pensacola Bay also shares similarities with other Florida estuaries such as Tampa and Florida Bays in having extensive shoals capable of supporting seagrasses and having clear waters with high light penetration. However, conspicuous features of Pensacola Bay shoals are the lack of macroalgae and low seagrass coverage. Hypoxia in Pensacola Bay varies from well-known estuaries such as the Baltic Sea, Chesapeake Bay, or the Neuse River, in that sub-pycnocline waters and

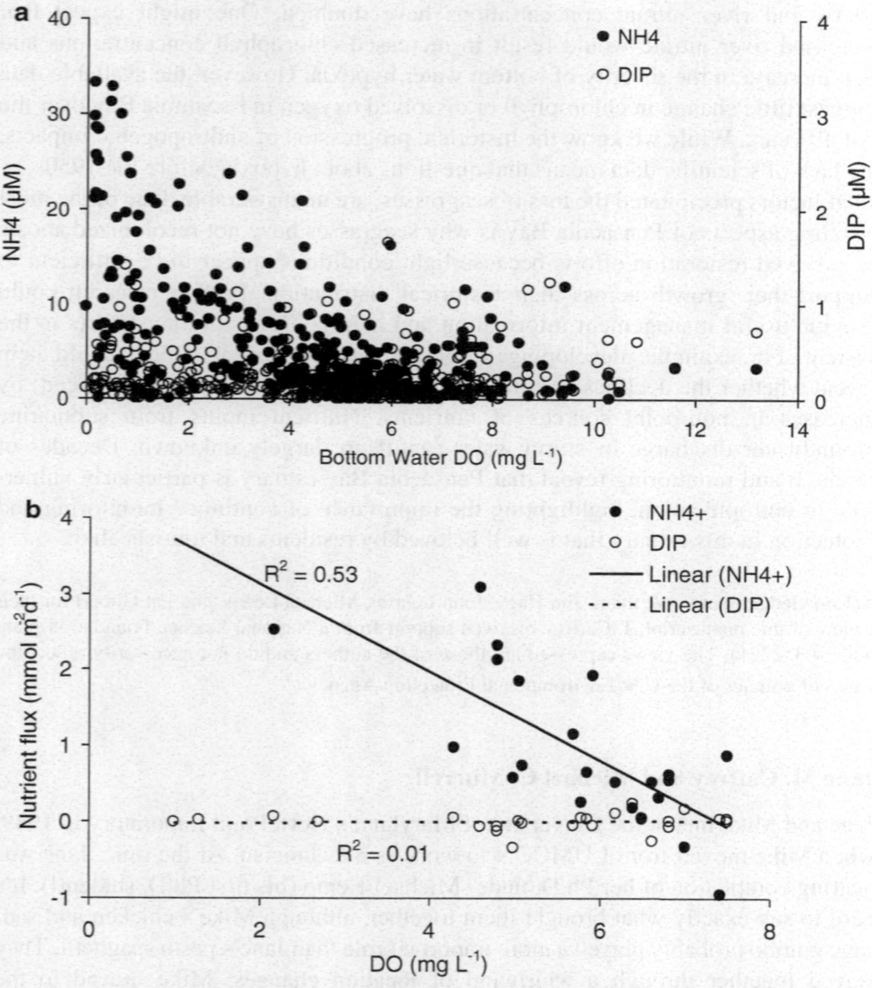

Fig. 5 (**a**) Relationship between NH_4^+ (*solid symbols*), DIP (*open symbols*), and dissolved oxygen concentrations in bottom water samples collected between 2002 and 2004 in Escambia Bay channel (Murrell unpublished). (**b**) Relationship between sediment-water fluxes of NH_4^+ (*solid symbols*) and DIP (*open symbols*) and the dissolved oxygen concentrations in the overlying water. Data sources are DiDonato et al. (2006), Murrell et al. (2009), and Smith and Caffrey (2009) from experiments conducted from 2000 to 2005. Positive fluxes indicate release from sediment

the benthos receive sufficient light to support photosynthesis. Thus, benthic oxygen production acts to offset hypoxia.

Pensacola Bay has a long history of anthropogenic impacts. Major point sources of nutrients from the 1950s and 1960s were greatly reduced by the 1980s and 1990s and many have been completely eliminated from surface discharges. At the same time, the population of watershed has increased fivefold between 1970s and

2010s and river nitrate concentrations have doubled. One might expect that increased river nitrate would result in increased chlorophyll concentrations and also increase in the severity of bottom water hypoxia. However, the available data suggest little change in chlorophyll or dissolved oxygen in Escambia Bay over the last 40 years. While we know the historical progression of anthropogenic impacts, the lack of scientific data means that questions about hypoxia before the 1950s, or what factors precipitated the loss of seagrasses, are unanswerable. One of the most puzzling aspects of Pensacola Bay is why seagrasses have not recolonized shoals or survived restoration efforts because light conditions appear to be sufficient to support their growth across their historical distribution. Further research could provide useful management information and help guide restoration efforts in the system. For example, developing a comprehensive nutrient budget could help reveal whether the declines in point sources have been balanced (or outpaced) by increases in non-point sources of nutrients. Nutrient inputs from submarine groundwater discharge or storm water runoff are largely unknown. Decades of research and monitoring reveal that Pensacola Bay estuary is particularly vulnerable to eutrophication, highlighting the importance of continued monitoring and protection in this estuary that is well beloved by residents and tourists alike.

Acknowledgements We thank Jim Hagy, John Lehrter, Michael Lewis, and Pat Glibert for their review of this manuscript. J. Caffrey received support from a National Science Foundation grant (OCE-0352221). The views expressed are those of the authors and do not necessarily reflect the views or policies of the U.S. Environmental Protection Agency.

Jane M. Caffrey and Michael C. Murrell

Jane and Mike met at the University of Maryland's Horn Point Laboratory in 1989, when Mike moved from LUMCON to work as a technician. At the time, Jane was nearing completion of her Ph.D. under Michael Kemp (his first Ph.D. student!). It's hard to say exactly what brought them together, although Mike's chicken and sausage gumbo probably played a more important role than Jane's pesto spaghetti. They stayed together through a whirlwind of location changes: Mike moved to the University of Washington to continue working as an Oceanographer for Evelyn Lessard and Jane took a Fulbright postdoc at the University of Aarhus under Henry Blackburn. After her exhilarating experience in Denmark, Jane moved to a postdoctoral position at the USGS in Menlo Park. So, while still separated by ~800 miles, at least they were the same time zone. Then, on Valentine's Day 1992, Mike moved to the San Francisco Bay area in preparation for starting his Ph.D. at the University of California Santa Cruz with Mary Silver as his academic advisor and Tim Hollibaugh (SFSU) as his research advisor. Jane and Mike worked side by side on the USGS vessel, the R/V Polaris (recently retired), as Mike examined bacterioplankton and microzooplankton dynamics in North San Francisco Bay and Jane studied patterns in benthic nutrient cycling as modulated by the spring phytoplankton bloom. After her postdoc ended, Jane became a Research Oceanographer at UCSC and also Research Coordinator at the Elkhorn Slough National Estuarine Research Reserve.

Following Mike's graduation, they all (baby Amelie makes 3) moved back East where Mike took a postdoc with the EPA at the Gulf Ecology Division in Gulf Breeze, FL, which evolved into a career position. Jane began collaborations with the University of West Florida, which also evolved into an academic faculty position. Over the years, Mike has helped to develop Jane's appreciation of microbes and plankton as discrete organisms, while Jane has helped to develop Mike's appreciation of biogeochemical cycling and playing in the mud.

References

Alexander RB, Smith RA (1990) County-level estimates of nitrogen and phosphorus fertilizer use in the United States, 1945–1985. U.S.G.S. Open-file report 90–130

Bianchi TS, Pennock JR, Twilley RR (eds) (1999) Biogeochemistry of Gulf of Mexico Estuaries, 1st edn. John Wiley and Sons, New York, NY

Bricker S, Longstaff B, Dennison W, Jones A, Boicourt K, Wicks C, Woerner J (2007) Effects of nutrient enrichment in the nation's estuaries: a decade of change. NOAA, coastal ocean program, decision analysis series no 26. National Centers for Coastal Ocean Science, Silver Spring, MD, p 322

Cloern JE, Foster SQ, Kleckner AE (2014) Phytoplankton primary production in the world's estuarine-coastal ecosystems. Biogeosciences 11:2477–2501

Conley DJ, Humborg C, Rahm L, Savchuk OP, Wulff F (2002) Hypoxia in the Baltic Sea and basin-scale changes in phosphorus biogeochemistry. Environ Sci Technol 36:5315–5320

Conley DJ, Björck S, Bonsdorff E, Carstensen J, Destouni G, Gustafsson BG, Hietanen S, Kortekaas M, Kuosa H, Markus Meier H (2009) Hypoxia-related processes in the Baltic Sea. Environ Sci Technol 43:3412–3420

Diaz RJ, Rosenberg R (2008) Spreading dead zones and consequences for marine ecosystems. Science 321:926–929

DiDonato GT, Lores EM, Murrell MC, Smith LM, Caffrey JM (2006) Benthic nutrient flux in a small estuary in Northwestern Florida (USA). Gulf Caribb Res 18:15–25

Dunn WE (1916) The Spanish search for La Salle's colony on the Bay of Espiritu Santo, 1685–1689. Southwest Hist Q 19:323–369

Engle VD, Kurtz JC, Smith LM, Chancy C, Bourgeois P (2007) A classification of U.S. estuaries based on physical and hydrologic attributes. Environ Monit Assess 129:397–412

George SM, Isphording WC, Meylan MA (1988) Recent sedimentary environment of Pensacola Bay, Florida. Gulf Coast Assoc Geol Soc Trans 38:581

Glibert PM, Wilkerson FP, Dugdale RC, Raven JA, Dupont CL, Leavitt PR, Parker AE, Burkholder JM, Kana TM (2016) Pluses and minuses of ammonium and nitrate uptake and assimilation by phytoplankton and implications for productivity and community composition, with emphasis on nitrogen-enriched conditions. Limnol Oceanogr 61:165–197

Hagy JD III, Murrell MC (2007) Susceptibility of a Gulf of Mexico estuary to hypoxia: an analysis using box models. Estuar Coast Shelf Sci 74:239–253

Handley L, Altsman D, DeMay R (2007) Seagrass status and trends in the northern Gulf of Mexico: 1940–2002. U.S. Geological Survey Scientific investigations report 2006–5287 and U.S. Environmental Protection Agency, 855-R-04-003, p 267

Hoos AB, McMahon G (2009) Spatial analysis of instream nitrogen loads and factors controlling nitrogen delivery to streams in the southeastern United States using spatially referenced regression on watershed attributes (SPARROW) and regional classification frameworks: hydrologic processes. doi:10.1002/hyp.7323

Isphording WC, Imsand FD, Flowers GC (1989) Physical characteristics and aging of Gulf coast estuaries. Am Assoc Petrol Geol Bull 73:9

Juhl AR, Murrell MC (2008) Nutrient limitation of phytoplankton growth and physiology in a subtropical estuary. Bull Mar Sci 82:59–82

Kemp WM, Sampou PS, Caffrey JM, Mayer MA, Henriksen K, Boynton WR (1990) Ammonium recycling versus denitrification in Chesapeake Bay sediments. Limnol Oceanogr 35:1545–1563

Kemp WM, Boynton WR, Adolf JE, Boesch DF, Boicourt WC, Brush G, Cornwell JC, Fisher TR, Glibert PM, Hagy JD, Harding LW, Houde ED, Kimmel DG, Miller WD, Newell RIE, Roman MR, Smith EM, Stevenson JC (2005) Eutrophication of Chesapeake Bay: historical trends and ecological interactions. Mar Ecol Prog Ser 303:1–29

Koop K, Boynton WR, Wulff F, Carman R (1990) Sediment-water oxygen and nutrient exchanges along a depth gradient in the Baltic Sea. Mar Ecol Prog Ser 63:65–77

Lewis FG (2010) East Bay/Blackwater Bay/Lower Yellow River preliminary baseline resource characterization: with a discussion of flow-dependent habitats and species. NWFWMD. Water resources special report 2010-02. http://www.nwfwmd.state.fl.us/system/assets/822/original/EastBayResourceCharacterization.pdf

Li W (1998) Annual average abundance of heterotrophic bacteria and *Synechococcus* in surface ocean waters. Limnol Oceanogr 43:1746–1753

Livingston RG (2001) Eutrophication processes in coastal systems: Origin and succession of plankton blooms and effects on secondary production in Gulf Coast Estuaries. CRC Press, Boca Raton, FL

May EB (1973) Extensive oxygen depletion in Mobile Bay, Alabama. Limnol Oceanogr 18:353–366

Murrell MC, Caffrey JM (2005) High cyanobacterial abundance in three northeastern Gulf of Mexico estuaries. Gulf Caribb Res 17:95–106

Murrell MC, Lores EM (2004) Phytoplankton and zooplankton seasonal dynamics in a subtropical estuary: importance of cyanobacteria. J Plankton Res 26:371–382

Murrell MC, Stanley RS, Lores EM, DiDonato GT, Smith LM, Flemer DA (2002) Evidence that phosphorus limits phytoplankton growth in a Gulf of Mexico estuary: Pensacola Bay, FL, USA. Bull Mar Sci 70:155–167

Murrell MC, Hagy JD, Lores EM, Greene RM (2007) Phytoplankton production and nutrient distributions in a sub-tropical estuary: importance of freshwater flow. Estuar Coast Shelf Sci 30:390–402

Murrell MC, Campbell JG, Hagy JD III, Caffrey JM (2009) Effects of irradiance on benthic and water column processes in a Gulf of Mexico estuary: Pensacola Bay, Florida, USA. Estuar Coast Shelf Sci 81:501–512

Nixon SW (1995) Coastal marine eutrophication: a definition, social causes, and future concerns. Ophelia 41:199–219

Olinger L, Rogers R, Fore P, Todd R, Mullins B, Bisterfield F, Wise L (1975) Environmental and recovery studies of Escambia Bay and the Pensacola Bay system. Florida US Environmental Protection Agency, Region IV, Gulf Breeze, FL, p 345

Pennock JR, Sharp JH, Schroeder WW (1994) What controls the expression of estuarine eutrophication? Case studies of nutrient enrichment in the Delaware Bay and Mobile Bay Estuaries, USA. In: Dyer KR, Orth RJ (eds) Changes in fluxes in estuaries: implications from science to management. Olsen and Olsen, Midvale, UT, pp 139–146

Phlips EJ, Badylak S, Lynch TC (1999) Blooms of the picoplanktonic cyanobacterium *Synechococcus* in Florida Bay, a subtropical inner-shelf lagoon. Limnol Oceanogr 44:1166–1175

Pinckney JL, Paerl HW, Harrington MB, Howe KE (1998) Annual cycles of phytoplankton community-structure and bloom dynamics in the Neuse River Estuary, North Carolina. Mar Biol 131:371–381

Smith KA, Caffrey JM (2009) The effects of human activities and extreme meteorological events on sediment nitrogen dynamics in an urban estuary, Escambia Bay, Florida, USA. Hydrobiologia 627:67–85

Stanley DW, Nixon SW (1992) Stratification and bottom-water hypoxia in the Pamlico River estuary. Estuaries 15:270–281

Testa JM, Kemp WM (2012) Hypoxia-induced shifts in nitrogen and phosphorus cycling in Chesapeake Bay. Limnol Oceanogr 57:835–850

Thompson VD, Worth JE (2011) Dwellers by the sea: Native American adaptations along the southern coasts of eastern North America. J Archaeol Res 19:51–101

Thorpe P, Bartel R, Ryan P, Albertson K, Pratt T, Cairns D, Sultana F, Macmillan T, Culbertson M, Montford H (1997) The Pensacola Bay system surface water improvement and management plan. Northwest Florida Water Management District, Crestview, FL, http://www.nwfwmd.state.fl.us/water-resources/swim/pensacola-bay/

USEPA (2004) The ecological condition of the Pensacola Bay system, northwest Florida (1994–2001) EPA 620-R-05-002. US Environmental Protection Agency, Office of Research and Development, National Health and Ecological Effects Research Laboratory, Gulf Ecology Division, Gulf Breeze, FL, p 38

Worth JE, Priestley HI (2010) The Luna papers, 1559–1561: volumes 1 and 2. University of Alabama Press, Tuscaloosa, AL, http://www.uapress.ua.edu/product/Luna-Papers-1559-1561,4754.aspx

Yarbro LA, Carlson PR (2013) Seagrass integrated mapping and monitoring program. Mapping and monitoring report no. 1. Florida Fish and Wildlife Conservation Commission. FWRI technical report TR-17. http://myfwc.com/research/habitat/seagrasses/projects/active/simm-report-1/

Unpublished Reports

Academy of Natural Sciences of Philadelphia (ANSP) (1953) Escambia River Florida, vicinity of the Chemstrand Corporation Plant Fall 1952-Spring 1953 Stream survey report for the E I. duPont deNemours and Company. Department of Limnology, Philadelphia, PA, p 165

Florida State Board of Health (FSBH), Bureau of Sanitary Engineering, Northwest Florida Regional Office (1969) Report of investigations into pollution of Pensacola area waters. Florida State Board of Health, Jacksonville, FL

Gallagher T, Thuman A, Blumberg A, Ahsan A (1999) Hydrodynamic and water quality modeling. Pensacola Bay System environmental study. Unpublished report to Champion International Corporation, p 98

Hopkins TS (1969) The Escambia River and Escambia Bay during summer 1969. University of West Florida, Pensacola, FL, p 54

Websites

National Water Quality Monitoring Council Water Quality Portal (WQP). http://www.waterqualitydata.us/

SPARROW model from USGS. http://water.usgs.gov/nawqa/pubs/nitrogen_loads/index.html

Meeting in the Middle: On the Interactions Between Microalgae and Their Predators or Zooplankton and Their Food

Karen H. Wiltshire and Maarten Boersma

Introduction

There are many ways to look at interactions within the plankton, and the angle with which investigators approach their research subjects depends on their background, education, and individual research culture. When a zooplankton ecologist and a phytoplankton hydrobiologist meet and decide to cooperate, this facilitates interesting discussions, especially when the persons involved also originally come from different countries. Whereas from a microalgal point of view zooplankters are dangerous predators, which need to be avoided, deceived, and combatted (Wiltshire et al. 2003), and which luckily are stupid enough to smell rather distinct (Wiltshire and Lampert 1999), zooplankters see just food. This food may vary in quality depending on its nutrient content (Boersma 2000), or biochemical composition (Boersma et al. 2001), so salad isn't always just salad. The fact that these food items may actually scavenge nutrients from the gut of the zooplankters (Boersma and Wiltshire 2006) is not of interest to the little animals. Of course, this is a simplified view of the world, and when we consider the whole food web, microalgae and zooplankters are tightly interconnected both serving as food of differing quality to higher trophic levels, depending on the nutrient availability to the primary producers (Boersma et al. 2008). Thus, when trying to understand the effects of anthropogenic stress on ecosystems it

K.H. Wiltshire (✉)
Alfred-Wegener-Institut Helmholtz-Zentrum für Polar- und Meeresforschung, Wattenmeerstation Sylt, Hafenstrasse 43, 25992 List/Sylt, Germany

Jacobs University Bremen, Bremen, Germany
e-mail: karen.wiltshire@awi.de

M. Boersma
Alfred-Wegener-Institut Helmholtz-Zentrum für Polar- und Meeresforschung, Biologische Anstalt Helgoland, Postfach 180, 27483 Helgoland, Germany

University of Bremen, Bremen, Germany

P.M. Glibert, T.M. Kana (eds.), *Aquatic Microbial Ecology and Biogeochemistry: A Dual Perspective*, DOI 10.1007/978-3-319-30259-1_17

is necessary to closely cooperate and investigate changes in nutrient dynamics, phytoplankton densities, and zooplankton distributions (Wiltshire and Manly 2004; Wiltshire et al. 2008; Boersma et al. 2015).

We were very fortunate to inherit databases (and the work) of the Helgoland Roads time series of the Biologische Anstalt Helgoland. This time series, with daily measurements of temperature, nutrients, transparency, and algal densities, was originally set up in 1962 (Wiltshire et al. 2010), and extended with zooplankton measurements in 1975 (Greve et al. 2004). The combination of high-frequency measurements has allowed us to investigate the interplay between phytoplankton, zooplankton, and nutrients (Wiltshire and Manly 2004; Wiltshire et al. 2008, 2015; Boersma et al. 2015; Meunier et al. 2015), and continues to be a rich resource for contemporary questions. Indeed, there are several aspects of the interactions in the lower trophic levels that remain somewhat elusive. Whereas it is generally accepted that heterotrophic poikilotherms are very responsive to changes in temperature, often the effect of changing light conditions rather than temperature is thought to be more important for phytoplankton (Sverdrup 1953; Sommer et al. 1986; Ellertsen 1993). Thus, we would expect in the currently changing environment that zooplankton dynamics should be rather responsive to the change in temperature and phytoplankton dynamics should not (Wiltshire et al. 2008). This difference could potentially lead to a mismatch between phytoplankton and zooplankton temporal occurrence with eventual consequence for herbivore secondary production (Cushing 1974) and potential release of predation pressure on phytoplankton. In fact, in an earlier study Wiltshire et al. (2008) indeed observed a change in the temporal dynamics of zooplankton, and almost none in the phytoplankton. Moreover, we are currently seeing an increase in the phytoplankton densities in the North Sea (Wiltshire et al. 2010; Alvarez-Fernandez et al. 2012). As we now have several additional years of data it is time for a revision of the current state of knowledge, especially as strong changes in temperature and changing hydrology (Callies and Scharfe 2015) have coincided with changes in light penetration depths (and consequently light availability for the algae).

Materials and Methods

In 1962, the Biologische Anstalt Helgoland initiated a long-term monitoring program at Helgoland Roads (54° 11, 3′N, 7° 54, 0′E). Apart from temperature recordings, this program involves monitoring nutrients (SiO_4, NO_2, NO_3, NH_4, PO_4), salinity, light penetration (measured as Secchi depths), and phytoplankton species composition on a work-daily basis. Phytoplankton are counted to species level whereby 370 species are recognized (Hickel et al. 1993; Wiltshire and Dürselen 2004; Wiltshire and Manly 2004). From 1975 onwards, zooplankton sampling was added to the time series at Helgoland Roads (Greve et al. 2004). This involves sampling three times a week, at the same time as the phytoplankton and nutrient samples of the daily time series. Several zooplankton taxa are identified to species level, but many have been counted to genus level only (Greve et al. 2004). Since especially with copepods counting to species is very time consuming and tedious for many genera, the standard protocol for most genera was originally established using specific taxonomical units (Greve et al. 2004).

Here, we were mainly interested in the changes in phenology in zooplankton and phytoplankton, and hence focusing on two major groups: diatoms and calanoid copepods. The start of the diatom and calanoid copepod growth season was defined as the week number where the cumulative density reached 10% of the total annual cumulative density (Mackas et al. 2012). Using 10% rather than 15 or 20% guaranteed that we concentrate on the beginning of the season only. In our analysis all algal densities were converted to carbon, using appropriate conversion factors based on average carbon contents for given species and the biovolume calculations by Hillebrand et al. (1999). Thus, we first established the total biovolume (in $\mu m^3\ l^{-1}$) for diatoms, and calculated biomasses using the conversion factor of Montagnes et al. (1994), 0.1 pg C μm^{-3}.

Furthermore, we computed the average annual temperatures and average annual Secchi depths, and related these with the measurements of start of season for phytoplankton and zooplankton.

Results

In contrast to what was observed before (Wiltshire et al. 2008), with the new data included and the slightly different metric used here (and in Wiltshire et al. 2015), we observed that both zooplankton and phytoplankton showed a significant movement to earlier occurrence with time. Both regressions were highly significant (Table 1), indicating that for both the diatoms and the copepods the start of the season has come significantly earlier. Interestingly, we observed no significant difference in the slope of the regression line between the two. Thus the change in timing of the diatoms happened in the same speed as in the copepods. In both cases we observed a shift of about 1 week per decade; thus in the total of the 40–50-year sampling period, there was a shift forward in the season of about a month. At the same time, we observed a strong shift in both the temperature, as was reported many times before, and the water transparency. In fact, after standardization of both temperature and the Secchi disk transparency, the rate of change was identical (Fig. 1b, c). However, we observed that no significant correlation could be found

Table 1 Summary table of the regression analyses relating temperature, Secchi disk transparency, and the week of 10% cumulative densities of zooplankton and phytoplankton

	n	int	s.e. int	*b*	s.e. *b*	r^2	*p*
Diatoms with year	50	235.5	60.4	−0.11	0.03	0.22	<0.001
Copepods with year	36	296.5	89.6	−0.14	0.045	0.22	<0.004
Temperature with year	50	−50.8	11.8	0.03	0.006	0.36	<0.001
Secchi with year	44	−35.7	8.94	0.02	0.004	0.31	<0.001
Secchi with temperature	44	2.98	0.97	0.05	0.095	0.01	0.58
Diatoms with temperature	50	21.2	6.71	−0.4	0.68	0.01	0.55
Copepods with temperature	36	37.3	6.79	−2.05	0.67	0.22	<0.005
Diatoms with Secchi	44	30.7	2.71	−3.99	0.77	0.39	<0.001
Copepods with Secchi	36	22.3	4.02	−1.64	1.12	0.06	0.15

Int intercept of the regression line, *b* slope, *s.e.* standard error

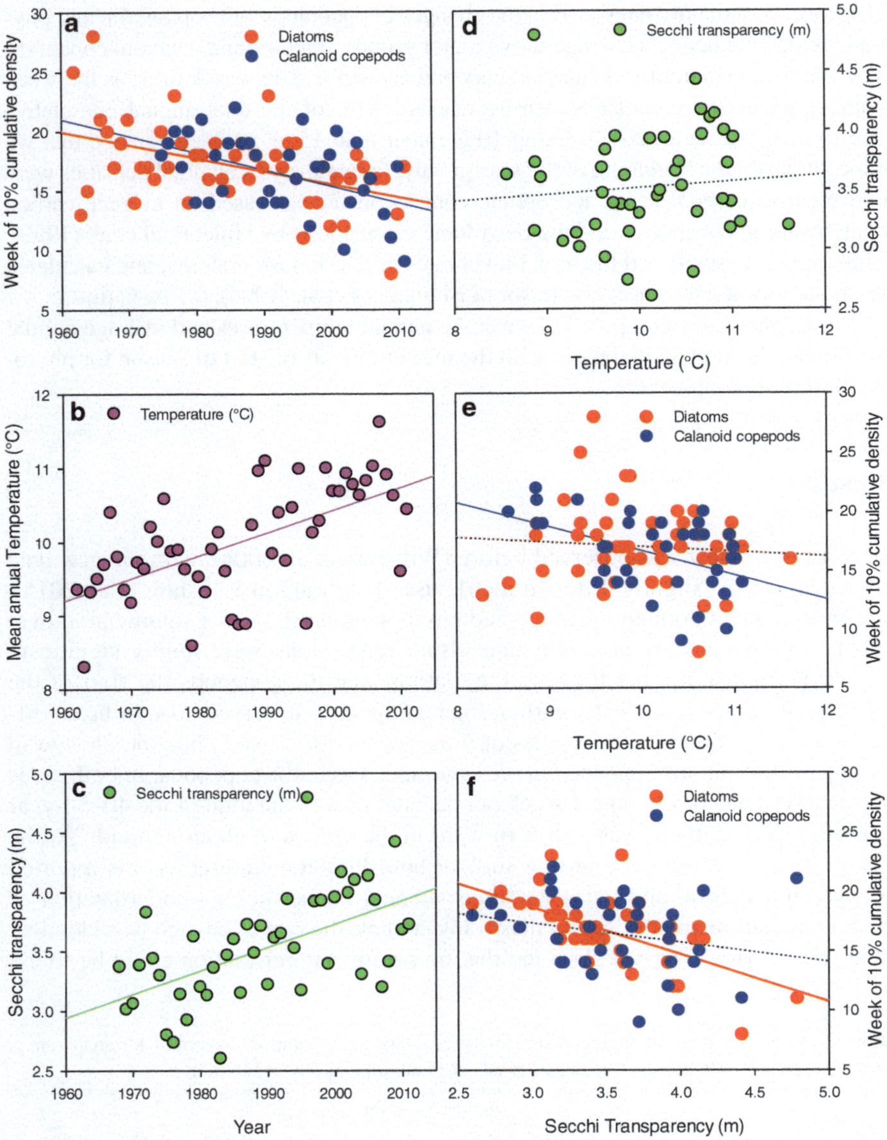

Fig. 1 Changes in the phenology of diatoms and calanoid copepods. Significant regression lines are represented by *solid lines*, *dotted regression lines* are not significant (see also Table 1). (**a**) Week of 10 % cumulative densities of calanoid copepods and diatoms with time. (**b**) Change in average annual temperature with time. (**c**) Change in water transparency with time. (**d**) Relationship between mean annual temperature and Secchi disk transparency. No significant correlation between the two could be observed. (**e**) Only the start of the season of the copepods showed a significant regression with temperature. (**f**) Only the start of the season of the diatoms showed a significant regression with Secchi disk transparency

between temperature and the water transparency (Table 1; Fig. 1d). A multiple regression with both factors, temperature and Secchi transparency as independent variables, and the start of season of the diatoms and copepods as the dependent ones, revealed that with the copepods only temperature was significantly related with the start of season, and not the water transparency, whereas the opposite was true for the diatoms (Fig. 1f; Table 1). Thus, for the Helgoland Roads series we conclude that through completely different and largely independent mechanisms we have seen a concurrent shift in phenology of phytoplankton and zooplankton over the last 50 years. Whereas the change in phytoplankton timing was largely related to an increase in water transparency (Wiltshire et al. 2015), the change in the occurrence of zooplankton could be attributed solely to changing temperatures. Thus, it is unlikely that mismatch phenomena have caused the changes in phytoplankton and zooplankton densities that have occurred in this area (Boersma et al. 2015).

Discussion

The current literature on the effects of global warming is full of examples where through differences in reactions of different players in the ecosystem mismatch situations now occur, where there used to be a match of herbivores with their food, or carnivores with herbivores previously (Beaugrand et al. 2003; Edwards and Richardson 2004). As the main changing driver, temperature, is thought to be more influential changing the phenology of heterotrophs than autotrophs (Ellertsen et al. 1995), the result of increasing temperatures is thought to be a larger movement forward in time for heterotrophs than for autotrophs. Here, we observed that at least for the Helgoland Roads time series this was not the case. The change in phenology over a 36-year period (1975–2011) for diatoms and calanoid copepods was identical. Both moved forward in the season by about a month over this time. However, the mechanisms by which they have moved are completely different. The change in timing of the diatom blooms seems to be completely driven by changes in the light conditions, as measured with a Secchi disk. In the southern North Sea, algal densities do not affect water transparency at all (Wiltshire et al. 2015), and the highest transparency is typically found in summer, when algal biomass is high. The reason for this is the strong effects of wind and suspension of sediments on water clarity (Wiltshire et al. 2008), and higher wind speeds are usually found in autumn to spring. Thus, even though the mechanism for the bloom is different when compared to for example those described by Sverdrup (1953), the causes are similar. Enough light needs to penetrate the water for the spring bloom to start, either by stratification of the water column or by the decrease in suspended material combined with the increase in solar radiation. With an average depth of around 5 m at the sampling station, this means that light penetrates to the bottom when the Secchi disk reading is around 2.5 m. In fact, the week where Secchi disk penetration depth exceeded 2.5 m has moved earlier ($r^2=0.13$; $p=0.02$) in the period of investigation; thus there is more light earlier at the Helgoland Roads sampling station. This has led to the advancement of the diatom bloom.

Heterotrophs do not respond to light, but to temperature. Congruent to other areas in the world, the North Sea has warmed significantly in the last 50 years. As a result, the occurrence of many zooplankton and ichthyoplankton species has moved forward in the year (Greve et al. 2005; Mackas et al. 2012). Many have speculated that global warming should affect zooplankton more than phytoplankton and hence global warming could potentially lead to a mismatch between resources and consumers. Although this has certainly occurred (Edwards and Richardson 2004), the situation in the Southern North Sea is different. As a result of the corresponding shift of both light conditions and temperature the observed trends for both food and predators are identical. Thus, although the mechanisms are completely different, the result is that everything has moved forward in time by about a month.

So, what have we learned? Is the southern North Sea an unusual place, where the global change processes that have been described in other areas work differently? To a certain degree, this is the case. The hydrography of the Southern North Sea has changed dramatically in the last 50 years (Scharfe 2013), especially around Helgoland, with a more prominent open-sea signal than was previously the case. This has led to an increase in salinity and to clearer water, and thus to the change in phenology of the phytoplankton (see also Wiltshire et al. 2015). We are aware of no other coastal seas where this has happened, but obviously temperatures have increased almost everywhere, but particularly strongly at Helgoland (Holt et al. 2012). Thus, we argue that the situation in the North Sea is special in the sense that phytoplankton and zooplankton have changed their phenology in a similar way. Despite this, we have observed a strong decrease in the densities of calanoid zooplankters (Boersma et al. 2015), and an increase in diatom biomass (Wiltshire et al. 2010). The reasons for this remain elusive. From a top-down perspective, lower herbivore densities of course would release predation pressure on the algae, and may lead to higher algal densities, but we would have to explain why there are less herbivores, something that can definitely not be done as fish or predatory zooplankton densities have not increased (Schlüter et al. 2010; Boersma et al. 2015). From a bottom-up perspective, more diatoms should mean more food for the zooplankters, suggesting that food should not be limiting. This may be the case quantitatively, but not qualitatively. One of the consequences of higher light penetration into the water column is that photosynthesis is increased, and more carbon is sequestered (Urabe and Sterner 1996; Sterner et al. 1998). This implies a higher carbon content relative to other nutrients such as nitrogen and phosphorus. Furthermore, one other peculiarity of the southern North Sea is that the nutrient conditions have changed dramatically. As a result of the large efforts in wastewater treatment in the major rivers, the inputs of nitrogen and phosphorus have plummeted (e.g., van Beusekom et al. 2009). If we consider that algal biomass is currently higher than previously this automatically implies that the nutrient content of the algae relative to the carbon is lowered. Our explanation for the reduction in zooplankton biomass is that this reduced relative nutrient content represents a resource of inferior quality for zooplankters (Boersma 2000; Malzahn and Boersma 2012). Thus, we need the combination of light, nutrients, phytoplankton, and zooplankton in combination with the changing abiotic factors to explain what has happened in the North Sea.

We expect that the link between the changing light conditions and the changing temperatures is going to break down sooner or later, which will lead to an additional difficulty for the herbivorous zooplankton, as then also the relatively good match in the timing of the animals and their food might break down. Hence, it could well be that the North Sea is continuing to become greener, despite the enormous reductions in nutrient inputs, with potential further decreases of the fishery yields as these depend on the secondary producers (Boersma et al. 2008).

Acknowledgements We thank the many mentors, supervisors, collaborators, and students that helped us on our scientific path. Without all of these people it would not have been possible to do the work that we have been so fortunate to be allowed to do. Mostly, it has been great fun with all of you, and we would not have wanted our careers any other way. Most importantly, we thank our two wonderful boys, Oisín and Malte, who certainly have been patient and loving through all the "I will be home in ten minutes ...." statements, but who always were proud of what we do.

Karen H. Wiltshire and Maarten Boersma

Conferences are highly useful professionally as well as rewarding private means to meet exciting new people. It only took us as long as the exchange of some coupons for alcoholic and nonalcoholic beverages to realize that a cooperation might be fruitful. Both Karen, who was doing her Ph.D. research in Germany, and Maarten, then a Ph.D. student in the Netherlands, worked on limnological themes, and after the "Fateful" meeting subsequently tried to find one common place of work. This was not as easy as it sounded, as junior positions were few and far between. As a result Maarten came to Germany at the exact time when Karen left for Scotland. However, we finally managed to work and live together in Plön, Germany, when we were offered a shared postdoctoral position by Winfried Lampert at the Max Planck Institute for Limnology. After that, the deal was that whoever was offered the first permanent position would take the other person with them. As a result, we moved to the Alfred-Wegener-Institut, Biologische Anstalt Helgoland in 2001, and never regretted this move to the "dark side" of salty water. We have been cooperating on plant-animal interactions ever since having met 26 years ago at a limnological conference.

References

Alvarez-Fernandez S, Lindeboom H, Meesters E (2012) Temporal changes in plankton of the North Sea: community shifts and environmental drivers. Mar Ecol Prog Ser 462:21–38

Beaugrand G, Brander KM, Lindley JA, Souissi S, Reid PC (2003) Plankton effect on cod recruitment in the North Sea. Nature 426:661–664

Boersma M (2000) The nutritional quality of P-limited algae for *Daphnia*. Limnol Oceanogr 45:1157–1161

Boersma M, Wiltshire KH (2006) Gut passage of phosphorus-limited algae through *Daphnia*: do they take up nutrients in the process? Arch Hydrobiol 167:489–500

Boersma M, Schöps C, McCauley E (2001) Nutritional quality of seston for the freshwater herbivore *Daphnia galeata x hyalina:* biochemical versus mineral limitations? Oecologia 129:342–348

Boersma M, Aberle N, Hantzsche FM, Schoo K, Wiltshire KH, Malzahn AM (2008) Nutritional limitation travels up the food chain. Int Rev Hydrobiol 93:479–488
Boersma M, Wiltshire KH, Kong S-M, Greve W, Renz J (2015) Long-term change in the copepod community in the southern German Bight. J Sea Res 101:41–50
Callies U, Scharfe M (2015) Mean spring conditions at Helgoland Roads, North Sea: graphical modeling of the influence of hydro-climatic forcing and Elbe River discharge. J Sea Res 101:1–11
Cushing DH (1974) The natural regulation of fish populations. In: Harden Jones FR (ed) Sea fisheries research. Paul Elek, London, pp 399–412
Edwards M, Richardson AJ (2004) Impact of climate change on marine pelagic phenology and trophic mismatch. Nature 430:881–884
Ellertsen HC (1993) Spring blooms and stratification. Nature 363:24
Ellertsen HC, Sandberg S, Tollefsen H (1995) Photoperiodic control of diatom spore growth – a theory to explain the onset of phytoplankton blooms. Mar Ecol Prog Ser 116:303–307
Greve W, Reiners F, Nast J, Hoffmann S (2004) Helgoland Roads meso- and macrozooplankton time-series 1974 to 2004: lessons from 30 years of single spot, high frequency sampling at the only off-shore island of the North Sea. Helgol Mar Res 58:274–288
Greve W, Prinage S, Zidowitz H, Nast J, Reiners F (2005) On the phenology of North Sea ichthyoplankton. ICES J Mar Sci 62:1216–1223
Hickel W, Mangelsdorf P, Berg J (1993) The human impact in the German Bight – eutrophication during 3 decades (1962–1991). Helgol Meeres 47:243–263
Hillebrand H, Durselen CD, Kirschtel D, Pollingher U, Zohary T (1999) Biovolume calculation for pelagic and benthic microalgae. J Phycol 35:403–424
Holt J, Hughes S, Hopkins J, Wakelin SL, Holliday NP, Dye S, Gonzalez-Pola C, Hjollo SS, Mork KA, Nolan G, Proctor R, Read J, Shammon T, Sherwin T, Smyth T, Tattersall G, Ward B, Wiltshire KH (2012) Multi-decadal variability and trends in the temperature of the northwest European continental shelf: a model-data synthesis. Prog Oceanogr 106:96–117
Mackas DL, Greve W, Edwards M, Chiba S, Tadokoro K, Eloire D, Mazzocchi MG, Batten S, Richardson AJ, Johnson C, Head E, Conversi A, Peluso T (2012) Changing zooplankton seasonality in a changing ocean: comparing time series of zooplankton phenology. Prog Oceanogr 97:31–62
Malzahn AM, Boersma M (2012) Effects of poor food quality on copepod growth are dose dependent and non-reversible. Oikos 121:1408–1416
Meunier CL, Boersma M, Wiltshire KH, Malzahn AM (2015) Zooplankton eat what they need: copepod selective feeding and potential consequences for marine systems. Oikos 125(1):50–58
Montagnes DJS, Berges JA, Harrison PJ, Taylor FJR (1994) Estimating carbon, nitrogen, protein, and chlorophyll-a from volume in marine-phytoplankton. Limnol Oceanogr 39:1044–1060
Scharfe M (2013) Analyse biologischer Langzeitveränderungen auf Basis hydroklimatischer parameter in der südlichen Nordsee (Helgoland). Ph.D. Hamburg, Hamburg
Schlüter MH, Merico A, Reginatto M, Boersma M, Wiltshire KH, Greve W (2010) Phenological shifts of three interacting zooplankton groups in relation to climate change. Glob Chang Biol 16:3144–3153
Sommer U, Gliwicz ZM, Lampert W, Duncan A (1986) The PEG model of seasonal succession of planktonic events in fresh waters. Arch Hydrobiol 106:433–471
Sterner RW, Clasen J, Lampert W, Weisse T (1998) Carbon: phosphorus stoichiometry and food chain production. Ecol Lett 1:146–150
Sverdrup HU (1953) On conditions for the vernal blooming of phytoplankton. J Conseil 18:287–295
Urabe J, Sterner RW (1996) Regulation of herbivore growth by the balance of light and nutrients. Proc Natl Acad Sci U S A 93:8465–8469
van Beusekom JEE, Loebl M, Martens P (2009) Distant riverine nutrient supply and local temperature drive the long-term phytoplankton development in a temperate coastal basin. J Sea Res 61:26–33

Wiltshire KH, Dürselen CD (2004) Revision and quality analyses of the Helgoland Reede long-term phytoplankton data archive. Helgol Mar Res 58:252–268
Wiltshire KH, Lampert W (1999) Urea excretion by *Daphnia*: a colony-inducing factor in *Scenedesmus*? Limnol Oceanogr 44:1894–1903
Wiltshire KH, Manly BFJ (2004) The warming trend at Helgoland Roads, North Sea: phytoplankton response. Helgol Mar Res 58:269–273
Wiltshire KH, Boersma M, Meyer B (2003) Grazer-induced changes in the desmid *Staurastrum*. Hydrobiologia 491:255–260
Wiltshire KH, Malzahn AM, Wirtz K, Greve W, Janisch S, Mangelsdorf P, Manly BFJ, Boersma M (2008) Resilience of North Sea phytoplankton spring blooms dynamics: an analysis of long term data at Helgoland Roads. Limnol Oceanogr 53:1294–1302
Wiltshire KH, Kraberg A, Bartsch I, Boersma M, Franke HD, Freund J, Gebuhr C, Gerdts G, Stockmann K, Wichels A (2010) Helgoland Roads, North Sea: 45 years of change. Estuar Coast Shelf Sci 33:295–310
Wiltshire KH, Boersma M, Carstens K, Kraberg AC, Peters S, Scharfe M (2015) Control of phytoplankton in a shelf sea: determination of the main drivers based on the Helgoland Roads Time Series. J Sea Res 105:42–52

Lake Transparency: A Window into Decadal Variations in Dissolved Organic Carbon Concentrations in Lakes of Acadia National Park, Maine

Collin Roesler and Charles Culbertson

Introduction

Recent decadal increases in dissolved organic carbon (DOC) concentrations in surface waters throughout Europe and the northeastern USA are well reported (De Wit et al. 2007; Driscoll et al. 2003; Evans et al. 2006; Hongve et al. 2004; Monteith and Evans 2005; Monteith et al. 2007; SanClements et al. 2012; Worrall and Burt 2007). A variety of drivers associated with global climate change are cited to explain rising trends in DOC, such as warming soils and increasing microbial activity to changing hydrology (Porcal et al. 2009); other investigations highlight decreasing atmospheric sulfur deposition over the last few decades as a more likely driver of rising DOC concentrations in surface waters (EkströM et al. 2011; Hruška et al. 2009; Oulehle and Hruška 2009; Strock et al. 2014; Tipping and Woof 1990). A prevailing hypothesis for this mechanism is that subsequent reduction in the acidity of both soil-solution and stream waters increases the net charge on humic substances, the dominant organic soil constituents that interact with the aqueous phase (Jaffé et al. 2008; Tipping and Hurley 1988). Consequently, the solubility of humic substances is increased, which facilitates their transport from terrestrial to aquatic environments. These humic substances comprise the largest pool of detrital organic carbon (Worrall et al. 2004) and are the major portion of the dissolved organic matter (DOM) pool (EkströM et al. 2011; Evans et al. 2006; Monteith et al. 2007; Oulehle and Hruška 2009; SanClements et al. 2012; Worrall and Burt 2007). Most of the reported increases in DOC occur in boreal forested catchments or those having

C. Roesler (✉)
Department of Earth and Oceanographic Science, Bowdoin College, Brunswick, ME, USA
e-mail: croesler@bowdoin.edu

C. Culbertson
U.S. Geological Survey, New England Water Science Center,
Maine Office, Augusta, ME, USA

P.M. Glibert, T.M. Kana (eds.), *Aquatic Microbial Ecology and Biogeochemistry: A Dual Perspective*, DOI 10.1007/978-3-319-30259-1_18

organic-rich soil (Driscoll et al. 2003; Evans et al. 2000; Hejzlar et al. 2003; Hruška et al. 2009; Monteith et al. 2007; Worrall and Burt 2007; Worrall et al. 2004).

A dominant characteristic of DOM in aquatic systems is the associated brown color. This is because a large portion of the DOM pool is chromophoric, colored DOM (CDOM), with strong absorption coefficients in the ultraviolet range of the spectrum, decreasing exponentially into the visible. As the CDOM concentration increases, visible light penetration decreases, reducing the euphotic depth and integrated primary productivity. Thus CDOM is one of the dominant constituents in aquatic systems that can impact transparency (EkströM et al. 2011; Jaffé et al. 2008; Miller and McKnight 2010; SanClements et al. 2012).

Among the most prominent landscape features at Acadia National Park (ANP), Maine, USA, are its 26 lakes and ponds that are host to a variety of recreational pursuits. Known for having exceptional water quality, some also serve as water supplies for towns on Mount Desert Island (MDI). The ANP Natural Resource Department operates a long-term lake monitoring network on Mount Desert Island (Fig. 1). Over the last 30 years, large variations in water transparency (clarity) as measured by Secchi depth have been observed. Similarly, transparency derived from Landsat imagery for other lakes in Maine suggests overall decrease in transparency (McCullough et al. 2013). These variations have been attributed to variations in DOM induced by both natural and anthropogenic factors, from changing acidity to development, respectively (Tipping and Woof 1990). In this study we investigate the relationship between Secchi depth and DOC as a means of expanding the record of DOC for future biogeochemical investigations (c.f., Boyce et al. 2010).

Methods

Twenty-five lakes comprise the lake-monitoring network at Acadia National Park in Maine (USA) (Fig. 1). The monitoring program has included Secchi depth estimates since 1975 and routine dissolved organic carbon concentrations since 1995. The data records of lake Secchi depth and DOC concentration were supplied by the ANP Natural Resources Department. Methods for determining Secchi depth and DOC concentrations are outlined in Gawley et al. (2014).

Model Description and Development

Light in aquatic systems is attenuated with depth by the absorption and scattering of photons by water itself and the particulate and dissolved materials in the water. Photons are removed by absorption and redirected by scattering, the latter ultimately being absorbed or backscattered out the surface. The materials responsible for light attenuation are as varied as aquatic systems are. Typically absorption is

Fig. 1 Lake monitoring network at Acadia National Park on Mount Desert Island, Maine, USA. *Symbol colored* by ANP designated numeric lake code, see following figures

dominated by particulate and dissolved organic material and algal pigments in the blue and green region of the spectrum and water in the red region. Scattering is dominated by organic particles such as living organisms and detrital particles, and inorganic particles such as suspended sediments. The more material in the water, generally the shallower the depth of light penetration. Thus the depth of light penetration can be used as a proxy for the concentration of materials in the water. When an aquatic system is dominated by one type of material (for example strongly absorbing DOM or CDOM), the depth of light penetration can serve as a proxy for that material. A model to estimate DOC concentration from Secchi depth observations is developed below based upon first principles of hydrologic optics and established optical proxies for biogeochemical parameters. The model assumes that CDOM is the dominant optical constituent in the lake system, thereby dominating lake transparency and the depth of light penetration, and that CDOM is a robust proxy for DOC. That CDOM does not scatter light in addition to absorbing it greatly simplifies the model assumptions. The model is validated with co-incident observations of DOC concentration and Secchi depth.

Light penetration in aquatic systems obeys Beer's law, decreasing exponentially with depth z (m):

$$E(z) = E_o \, e^{-k(z)*z} \tag{1}$$

where $E(z)$ is the irradiance (μmol quanta $m^{-2}\,s^{-1}$) at depth z, E_o is the incident irradiance, and $k(z)$ is the depth dependent diffuse attenuation coefficient (m^{-1}) which describes the rate of exponential light decay. Allowing the attenuation coefficient to be constant with depth and rearranging Eq. 1 permit the product of the attenuation coefficient, k (m^{-1}), and the Secchi depth, Z_S (m), to be expressed as

$$\ln \frac{E(Z_S)}{E_o} = -k \times Z_S \tag{2}$$

where $E(Z_s)/E_o$, when multiplied by 100, is the percent of surface light at the depth at which the Secchi disk disappears. Following Mankovsky (2014), the product of the diffuse attenuation and Secchi depth is given by the parameter, Ψ:

$$\Psi = k \times Z_S \tag{3}$$

which is mathematically expressed in terms of percent rather than fractional light levels. If the value of Ψ is constrained, with relatively small variations about an expected value, then there must exist a robust relationship between the attenuation coefficient and the Secchi depth; as the attenuation increases, the Secchi depth decreases. Thus, Secchi depth can take the place of measurements of absolute light levels.

Approximations to the radiative transfer equation yield a relationship between the diffuse attenuation coefficient, k, and the absorption coefficient, a (m^{-1}), via Gershun's equation (Gershun et al. 1939):

$$a = k \, \bar{\mu} \tag{4}$$

where $\bar{\mu}$ is the average cosine, a factor which describes the angularity of the radiance field via the average angle from nadir (overhead). This theoretical relationship states that the attenuation of light in the water column is linearly related to the absorption coefficient. The attenuation due to scattering is described not by the scattering coefficient but by the impact that scattering has on the angular distribution of light beams, which is parameterized as the cosine of that average angle of the light field. So if there is no light scattering by particles and the solar beam is directly overhead, the average cosine $\bar{\mu}$ of the photon path is 1 (cosine of 0° from nadir). If there are a lot of particles in the water and they scatter the solar beam so that the average angle is 60° from nadir, then $\bar{\mu}=0.5$ and the attenuation of light is two times the absorption coefficient. Because *k* is measured as a function of depth, photons traveling along a longer path (higher angle) have a greater likelihood of being absorbed compared to photons traveling vertically through the water column.

Combining Eqs. 3 and 4 provides a direct relationship between Secchi depth and the absorption coefficient:

$$Z_S = \Psi \times \bar{\mu} / a \quad (5)$$

At the wavelengths predominantly transparent for the Secchi disk, blue in clear waters, and green in more productive or inland waters, constituents other than water dominate the absorption. For the inland lakes and ponds, CDOM is the predominant absorber, with a spectrum well described by an exponential function:

$$a_{\mathrm{CDOM}}(\lambda) = a_{\mathrm{CDOM}}(\lambda_o) \times \mathrm{e}^{-S_{\mathrm{CDOM}}(\lambda-\lambda_o)} \quad (6)$$

where S_{CDOM} is the spectral slope describing how steep the spectrum decays from ultraviolet to visible wavelengths. The range of S_{CDOM} values is small globally (Babin et al. 2003; Roesler et al. 1989) with median value of 0.016 ± 0.002, so that variations in CDOM absorption are primarily determined by the magnitude of the absorption, $a_{\mathrm{CDOM}}(\lambda_o)$, rather than the spectral shape, and thus estimating the magnitude at any reference wavelength, λ_o, allows the whole spectrum to be computed analytically.

CDOM absorption at 254 nm has become a robust proxy for the concentration of DOC, and the ratio (of absorbance to DOC concentration) is called the carbon-specific ultraviolet absorption, or SUVA (Weishaar et al. 2003). Although the value of SUVA does vary with the composition of the DOM pool (specifically with the degree of aromaticity), due to such processes as dilution, photodegradation, and flocculation (Chen et al. 2004; Loiselle et al. 2010), the ranges are constrained. Thus the link between the optical properties and the biogeochemical properties of the DOM pool is established:

$$a_{\mathrm{CDOM}}(\lambda) = \mathrm{SUVA} \times [\mathrm{DOC}] \times \mathrm{e}^{-S_{\mathrm{CDOM}}(\lambda-\lambda_o)} \quad (7)$$

where $\mathrm{SUVA} = a_{\mathrm{CDOM}}(254) / \mathrm{DOC}$ (l mg^{-1} m^{-1}), and [DOC] is the dissolved organic carbon concentration (mg l^{-1}). Combining Eqs. 5 and 7 yields the final forward model for Secchi depth as a function of the optical and biogeochemical properties of DOM at the wavelengths of maximal transparency (e.g., $\lambda = 500$ nm):

$$Z_S = \Psi \times \bar{\mu} / \left(\text{SUVA} \times [\text{DOC}] \times e^{-S_{\text{CDOM}}(500-254)} \right) \quad (8)$$

The inversion of this equation yields an estimate of DOC from Secchi depth using typical values for the parameters Ψ, $\bar{\mu}$, SUVA, and S_{CDOM}:

$$[\text{DOC}] = \Psi \times \bar{\mu} / \left(Z_S \times \text{SUVA} \times e^{-S_{\text{CDOM}}(500-254)} \right) \quad (9)$$

Using published ranges for these four parameters, model estimates of DOC were computed for the range of observed Secchi depths in the 40-year data set. The range of values for the model parameters were Ψ varying from 1.5 to 2.2 (Armengol et al. 2003; Gallegos et al. 2011; Murray and Markager 2011), $\bar{\mu}$ varying from 0.7 to 0.9 (Kirk 1991), SUVA varying from 2.5 to 7.1 (l mg^{-1} m^{-1}), and S_{CDOM} varying from 0.013 to 0.018 nm^{-1} (Roesler et al. 2006 and other unpublished data from Roesler for Maine rivers).

Results

The time series observations of Secchi depth for the lake network indicate large variations in lake transparency both between lakes and within lakes over the 40-year record (Fig. 2a). Secchi depth ranges from approximately 0.5 to 22 m across the lake network. The seasonal variations (Fig. 2b) are relatively small within each lake, particularly when compared with the interannual variations, which demonstrate an approximate decadal cycle. This is particularly pronounced in the clearest of the lakes. That the seasonality of lake transparency is relatively small compared to the interannual variations suggest that larger scale factors control temporal variations in transparency. However, the significant differences between lakes suggest

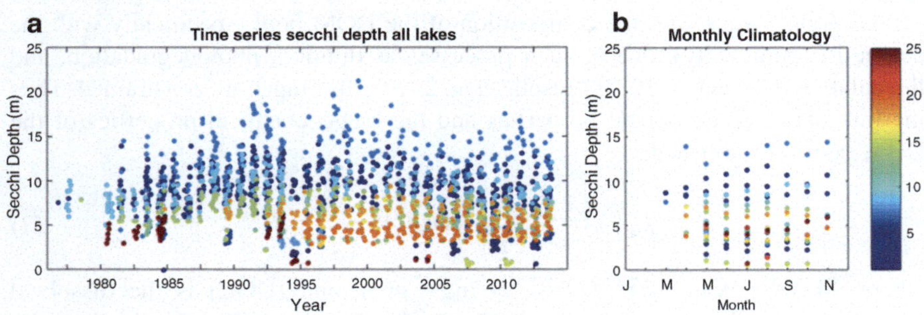

Fig. 2 (**a**) Time series observations of lake transparency as quantified by Secchi depth (m) and (**b**) monthly climatology of Secchi depth for each of the monitored lakes (*colorbar* is ANP designated lake code, see Fig. 1)

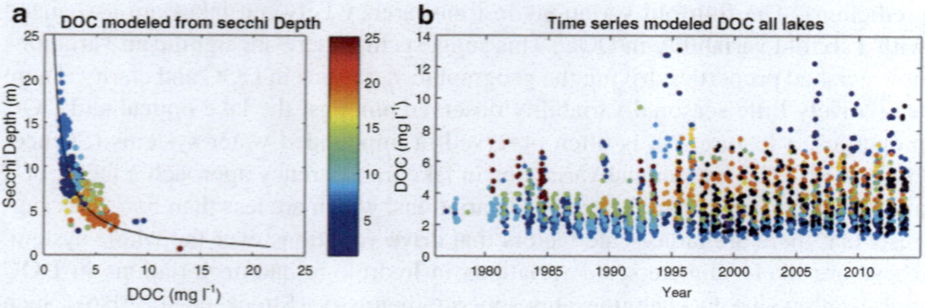

Fig. 3 (**a**) Dependence of Secchi depth on measured DOC concentration with best fit analytic model shown by the *solid black line*; see text for details. (**b**) Time series of DOC concentrations estimated from Secchi depth observations using analytic model. *Colorbar* of ANP designated lake code, as in Fig. 2. Measured DOC concentrations in *black symbols* for comparison in (**b**)

that there are important geographical forces (land cover/land use, watershed morphology and hydrology, lake depth, and residence time) that impact on the scale of the watersheds between lakes.

The Secchi depth in these lakes is strongly correlated to the DOC concentration measured over all the lakes in the sampling network and over the 20-year sampling interval. The dependence is nonlinear (Fig. 3a), with a rapid decline in transparency as the DOC concentration increases from lowest values. A suite of curves for the analytic model in Eq. 9 were computed for all combinations of parameter ranges. The best fit functions to the data (e.g., solid line in Fig. 3a) were obtained for parameter values of $\Psi = 2.0$–2.2, $\bar{\mu} = 0.8$–0.9, SUVA varying from 2.7 to 3.0 (l mg^{-1} m^{-1}), and S_{CDOM} varying from 0.0145 to 0.015 nm^{-1}. The curve shown in Fig. 3a is that computed from the mean of these ranges.

The analytic model was used to estimate [DOC] from the complete time series observations of Secchi depth (Fig. 3b). The observed [DOC] values are superimposed to demonstrate both the reasonableness of the model results and the limited spatial and temporal extent of [DOC] observations, even after the onset of routine sampling in 1995. Measured and modeled [DOC] values range from approximately 1 to 20 (mg l^{-1}), with the majority of the observations between 2 and 6 (mg l^{-1}). The RMS error between measured and modeled was 1.29 (mg l^{-1}).

Discussion

It appears that the dominant source of variations in lake transparency in the ANP lake system is caused by variations in CDOM and is strongly related to lake DOC. Although the ANP lakes certainly have phytoplankton and other particulate matter that would both absorb and scatter light, hence decreasing transparency, they either are not significant compared to CDOM (and thus may be the source of variability about the model) or covary with CDOM (and thus do not impact the model prediction). The fourfold variations in transparency between lakes are associated with a sixfold variability in DOC. This suggests that there are significant variations in watershed properties driving the geographic variations in DOC and clarity. There is relatively little seasonal variability observed amongst the lake optical and DOC properties, a feature that is often observed in impounded water systems (Spencer et al. 2012). The interannual variations in lake transparency approach a factor of 2 and are much larger than the seasonal variations, which are less than 5 %. This suggests that there are large-scale factors that drive variations over the whole system. These are likely climate-scale variations in hydrology and/or variations in DOC mobilization due to changing atmospheric chemistry (Strock et al. 2016). Such large-scale processes induce coherent variations over the entire lake network. This is in contrast to the lake-to-lake variations due to local processes and landscape characteristics.

A simple analytic model for lake DOC based upon Secchi depth observations was derived from first principles using published values for model parameters. The goodness of fit of the model to the data suggests that the model assumptions are robust for this environment.

Idso and Gilbert (1974) suggest that the depth at which the Secchi disk disappears coincides with the 18 % light level and thus Ψ would equal 1.7, although the exact value will vary depending upon the wavelength range of k (narrow band or PAR), and on the optical classification of the water. The clear blue waters of the Mediterranean, where the Secchi disk measurement originated (Wernand 2010), will necessarily have a different relationship than the turbid green waters of lakes due to the differing contributions to attenuation by absorbing and scattering constituents in the blue versus green wavebands and the relative contributions of absorption and scattering to the water. Recent studies into the exact product of attenuation and Secchi depth show variability of order 10 %. For example Murray and Markager (2011) presented a range of 1.85–2.15 for Danish estuaries, while Armengol et al. (2003) found 1.49–1.98 for a Spanish reservoir, and Gallegos et al. (2011) observed a range of 1.0–2.2 for Chesapeake Bay. Values in the range of 2–2.2 resulted in the strongest model for these data which is typical for greener waters and indicates that the Secchi depth in these waters represents the 11–13 % light level.

The average cosine, $\bar{\mu}$, is strongly dependent upon the solar angle of incident irradiance and to second order on the scattering-to-absorption ratio and the shape of the volume scattering function. Secchi depth observations are generally collected between 10 a.m. and 2 p.m. to maximize the solar zenith angle. In addition, if CDOM absorption is very strong, the reduction in average cosine due to scattering is minimized. Thus values between 0.7 and 0.9 are expected (Kirk 1991). Indeed the higher values of 0.8–0.9 yielded the strongest model fit and are consistent with lower scattering waters and more direct solar beams that would typically be observed when the Secchi depth measurements were taken.

The assumptions were that the optical proxies for SUVA and the slope of CDOM absorption, S_{CDOM}, observed over the extent of the four largest Maine watersheds on the mainland during 2004–2014 (Roesler et al. 2006 and Roesler, unpubl. data) were robust for those on Mount Desert Island over the last three decades. Typical observed SUVA values were 3.93 ± 1.22 ($l\ mg^{-1}\ m^{-1}$) and typical observed spectral slopes, S_{CDOM}, were −0.0143 ± 0.0008 nm^{-1}. The values that provided the strongest model fit were 2.7–3.0 for SUVA and moderate absorption slopes of 0.0145–0.0150, respectively. These values for SUVA suggest approximately 20 % aromaticity and are at the lower end of the range observed in a multi-year time series of 60 stations sampled in the four major river systems in Maine (Roesler et al. 2006 and unpublished data). This is consistent with observations of impounded systems compared to riverine systems dominated by runoff (Spencer et al. 2012).

Despite the good fit of the analytic model to the observations, there are still significant variations in the distribution of the observations. Little information is available regarding whether there has been a concomitant change in DOM composition along with its increased quantity (EkströM et al. 2011), although studies have indicated a shift from autochthonous (microbial) to allochthonous (terrestrial) sources (Jaffé et al. 2008; Monteith et al. 2007; Oulehle and Hruška 2009; SanClements et al. 2012). In a seminal study of 28 catchments in northern Europe having 35 years of monthly measurements, Erlandsson et al. (2008) reported that the increased DOM concentration *alone* could not account for the observed increases in water color (CDOM), suggesting that other factors such as photochemical transformation and/or altered DOM sources might be responsible. Such changes in the chemical quality of DOM (color, content, and reactivity) can have important ecological and societal consequences including increased transport of associated contaminants to aquatic environments, decreased primary production due to light attenuation, loss of recreational value as a result of water brownification, and increased water treatment costs for drinking water supplies impacted by high DOM content (De Wit et al. 2007; EkströM et al. 2011; Jaffé et al. 2008; Miller and McKnight 2010). Investigations linking DOM mobilization, transport, and transformation in study sites having long-term DOC records and well-established land coverage/land-use patterns are needed to resolve these uncertainties. Extending the DOC records via such optical models as presented here provides an avenue for such investigations.

Acknowledgements We thank William Gawley at Acadia National Park and the National Park Service Northeast Temperate Network (NETN) Monitoring Program for providing all the Secchi depth and DOC data used to develop the model presented in this chapter. These data are available via the USEPA Storage and Retrieval (STORET) water quality inventory database. We also thank Tom Huntington (USGS, New England Water Science Center) for his helpful review comments and Leslie Desimone (USGS, New England Water Science Center) for model verification.

Collin Roesler and Charles Culbertson

Charlie and Collin met in 1996 at a dinner party in Connecticut hosted by Pieter Visscher from the University of Connecticut. It was love at first site but they lived over 3000 miles apart, Charlie in California and Collin in Connecticut. It was their common interest in aquatic systems and microorganisms that made their life together possible. Charlie and his mentor, Ron Oremland at USGS, Menlo Park, CA, had worked extensively in Mono Lake, California. They had discovered a strange photosynthetic eukaryote that made its living under extreme conditions of low light and oxygen, and high pH and salinity. Ron turned over the project for them to find the story of this organism in the years of field data that had been collected. This allowed Collin to live for a few months in California to work with Charlie at USGS. When it became clear that more work was needed to understand the environmental tolerances of the organism, Ron enabled Charlie to move to Connecticut for a few months to conduct experiments in Collin's lab at the University of Connecticut. Collin's student Stacey Etheridge (now DeGrasse, and another member of the dual-career club) participated in these experiments with them. Putting the field and lab work together resulted in their first joint publication (Roesler, Collin S., Charles W. Culbertson, Stacey M. Etheridge, Ralf Goericke, Ronald P. Kiene, Laurence G. Miller, and Ronald S. Oremland. "Distribution, production, and ecophysiology of Picocystis strain ML in Mono Lake, California." *Limnology and Oceanography* 47, no. 2 (2002): 440–452). They joke that Mono Lake is one of the unique places that could have brought together an oceanographer and a limnological microbial ecologist.

After 3 years of across the country courtship, they made the decision to leave all their friends, two amazing jobs (permanent position at USGS and tenure-track position at the University of Connecticut), sell two beautiful houses, and move to coastal Maine. Collin was recruited to Bigelow Laboratory for Ocean Sciences by Barney Balch (who, as another member of the dual-career club, assured them that "miracles happen in Maine") and Charlie eventually was able to transfer to the USGS Water Science Center in Augusta. Collin has since moved to Bowdoin College, where she teaches oceanography. So miracles do happen in Maine! They have good jobs which provide lots of interesting science discussions. They live with their beautiful twins, Jack and Maeva, on a lovely farm on the Sheepscot River, where the ocean and the freshwater co-mingle.

References

Armengol J, Caputo L, Comerma M, Feijoó C, García JC, Marcé R, Navarro E, Ordoñez J (2003) Sau reservoir's light climate: relationships between Secchi depth and light extinction coefficient. Limnetica 22:195–210

Babin M, Stramski D, Ferrari GM, Claustre H, Bricaud A, Obolensky G, Hoepffner N (2003) Variations in the light absorption coefficients of phytoplankton, non-algal particles, and dissolved organic matter in coastal waters around Europe. J Geophys Res 108:C7

Boyce DG, Lewis MR, Worm B (2010) Global phytoplankton decline over the past century. Nature 466:591–596

Chen Z, Li Y, Pan J (2004) Distributions of colored dissolved organic matter and dissolved organic carbon in the Pearl River Estuary, China. Cont Shelf Res 24:1845–1856

De Wit HA, Mulder J, Hindar A, Hole L (2007) Long-term increase in dissolved organic carbon in streamwaters in Norway is response to reduced acid deposition. Environ Sci Technol 41:7706–7713

Driscoll CT, Driscoll KM, Roy KM, Mitchell MJ (2003) Chemical response of lakes in the Adirondack region of New York to declines in acidic deposition. Environ Sci Technol 37:2036–2042

EkströM SM, Kritzberg ES, Kleja DB, Larsson N, Nilsson PA, Graneli W, Bergkvist B (2011) Effect of acid deposition on quantity and quality of dissolved organic matter in soil–water. Environ Sci Technol 45:4733–4739

Erlandsson M, Buffam I, Fölster J, Laudon H, Temnerud J, Weyhenmeyer GA, Bishop K (2008) Thirty-five years of synchrony in the organic matter concentrations of Swedish rivers explained by variation in flow and sulphate. Glob Chang Biol 14:1191–1198

Evans M, Kaplan A, Cane M, Villalba R (2000) Globality and optimality in climate field reconstructions from proxy data. In: Markgraf V (ed) Present and past interhemispheric climate linkages in the americas and their societal effects. Academic, Cambridge, pp 53–72

Evans CD, Chapman PJ, Clark JM, Monteith DT, Cresser MS (2006) Alternative explanations for rising dissolved organic carbon export from organic soils. Glob Chang Biol 12:2044–2053

Gallegos CL, Werdell PJ, Mcclain CR (2011) Long-term changes in light scattering in Chesapeake Bay inferred from Secchi depth, light attenuation, and remote sensing measurements. J Geophys Res Ocean 1978–2012:116

Gawley WG, Mitchell BR, Arsenault EA (2014) Northeast Temperate Network lakes, ponds, and streams monitoring protocol: 2014 revision. In: NP Service (ed) Natural resource report NPS/NETN/NRR

Gershun A, Moon PH, Timoshenko G (1939) The light field. Massachusetts Institute of Technology, Cambridge, MA

Hejzlar J, Dubrovský M, Buchtele J, Růžička M (2003) The apparent and potential effects of climate change on the inferred concentration of dissolved organic matter in a temperate stream (the Malše River, South Bohemia). Sci Total Environ 310:143–152

Hongve D, Riise G, Kristiansen J (2004) Increased colour and organic acid concentrations in Norwegian forest lakes and drinking water – a result of increased precipitation? Aquat Sci 66:231–238

Hruška J, Krám P, Mcdowell WH, Oulehle F (2009) Increased dissolved organic carbon (DOC) in Central European streams is driven by reductions in ionic strength rather than climate change or decreasing acidity. Environ Sci Technol 43:4320–4326

Idso SB, Gilbert RG (1974) On the universality of the Poole and Atkins Secchi disk-light extinction equation. J Appl Ecol 11:399–401

Jaffé R, McKnight D, Maie N, Cory R, Mcdowell WH, Campbell JL (2008) Spatial and temporal variations in DOM composition in ecosystems: the importance of long-term monitoring of optical properties. J Geophys Res Biogeosci 113, G04032

Kirk JT (1991) Volume scattering function, average cosines, and the underwater light field. Limnol Oceanogr 36:455–467

Loiselle SA, Azza N, Gichuki J, Bracchini L, Tognazzi A, Dattilo AM, Rossi C, Cozar A (2010) Spatial dynamics of chromophoric dissolved organic matter in nearshore waters of Lake Victoria. Aquat Ecosyst Health Manag 13:185–195

Mankovsky VI (2014) Relation between the diffuse attenuation coefficient and the Secchi depth. Oceanology 54:32–37

McCullough IM, Loftin CS, Sader SA (2013) Landsat imagery reveals declining clarity of Maine's lakes during 1995-2010. Freshwater Sci 32:741–752

Miller MP, McKnight DM (2010) Comparison of seasonal changes in fluorescent dissolved organic matter among aquatic lake and stream sites in the Green Lakes Valley. J Geophys Res Biogeosci 2005–2012:115

Monteith D, Evans C (2005) The United Kingdom Acid Waters Monitoring Network: a review of the first 15 years and introduction to the special issue. Environ Pollut 137:3–13

Monteith DT et al (2007) Dissolved organic carbon trends resulting from changes in atmospheric deposition chemistry. Nature 450:537–540

Murray C, Markager S (2011) Relationship between Secchi depth and the diffuse light attenuation coefficient in Danish estuaries. NordAquaRemS Workshop

Oulehle F, Hruška J (2009) Rising trends of dissolved organic matter in drinking-water reservoirs as a result of recovery from acidification in the Ore Mts., Czech Republic. Environ Pollut 157:3433–3439

Porcal P, Koprivnjak J-F, Molot LA, Dillon PJ (2009) Humic substances—part 7: the biogeochemistry of dissolved organic carbon and its interactions with climate change. Environ Sci Pollut Res 16:714–726

Roesler CS, Perry MJ, Carder KL (1989) Modeling in situ phytoplankton absorption from total absorption spectra in productive inland marine waters. Limnol Oceanogr 34:1510–1523

Roesler CS, Abraham D, Franklin HM, Hoering E, Langner MR, Sauer MJ, Proctor CW, Barnard AH, Orico C, Aiken GR, Butler KD, Huntington T, Balch WB, Xue H (2006) Using optical proxies for biogeochemical properties to study land coverage and terrestrial inputs of organic carbon into coastal waters from the Penobscot Watershed to the Gulf of Maine. In: Paula Bontempi SA, Joan Cleveland (ed) Ocean Optics XVIII, p 12

SanClements MD, Oelsner GP, McKnight DM, Stoddard JL, Nelson SJ (2012) New insights into the source of decadal increases of dissolved organic matter in acid-sensitive lakes of the Northeastern United States. Environ Sci Technol 46:3212–3219

Spencer RG, Butler KD, Aiken GR (2012) Dissolved organic carbon and chromophoric dissolved organic matter properties of rivers in the USA. J Geophys Res Biogeosci 2005–2012:117

Strock KE, Nelson SJ, Kahl JS, Saros JE, Mcdowell WH (2014) Decadal trends reveal recent acceleration in the rate of recovery from acidification in the Northeastern US. Environ Sci Technol 48:4681–4689

Strock KE, Saros JE, Nelson SJ, Birkel SD, Kahl JS, McDowell WH (2016) Extreme weather years drive episodic changes in lake chemistry: implications for recovery from sulfate deposition and long-term trends in dissolved organic carbon. Biogeochemistry 127:353–365

Tipping E, Hurley M (1988) A model of solid-solution interactions in acid organic soils, based on the complexation properties of humic substances. J Soil Sci 39:505–519

Tipping E, Woof C (1990) Humic substances in acid organic soils: modelling their release to the soil solution in terms of humic charge. J Soil Sci 41:573–586

Weishaar JL, Aiken GR, Bergamaschi BA, Fram MS, Fujii R, Mopper K (2003) Evaluation of specific ultraviolet absorbance as an indicator of the chemical composition and reactivity of dissolved organic carbon. Environ Sci Technol 37:4702–4708

Wernand MR (2010) On the history of the Secchi disc. J Eur Opt Soc Rapid Publ 5:6

Worrall F, Burt T (2007) Flux of dissolved organic carbon from UK rivers. Global Biogeochem Cycles 21:GB1013

Worrall F, Harriman R, Evans CD, Watts CD, Adamson J, Neal C, Tipping E, Burt T, Grieve I, Monteith D (2004) Trends in dissolved organic carbon in UK rivers and lakes. Biogeochemistry 70:369–402

Part V
Focusing on Unique Systems, Processes and Dynamics

Phytoplankton Biodiversity in the Oligotrophic Northwestern Sargasso Sea

James L. Pinckney and Tammi L. Richardson

Introduction

One of the fundamental properties of ecosystems is the relationship between diversity and biomass. Phytoplankton are the major primary producers in oceanic systems and both their biodiversity and productivity determine the biomass and productivity of higher trophic levels (Irigoien et al. 2004; Vallina et al. 2014). Phytoplankton communities can be assembled by many co-occurring processes such as nutrient limitation, stoichiometry, top-down predation, turbulent mixing, mixotrophy, and taxon-specific life cycles, which vary in heterogeneous systems and also change in response to global climate cycles (Cloern and Jassby 2010; Finkel et al. 2010; Hillebrand et al. 2013). Reliable and repeatable measures of phytoplankton diversity are essential for understanding the relative importance of these processes on community assembly, structure, and function over time. Recent studies of oceanic phytoplankton diversity and biomass suggest that the relationship of biodiversity vs. biomass exhibits a unimodal function, similar to terrestrial communities, with maximum diversity occurring at intermediate levels of phytoplankton biomass (Irigoien et al. 2004; Vallina et al. 2014). Furthermore, the productivity and stability of phytoplankton communities increase with biodiversity at the community level (Tilman 1999; McCann 2000; Corcoran and Boeing 2012). Diversity indices (and associated values such as evenness, richness) may be used to provide a measure of biodiversity and as a bioindicator for ecosystem properties. For example, a high and stable diversity index value over time is consistent with high resource use efficiency and stable ecosystem functions (Ives and Carpenter 2007). Alternatively, a decrease in the diversity index at high biomass levels may signal monospecific phytoplankton blooms.

J.L. Pinckney (✉) • T.L. Richardson
Marine Science Program and Department of Biological Sciences,
University of South Carolina, Columbia, SC 29205, USA
e-mail: pinckney@sc.edu

P.M. Glibert, T.M. Kana (eds.), *Aquatic Microbial Ecology and Biogeochemistry: A Dual Perspective*, DOI 10.1007/978-3-319-30259-1_19

Traditional diversity indices are usually based on species composition and population abundances (Krebs 1999). The variety and phylogenetic associations of specific photosynthetic accessory pigments (chlorophylls and carotenoids) with different algal groups provide diagnostic biomarker compounds for differentiating the relative abundance of phytoplankton groups in mixed species assemblages (Roy et al. 2011). Photopigments and their concentrations are analogous to species and abundances, and provide the ability to assess phytoplankton diversity at the community level using common diversity indices (Noble et al. 2003; Sherrard et al. 2006; Pinckney et al. 2015). Thus, photopigment-based measures offer an efficient way to quantify community or functional diversity at the level of major algal groups. Furthermore, microscopy and direct count preservation methods to enumerate species may underestimate phytoplankton species richness while larger sample volumes for high-performance liquid chromatography (HPLC) photopigment analysis are more likely representative "true" phytoplankton community composition (Cermeño et al. 2014).

Numerous studies of the phytoplankton communities in the northwestern Sargasso Sea near Bermuda have been reported, especially associated with the Bermuda Atlantic Time-Series Study (BATS) (Bidigare et al. 1990; Steinberg et al. 2001; Not et al. 2007; Lomas et al. 2010; Treusch et al. 2012). However, none of these studies attempted to quantify the phytoplankton diversity-biomass relationship. The oligotrophic Sargasso Sea provides an ideal environment for examining the characteristics of the upward slope of the diversity-biomass curve because very few natural systems (both terrestrial and aquatic) exhibit such extreme low abundances of autotrophs. Measures of the phytoplankton diversity-biomass relationship, especially for low biomass concentrations, are essential for determining community assembly characteristics under oligotrophic conditions, which characterize much of the world's oceans. The purpose of this study was to use photopigment concentrations to quantify the relationship between diversity measures and biomass for Sargasso Sea phytoplankton. We further apply biodiversity indices to determine if photopigment-based diversity differs between major oceanographic features such as cyclonic and anticyclonic eddies, and among stations within these eddies.

Materials and Methods

Data were collected on four cruises in the Sargasso Sea (Table 1). On all but one cruise (1206), sampling was conducted at three stations: at the center of a mesoscale eddy, the approximate edge of an eddy, and the Bermuda Atlantic Time Series (BATS) site (31.667° N, 64.167° W) (Cotti-Rausch et al. 2016). Hydrocasts were performed before dawn. Samples (1–2 l) for determination of diagnostic photopigments by HPLC were collected from discrete depths (0–100 m) on all casts and filtered under gentle vacuum onto Whatman GF/F filters (Pinckney et al. 1996). Each filter was folded and placed into a 1.5 ml cryotube and immediately frozen at

Table 1 Dates and locations of sampling in the Sargasso Sea in 2011 and 2012

Cruise number	Dates	Season	Eddy type	Location, eddy center	Location, eddy edge	Eddy age (months)	Transect length (km)
AE1102	24 February to 5 March 2011	Spring	Anticyclone (AC1)	29° 42′ N, 64° 06′ W	30° 30′ N, 64° 08′ W	6	215
AE1118	22 July to 5 Aug 2011	Summer	Cyclone (C1)	30° 49′ N, 65° 47′ W	31° 17′ N, 64° 55′ W	2	178
AE1206	15–23 March 2012	Spring	Cyclone (C2)	32° 50′ N, 63° 29′ W	Not sampled	6	145
AE1219	19–31 July 2012	Summer	Anticyclone (AC2)	33° 30′ N, 64° 27′ W	32° 23.0′ N, 64° 22.0′ W	1	205

The abbreviations AC1, C1, C2, and AC2 are used in the text to refer to eddies sampled on cruises AE1102, AE1118, AE1206, and AE1219, respectively

−80 °C. At the end of each cruise, filters were shipped in liquid nitrogen to the University of South Carolina where they were stored at −80 °C until analysis.

Phytoplankton photopigment concentrations were measured using HPLC (Roy et al. 2011). Filters were first lyophilized for 18–24 h at −50 °C. Photopigments were then extracted by adding 750 μl of 90 % aqueous acetone solvent followed by storage for 12–20 h at −20 °C. Filtered extracts (250 μl) were injected into a Shimadzu HPLC with a single monomeric column (Rainin Microsorb, 0.46 × 1.5 cm, 3 μm packing) and a polymeric (Vydac 201TP54, 0.46 × 25 cm, 5 μm packing) reverse-phase C18 column in series. A nonlinear binary gradient consisting of solvent A (80 % methanol: 20 % 0.5 M ammonium acetate) and solvent B (80 % methanol: 20 % acetone) was used for the mobile phase (Pinckney et al. 2001). Absorption spectra and chromatograms (440 ± 4 nm) were obtained using a Shimadzu SPD-M10av photodiode array detector and pigment peaks were identified by comparing retention times and absorption spectra with pure standards (DHI, Denmark). The synthetic carotenoid β-apo-8′-carotenal (Sigma) was used as an internal standard. The effective limit of detection for all pigments was <0.004 μg l^{-1} (Hooker et al. 2010).

Measures of photopigment diversity for each sample were calculated using the individual pigments and concentrations as variables in standard diversity indices (Primer v. 5) (Krebs 1999; Noble et al. 2003; Sherrard et al. 2006; Pinckney et al. 2015). Pigment concentrations (in units of μg l^{-1}) were multiplied by 1000 for diversity calculations, which included Margalef's richness (*d*; an index based on the number of different photopigments in a sample):

$$d = \frac{(S-1)}{\log N}$$

where S is the total number of pigments in the sample and N is the total number of possible pigments (Krebs 1999). Pielou's evenness (j') is a measure of the relative concentrations for each photopigment and quantifies the equality of the concentrations of pigment for each photopigment (Krebs 1999):

$$j' = \frac{H'}{\log S}$$

where H' is the Shannon-Wiener diversity index which is calculated by (Krebs 1999):

$$H' = -\sum p_i \log_e (p_i)$$

where p_i is the proportion of the total count from the *i*th pigment. Pielou's evenness index ranges from 0 to 1 with higher values indicating more even distributions. The Shannon-Wiener index and Margalef's evenness index usually range between 1.5 and 3.5.

The 11 photopigments used for the diversity calculations were alloxanthin (cryptophytes), chlorophyll *b* (chlorophytes), chlorophyll c_3 (haptophytes), fucoxanthin (diatoms), lutein (chlorophytes), peridinin (dinoflagellates), prasinoxanthin (prasinophytes), zeaxanthin (cyanobacteria), 9′ cis-neoxanthin (chlorophytes, prasinophytes), 19′ butanoyloxyfucoxanthin (haptophytes, pelagophytes), and 19′ hexanoyloxyfucoxanthin (haptophytes, coccolithophytes). Chl *a* was excluded because it is common to all phytoplankton groups. The HPLC method used in this study could not fully resolve di-vinyl chl *a* from mono-vinyl chl *a*. Thus, chl *a* as reported in this chapter represents the sum of the weights of mono-vinyl chl *a*, di-vinyl chl *a*, as well as epimers and allomers of chl *a*. Thus we were unable to separate prochlorophytes from the total phytoplankton community. The combined data were used to determine sample sizes necessary for accurate estimates of mean values for richness, evenness, and diversity based on pooled sample variance (Eckblad 1991).

Data were tested for normality using a Kolmogorov-Smirnov test. As data were non-normal, they were then compared among the four different cruises and 11 stations using a nonparametric Kruskal-Wallis test. Post hoc means comparisons were done using a Mann–Whitney *U* test with a Bonferroni correction for multiple pairwise comparisons (Dytham 2011).

Results

The combined data set included 1642 measurements of phytoplankton photopigments in samples collected from a range of water mass types and depths to 100 m. The pooled variances for richness, evenness, and diversity were 0.140, 0.005, and 0.193, respectively. Estimates of the mean within ±10 % of the true mean at a 0.05 level of significance would require sample sizes of 50, 3, and 36 for richness, evenness, and diversity, respectively. Thus, our sample size, even if divided by 11 for among-station comparisons, was sufficiently large enough for robust calculations.

The average chl *a* concentration was 0.195 (±0.151 SD) and ranged from 0.008 to 0.905 μg l^{-1}. Margalef's richness index averaged 1.070 (±0.374 SD) and generally increased with an increase in chl *a* concentration, while the variation in richness was highest at low biomass values (Fig. 1). Pielou's evenness index averaged 0.837 (±0.068 SD) and chl *a* concentrations showed no obvious pattern other than high variation at low chl *a* concentrations (Fig. 1). A plot of richness vs. evenness indicated that the two indices were not correlated (Fig. 2).

The Shannon-Wiener photopigment diversity index averaged 1.484 (±0.440 SD) and ranged from 0.323 to 2.147 (Fig. 3). H' values increased sharply with biomass until reaching a chl *a* concentration of approximately 0.2 μg l^{-1}. The data were fit to a saturating hyperbolic type equation to yield the following relationship:

$$\text{Photopigment diversity}\left(H'\right) = 1.93 \times \left(\frac{\text{chl}\,a}{\left(0.033 + \text{chl}\,a\right)}\right)$$

where chl *a* is the concentration of chl *a* (μg l^{-1}) for each respective sample, 1.93 (±0.13 SE) is the maximum diversity, and 0.033 (±0.001 SE) is the half-saturation constant for diversity (adj $r^2=0.62$, $p<0.001$). The maximum possible diversity,

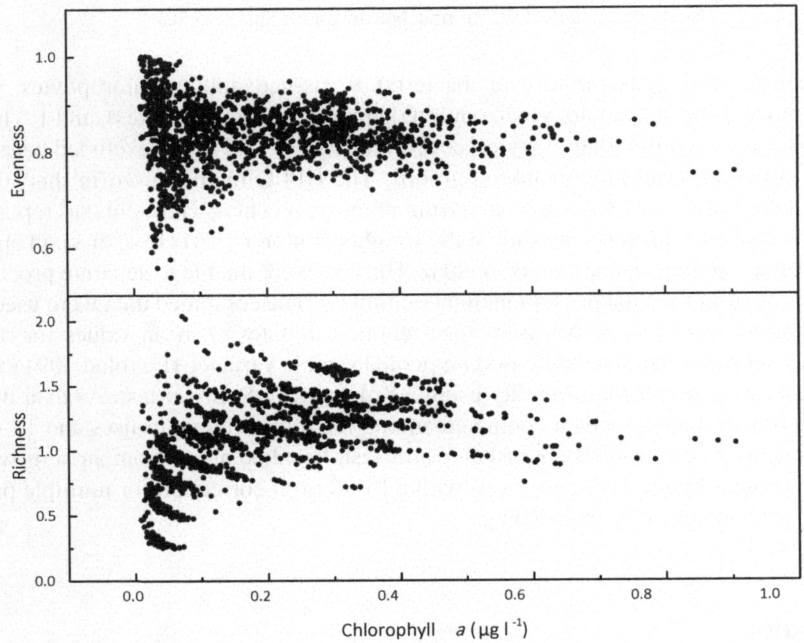

Fig. 1 Scatterplots of photopigment richness index (*d*) and evenness index (*j′*) vs. chlorophyll *a*

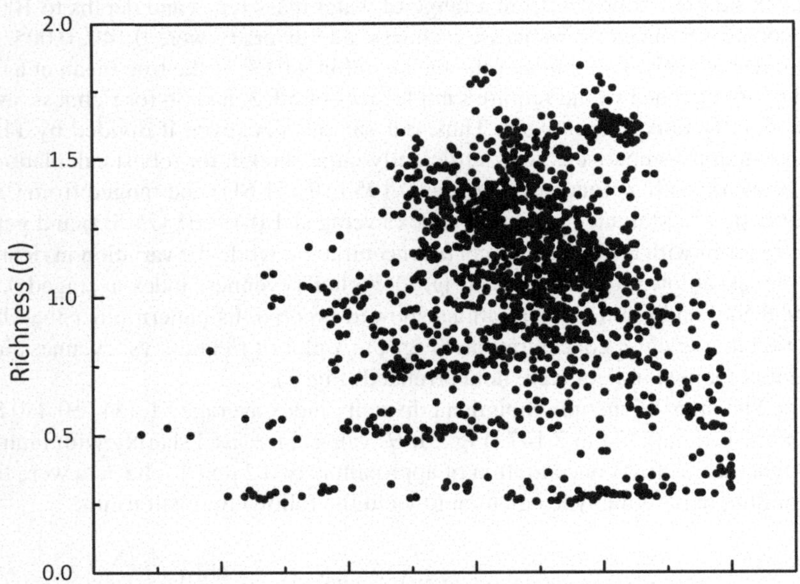

Fig. 2 Scatterplot of photopigment group Margalef's richness index (*d*) and Pielou's evenness index (*j′*)

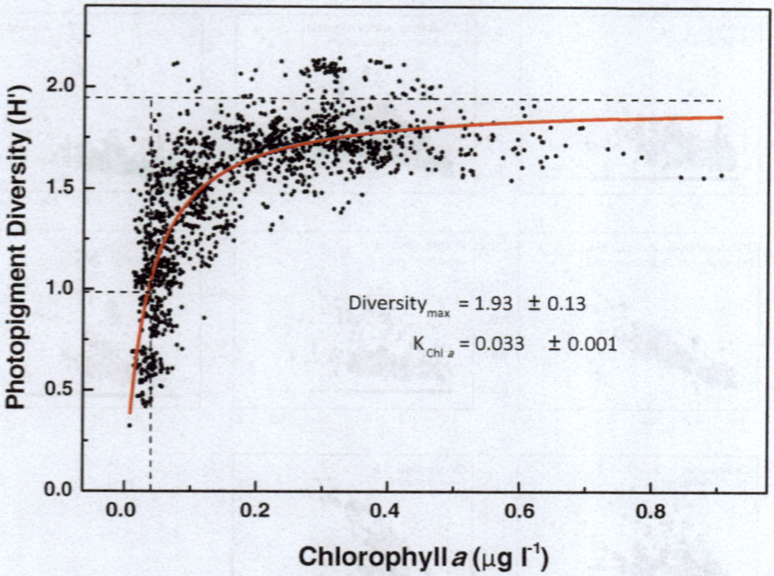

Fig. 3 Scatterplots of photopigment diversity index (*H′*) vs. chlorophyll *a*. Parameters for the fitted Michaelis-Menten curve are shown on the graph (adj $r^2=0.62$). *Dashed lines* illustrate maximum photopigment diversity ($Diversity_{max}$) and the half-saturation constant (K_{chla}). *Shading* around the curve indicates the 95 % confidence interval for the regression

assuming equal concentrations of all 11 photopigments, could be as high as 2.40. Thus the measured maximum diversity (1.93) is lower than the theoretical maximum (2.40). These results suggest that phytoplankton diversity increases with biomass, but reaches a maximum value of 1.93 for chl *a* concentrations >1.0 $\mu g\ l^{-1}$. The concentrations of individual diagnostic pigments were plotted as a function of chl *a* to determine if there were any consistent trends between low diversity and certain algal groups (Fig. 4). Although pigment and chl *a* concentrations are autocorrelated, there were no consistent trends between low biomass and specific algal groups. Thus there does not appear be a particular algal group assemblage characteristic of lower diversity. A frequency distribution was constructed for the photopigment diversity index (*H′*) values to further examine the range in *H′* over the study area (Fig. 5). The plot illustrates a number of occurrences of low diversity (i.e., <1.5) and suggests a mosaic pattern of patches within which diversity ranges over a gradient of values.

Diversity showed a significant difference between stations ($\chi^2=370$, $p<0.001$) (Fig. 3). Post hoc comparisons showed that *H′* at the BATS station sampled in February of 2011 was significantly higher than diversity at all other stations ($p<0.05$) (Fig. 3). Diversity was also high at the edge of the anticyclone on that same cruise, but otherwise the data do not show a clear pattern in the relative diversity for cyclonic vs. anticyclonic eddies or for stations within an eddy (Fig. 6).

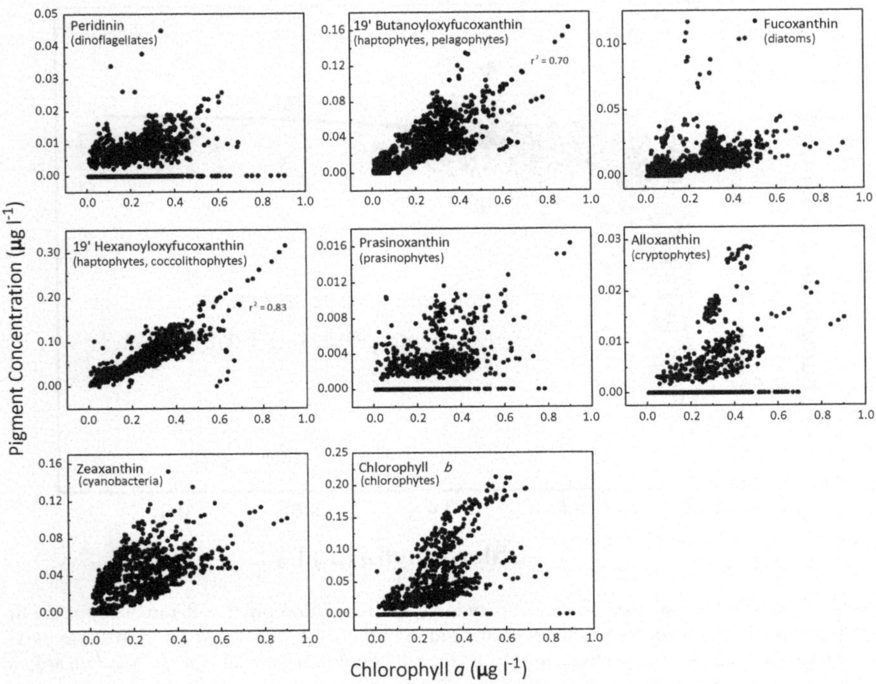

Fig. 4 Scatterplots of individual photopigments vs. phytoplankton biomass (chl *a*). The major algal groups associated with each pigment are indicated on each plot

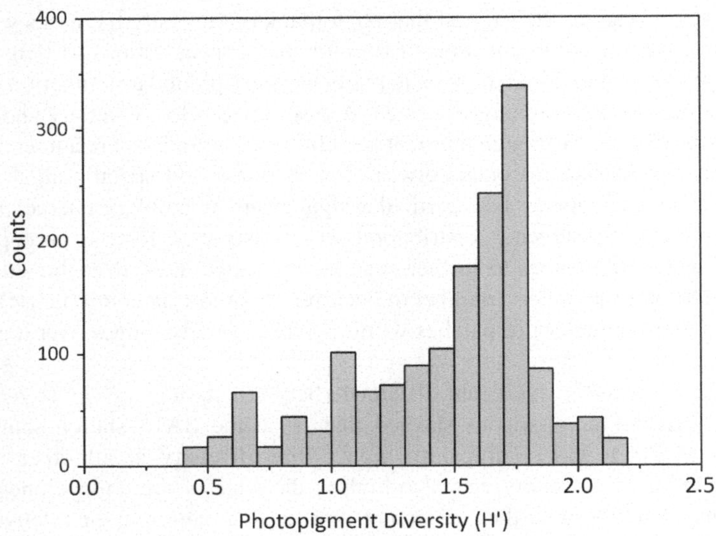

Fig. 5 Frequency distribution of photopigment diversity (H') measures from individual samples

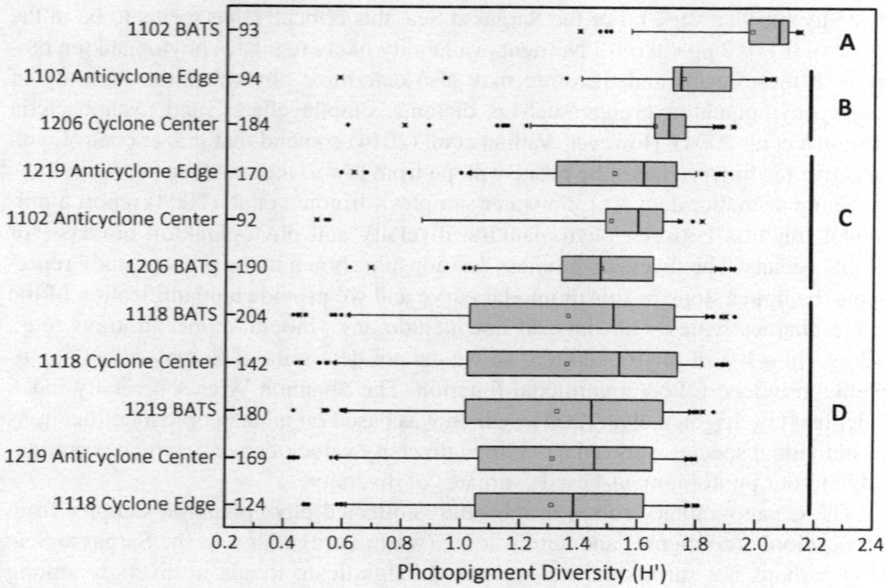

Fig. 6 Boxplots of photopigment diversity index (H') for each of the water mass types sampled in this study. Sample sizes for each boxplot are indicated along the *left y*-axis. Locations that were not significantly different ($p < 0.05$) are shown by the *vertical lines* and *bold letters* on the *right*. Boxplots illustrate the 25th and 75th quartiles (the *box*), the *line* in the *box* is the median, the *dot* in the *box* is the mean, and the whiskers show the 5th and 95th percentiles for the data

Discussion

The data used in our analysis were obtained from a series of oceanographic cruises that covered a wide area (ca. 860×10^3 km^2) of the Sargasso Sea over a 2-year time span. The large sample size (1642) constitutes a significant survey of the variation in phytoplankton community composition in the upper 100 m of the water column for this region.

Values for Margalef's richness and Pielou's evenness fall within the bounds calculated for phytoplankton from oceanic ecosystems (Irigoien et al. 2004; Vallina et al. 2014). The most striking feature for the richness and evenness data is the relatively high variability in these indices relative to phytoplankton biomass. These results, combined with the lack of a correlation between richness and evenness, suggest that, although the Sargasso Sea is an oligotrophic system, phytoplankton community diversity is quite heterogeneous. The phytoplankton assemblages within the low biomass patches are also quite variable, with a range of algal groups being represented in even the lowest biomass patches.

The Sargasso Sea typifies the paradigm of high diversity and low biomass for oligotrophic waters. Our results illustrate this paradigm but suggest that there is a critical value at which a reduction in phytoplankton biomass is related to a rapid decrease in diversity (or vice versa). For the Sargasso Sea, this critical value seems to be in the range of 0.1–0.2 μg chl *a* l^{-1}. Nutrient availability likely regulates phytoplankton biomass at these levels and therefore may also determine phytoplankton diversity of major phytoplankton groups such as diatoms, dinoflagellates, and cyanobacteria (Irigoien et al. 2004). However, Vallina et al. (2014) contend that grazer control with selective feeding produces the positive slope from low to intermediate productivity.

Using a smaller data set (353 water samples), Irigoien et al. (2004) report a unimodal function between phytoplankton diversity and phytoplankton biomass for global oceans. The diversity-biomass relationship shown in the present study represents the upper slope of this unimodal curve and we provide a quantification of the curve characteristics. Our data do not include any "bloom" concentrations (e.g., >2 μg chl *a* l^{-1}) of phytoplankton, so we cannot determine if Sargasso Sea phytoplankton indeed follow a unimodal function. The Shannon-Wiener diversity index calculated by Irigoien et al. (2004), which was based on microscopic identifications of individual species, showed a maximal diversity value of ca. 2.5, which is comparable to our photopigment-based estimates of diversity.

The oceanographic cruises in this study collected phytoplankton samples from two cyclonic (cold-core) and anticyclonic (warm-core) eddies in the Sargasso Sea. It is perhaps not surprising that we saw no significant trends in diversity among cruises or among stations within a cruise, because of the high degree of submesoscale (1–10 km scale) variability that results from the physics of eddy circulation (e.g., McGillicuddy et al. 1998). At even smaller scales (1–100 m) we observed significant variations in phytoplankton community composition (estimated by CHEMTAX analysis of the HPLC-derived photopigment concentrations used in this paper) with depth (Cotti-Rausch et al. 2016). In some cases, diversity of phytoplankton collected at the chlorophyll maximum was much higher than in populations nearer to the surface. This follows the general trend of increased H' with chl *a* concentration that we saw in the aggregated data set.

The characteristics of eddy circulation and associated nutrient fluxes likely play a major role in maintaining phytoplankton diversity in the Sargasso Sea. These features produce a range of habitats supporting a wide range in phytoplankton biomass. The significant positive relationship between diversity and biomass illustrated in this study shows that nutrient inputs associated with eddies likely foster phytoplankton diversity. As nutrients become depleted, diversity also declines, but the algal group assemblages vary in the nutrient-depleted patches. The implication of this study is that phytoplankton biomass levels above ca. 0.2 μg chl *a* l^{-1} are necessary to maintain biodiversity for this region of the Sargasso Sea. Increases in biomass above this threshold do not result in substantial changes in diversity. Since phytoplankton biomass is likely regulated by nutrient availability at these low concentrations, nutrient supply may be the key regulator for diversity in this system. Unlike terrestrial systems, extremely low concentrations of phytoplankton biomass can result in a variety of algal group assemblages.

Acknowledgements Funding for this research was provided by the National Science Foundation (OCE-1030345 to T.L.R.). We thank Bridget Bachman, Emily Goldman, and Eric Lachenmeyer for sample collection, processing, and HPLC analyses.

James L. Pinckney and Tammi L. Richardson

Tammi and Jay first met when Tammi was being interviewed for a postdoc position in Hans Paerl's lab at the UNC Institute of Marine Sciences (IMS). The opening was to replace a guy named Jay Pinckney, who was leaving for a tenure-track position in the Oceanography department at Texas A&M. The project focused on the effects of different forms of nitrogen on phytoplankton community composition in the Neuse River Estuary. Tammi got the job, and since Jay was a PI on the project he traveled back and forth to Morehead City during the first year to participate in the microcosm experiments (that was his story anyway). This project resulted in their first publication together (Richardson et al. 2001). After a year or so apart, and when her postdoc ended, Jay likes to say that Tammi "followed him to Texas." Really, she was incredibly fortunate to land a soft-money position at Texas A&M, working with George Jackson on inverse modeling of marine food webs. Jay was well into his Assistant Professor position by this time and was working with his first graduate students on phytoplankton communities in Galveston Bay. Tammi and Jay lived in College Station for about 5 years, during which they got married, built a house, and had a baby (pretty much in that order). Before the arrival of said baby (Jacob Linehan Pinckney), however, Tammi saw an ad for a tenure-track Assistant Professor position in Biological Oceanography at the University of South Carolina. She applied, interviewed at 12 weeks gestation, and was offered the position. The clever folks at U-SC knew that they were dealing with a dual-career couple and they successfully convinced their Provost to create a second position for Jay. Tammi and Jay moved to Columbia and started their new jobs as Assistant and Associate Professor, respectively, in January of 2005. Jay is amazed and/or dismayed by the fact that he went to Texas as a bachelor, with all of his worldly belongings in the back of his Isuzu pickup truck, but returned to his home state of South Carolina with a wife, a baby, two cats, a graduate student, the graduate student's girlfriend (now wife), and the graduate student's turtle in a convoy of two cars and a moving van that contained the contents of two laboratories and a house. Tammi doesn't necessarily recommend starting a tenure-track job with a 4-month-old baby, but it clearly worked out ok. Tammi and Jay were both promoted to Professor at U-SC in January of 2015 and continue to enjoy teaching, advising graduate students, and working in the field whenever possible.

References

Bidigare RR, Marra J, Dickey TD, Iturriaga R, Baker KS, Smith RC, Pak H (1990) Evidence for phytoplankton succession and chromatic adaptation in the Sargasso Sea during spring 1985. Mar Ecol Prog Ser 60:113–122

Cermeño P, Teixeira IG, Branco M, Figueiras FG, Marañon E (2014) Sampling the limits of species richness in marine phytoplankton communities. J Plankton Res 36:1135–1139

Cloern JE, Jassby AD (2010) Patterns and scales of phytoplankton variability in estuarine-coastal ecosystems. Estuar Coast Shelf Sci 33:230–241

Corcoran AA, Boeing WJ (2012) Biodiversity increases the productivity and stability of phytoplankton communities. PLoS One 7, e49397. doi:10.1371/journal.pone.0049397

Cotti-Rausch BE, Lomas MW, Lachenmyer EM, Goldman EA, Bell DW, Goldberg SR, Richardson TL (2016) Mesoscale and sub-mesoscale variability in phytoplankton community composition in the Sargasso Sea. Deep Sea Res Part I 110:106–122

Dytham C (2011) Choosing and using statistics. Wiley-Blackwell, Chichester

Eckblad JW (1991) How many samples should be taken? Bioscience 41:346–348

Finkel Z, Beardall J, Flynn K, Quigg A, Rees T, Raven J (2010) Phytoplankton in a changing world: cell size and elemental stoichiometry. J Plankton Res 32:119–137

Hillebrand H, Steinert G, Boersma M, Malzahn A, Meunier C, Plum C, Ptacnik R (2013) Goldman revisited: faster-growing phytoplankton has lower N:P and lower stoichiometric flexibility. Limnol Oceanogr 58:2076–2088

Hooker SB, Thomas CS, Van Heukelem L, Schlüter L, Russ ME, Ras J, Claustre H, Clementson L, Canuti E, Berthon J-F, Perl J, Normandeau C, Cullen J, Kienast M, Pinckney JL (2010) The fourth SeaWiFS HPLC analysis round-Robin experiment (SeaHARRE-4). NASA/TM–2010–215857. National Aeronautics and Space Administration, Goddard Space Flight Center, Greenbelt, MD

Irigoien X, Huisman J, Harris RP (2004) Global biodiversity patterns of marine phytoplankton and zooplankton. Nature 429:863–867

Ives AR, Carpenter SR (2007) Stability and diversity of ecosystems. Science 317:58–62

Krebs CJ (1999) Ecological methodology. Harper & Row, New York, NY

Lomas MW, Steinberg DK, Dickey T, Carlson CA, Nelson NB, Condon RH et al (2010) Increased ocean carbon export in the Sargasso Sea linked to climate variability is countered by its enhanced mesopelagic attenuation. Biogeosciences 7:57–70

McCann RS (2000) The diversity-stability debate. Nature 405:228–233

McGillicuddy DJ Jr, Robinson AR, Siegel DA, Jannasch HW, Johnson R, Dickey TD, McNeil J, Michaels AF, Knap AH (1998) Influence of mesoscale eddies on new production in the Sargasso Sea. Nature 394:263–266

Noble PA, Tymowski RG, Fletcher M, Morris JT, Lewitus AJ (2003) Contrasting patterns of phytoplankton community pigment composition in two salt marsh estuaries in Southeastern United States. Appl Environ Microbiol 69:4129–4143

Not F, Gausling R, Azam F, Deidelberg JF, Worden A (2007) Vertical distribution of picoeukaryotic diversity in the Sargasso Sea. Environ Microbiol 9:1233–1252

Pinckney, JL, D Millie, K Howe, H Paerl, J Hurley. 1996. Flow scintillation counting of C-labeled microalgal photosynthetic pigments. Journal of Plankton Research 18:1867–1880

Pinckney JL, Richardson T, Millie D, Paerl H (2001) Application of photopigment biomarkers for quantifying microalgal community composition and *in situ* growth rates. Org Geochem 32:585–595

Pinckney JL, Millie D, Howe K, Paerl H, Hurley J (1996) Flow scintillation counting of C-labeled microalgal photosynthetic pigments. J Plankton Res 18:1867–1880

Roy S, Llewellyn CA, Egeland ES, Johnsen G (2011) Phytoplankton pigments. Cambridge University Press, Cambridge

Sherrard NJ, Nimmo M, Llewellyn CA (2006) Combining HPLC pigment markers and ecological similarity indices to assess phytoplankton community structure: an environmental tool for eutrophication? Sci Total Environ 361:97–110

Steinberg DK, Carlson CA, Bates NR, Johnson RH, Michaels AF, Knap AH (2001) Overview of the US JGOFS Bermuda Atlantic Time-series Study (BATS): a decade-scale look at ocean biology and biochemistry. Deep Sea Res II 48:1405–1447

Tilman D (1999) The ecological consequences of changes in biodiversity: a search for general principles. Ecology 80:1455–1475

Treusch AH, Demir-Hilton E, Vergin KL, Worden AZ, Carlson CA, Donatz MG, Burton RM, Giovannoni SJ (2012) Phytoplankton distribution patterns in the northwestern Sargasso Sea revealed by small subunit rRNA genes from plastids. ISME J 6:481–492

Vallina SM, Follows MJ, Dutkiewics S, Montoya JM, Cermeno P, Loreau M (2014) Global relationship between phytoplankton diversity and productivity in the ocean. Nat Commun 5:4299

Biological Oceanography of the Gulf of Carpentaria, Australia: A Review

Peter C. Rothlisberg and Michele A. Burford

Introduction

The world's shallow shelf waters are typically well studied, by virtue of being close to much of the world's human population which live on or near the coast. The Gulf of Carpentaria (GoC) in tropical northern Australia is an exception to this, despite being a similar size to the North Sea. This remote environment has a low human population inhabiting coastal areas (ca. 30,000 permanent residents as of June 2013, Australian Bureau of Statistics), and has been recognised as one of the world's least impacted marine ecosystems (Halpern et al. 2008). Tropical northern Australia (i.e. Northern Region) is viewed as a region of high ecological value (DEWHA 2008); however, the biodiversity of the pelagic fauna is poorly understood (Rothlisberg et al. 2005).

Study Area

The Gulf of Carpentaria and adjacent Arafura Sea form a low-latitude epicontinental sea between Australia and Papua New Guinea which have fluctuated between open-ocean, estuarine and lacustrine (Lake Carpentaria) environments during the sea-level changes of the Late Quaternary (Chivas et al. 2001) (Fig. 1). The final

P.C. Rothlisberg
CSIRO Oceans and Atmosphere Flagship, GPO Box 2583, Brisbane, 4001 QLD, Australia

M.A. Burford (✉)
Australian Rivers Institute, Griffith University, Nathan 4111, QLD, Australia
e-mail: m.burford@griffith.edu.au

P.M. Glibert, T.M. Kana (eds.), *Aquatic Microbial Ecology and Biogeochemistry: A Dual Perspective*, DOI 10.1007/978-3-319-30259-1_20

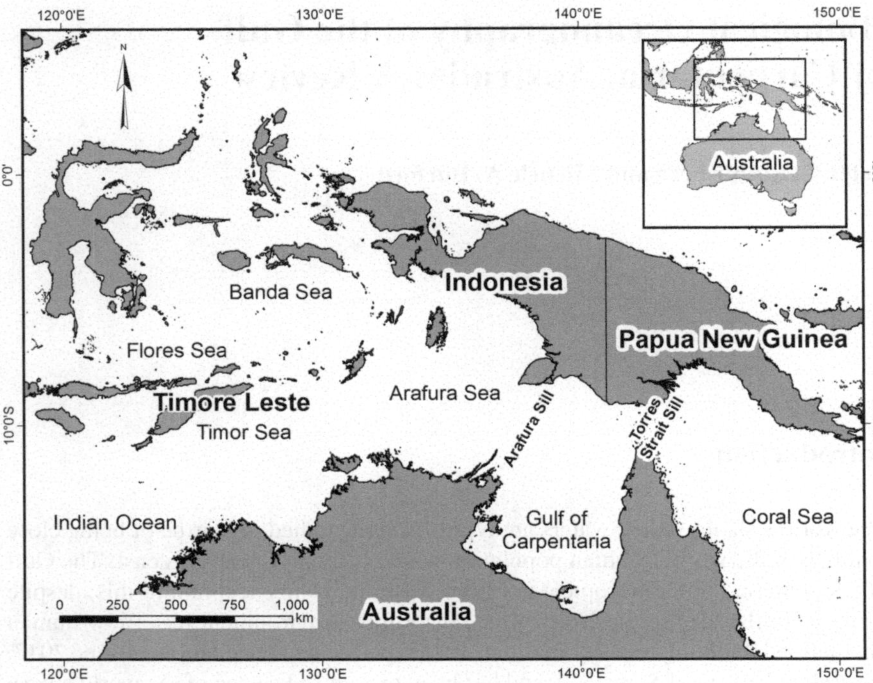

Fig. 1 Gulf of Carpentaria, Australia and adjacent seas

marine transgression occurred about 9.7 ka (Chivas et al. 2001). The area is bounded in the west by the Arafura Sill (53 m [=Wessel Rise]) and on east by the Torres Straight Sill (12 m). These sills limit water exchange with the Coral Sea to the east and the Timor Sea and Indian Ocean to the west (Forbes 1984; Condie and Dunn 2006; Condie 2011).

The present Gulf is large (ca. 3.5×10^5 km^2) and shallow (mean depth 50 m, maximum depth 70 m) and lies between 12 and 17.5°S latitude. The seasonal winds and mixing are dominated by two monsoons. During the summer (October–March) wet season, winds are from the northwest. Studies by Rothlisberg and his colleagues showed that the winds are variable in strength allowing stratification in the deeper waters of the Gulf (Rothlisberg et al. 1989). Rainfall is also highly seasonal, with 80 % of tropical Australia's annual mean precipitation falling from December to March (Bureau of Meteorology 1988). In contrast, in the winter dry season (April–September), winds are persistently from the southeast which mixes the entire water column (Burford and Rothlisberg 1999).

Currents and Hydrography

Our understanding of the Gulf's current regimes relies heavily on modelling studies with limited tidal and current observations (Forbes and Church 1983; Rothlisberg et al. 1983a, 1996; Wolanski 1993; Condie et al. 1999, and Condie 2011). Within the Gulf there are two distinct zones: a turbid, well-mixed, nearshore zone with a depth to a maximum of 15–20 m, and deeper waters separated from the coastal zone by a boundary (buoyancy) current (Wolanski and Ridd 1990; Wolanski 1993). River run-off and nutrients are effectively trapped within the coastal zone by the boundary current (Rothlisberg et al. 1994; Burford et al. 2009). The Gulf's high tidal and wind mixing dominates over advective processes and there is limited water exchange with adjacent seas (Forbes 1984; Condie 2011). Condie (2011) estimated a flushing timescale for the entire Gulf to be in excess of 2 years.

A hydrographic atlas has been published (Rothlisberg et al. 1989). The Atlas plots and contours the surface, bottom and difference in temperature, salinity, sigma-t, dissolved oxygen, nitrate + nitrite and silicate values at all stations sampled. A link to the printed report is in the references and data files can be retrieved via http://dx.doi.org/10.4225/08/5593848D9FE36.

Phytoplankton and the Role of Nutrients

The Gulf has a more diverse net phytoplankton community with elaborate morphologies compared with Australian subtropical and temperate waters (Hallegraeff and Jeffrey 1984). Morphologies include large spines, setae and wing-like structures. The benefits of this morphology are unclear but may be related to improving buoyancy in the warm tropical waters, protection against predators and/or increased nutrient uptake. Analysis of net samples revealed 72 genera with a dominance of diatom species (*Rhizosolenia, Chaetoceros,* and *Thalassionema*) as well as the filamentous nitrogen-fixing cyanobacterium, *Trichodesmium* (Rothlisberg et al. 1994; Burford et al. 1995). A number of the diatom species contained the symbiotic nitrogen-fixing cyanobacterium, *Richelia,* presumably to provide an additional source of nitrogen for the diatom. Previous studies had also highlighted the importance of nanoplankton (70–95 % of total chlorophyll *a*) in the central Gulf in summer (Hallegraeff and Jeffrey 1984; Rothlisberg et al. 1994). The species are similar to those found in Australian subtropical and temperate waters.

The Gulf of Carpentaria has some of the highest depth-integrated primary productivity rates in Australia's coastal waters (Condie and Dunn 2006) but comparable with other tropic regions, e.g. Furnas and Mitchell (1996). The first study of primary productivity in the Gulf of Carpentaria was by Motoda et al. (1978), but they only undertook their incubations at one light level. Since then Burford and Rothlisberg have undertaken a number of cruises to measure primary productivity in both the deep and shallow regions of the Gulf (Rothlisberg et al. 1994; Burford

Table 1 Seasonal and spatial variation in primary productivity in the Gulf of Carpentaria (mean (SD of stations), mg C m^{-2} day^{-1})

Season	Shallow (<20 m)	Deep (>20 m)	Source
Summer	1430 (400)	660 (108)	Rothlisberg et al. (1994)
Summer	950 (320)	950 (130)	Burford and Rothlisberg (1999)
Winter	750 (380)	560 (350)	Burford and Rothlisberg (1999)

and Rothlisberg 1999). The studies showed two key features: productivity was higher in summer (660–1430 mg C m^{-2} day^{-1}) than winter (560–750 mg C m^{-2} day^{-1}) due to calmer, stratified conditions in the summer period, compared with the wind-mixed waters in winter (Table 1). However, cyclones [=hurricanes] in the summer monsoon period can also cause periodic mixing of the whole water column (Burford et al. 2009).

The high productivity in offshore waters in summer months is despite surface nitrate/nitrite and phosphate concentrations being near detection limits (<0.3 μmol l^{-1}) (Burford and Rothlisberg 1999). The low nutrient levels combined with molar nitrogen:phosphorus ratios of 0.8–8.2 suggest that the Gulf is likely to be nitrogen limited. In this environment, nitrogen-fixing cyanobacteria dominate, i.e. *Trichodesmium* spp. (Burford et al. 1995). It may also explain why so many diatom species have symbiotic nitrogen-fixing cyanobacteria, such as *Richelia.*

During summer, the chlorophyll maximum zone is relatively deep in the water column and the main source of nitrogen is via benthic mineralization of nitrogen fixed by cyanobacteria, i.e. *Trichodesmium* (Burford et al. 2009). Nitrogen fixation rates in blooms in this study were as high as 76 μmol N m^{-2} h^{-1}. In winter, primary productivity is fuelled by wind-driven mixing bringing accumulated nitrogen from the bottom waters to the surface, but light limited because of the re-suspension of fine sediments, as well as the calcareous coccoliths from previous coccolithophorid blooms (Brown and Yoder 1994; Burford and Rothlisberg 1999).

While understanding of the deeper waters of GoC is based on a limited number of cruises, studies of the coastal waters (<20 m deep) are even more limited. A significant proportion of the area of the GoC is within the coastal boundary layer and includes large areas of seagrass beds (11 species, 906.4 km^2, in 671.1 km of fringing coastline) (Poiner et al. 1987; Condie and Dunn 2006). These shallow environments have considerable sediment re-suspension due to tidal mixing, and at times, wind mixing, resulting in high light attenuation, reducing primary productivity rates. Additionally these environments are on the receiving end of sediment and nutrient loads from freshwater runoff during the wet season (Rothlisberg et al. 1994; Burford et al. 2012). Freshwater runoff triggers recruitment of estuarine shrimp into the offshore fishery as the low salinity drives the shrimp from their estuarine nursery grounds to higher salinity waters (Vance et al. 1985); however, the role of the freshwater runoff on coastal productivity remains poorly understood.

Zooplankton

There are only two published studies specifically on the zooplankton community of the GoC: zooplankton biomass estimates by Rothlisberg and Jackson (1982), and a copepod species list and associated biogeographic affinities with Southeast Asian seas (Othman et al. 1990).

Zooplankton biomass was measured at up to 72 stations on ten Gulf-wide cruises over a 20-month period from August 1975 to May 1977 (Rothlisberg and Jackson 1982). Stepped oblique tows from surface to near-bottom with paired plankton nets with both 142 and 500 μm mesh were made. The mean biomass estimate (142 μm net) over all cruises and all stations was 77 mg m^{-3} (range 50–123 mg m^{-3}) and compares with the very high abundances found only in seasonal upwelling areas south of Java and off the northwest shelf of Australia.

Othman et al. 1990 used the 142 μm mesh samples from a 23-station subset from the same ten cruises to characterize the copepod community and found 102 species: 68 from the sub-order Calanaoida, 30 Cyclopoida and 4 Harpacticoida—13 of these species were new to science and 23 were new records for Australia.

Mean copepod numbers were higher in the inshore waters (<20–30 m) especially on the mid-eastern shore (Othman 1986). Abundances dropped in deeper waters in the central and northern Gulf. Conversely, inshore stations generally had fewer species. Numbers increased with depth and distance from shore. The top ten species accounted for ca. 60 % of the total numbers of copepods sampled (Table 2). The calanoid copepod *Parvocalanus crassirostris* was the most abundant species accounting for almost 17 % of the copepod abundance. The most abundant harpacticoid and second most abundant species was *Euterpina acutifrons*. Five cyclopoid species from the genera *Oithona* and *Oncaea* were in the top ten—*Oncaea clevei* was the most abundant.

Table 2 Top ten species of copepod based on total abundance (No. m^{-3}) over ten cruises in the Gulf of Carpentaria (modified from Othman (1986))

Rank	Species	Total abundance (no. m^{-3})	Cumulative percent abundance
1.	*Parvocalanus crassirostris*	19,111	16.7
2.	*Euterpina acutifrons*	11,248	26.5
3.	*Paracalanus aculeatus*	11,074	36.2
4.	*Oncaea clevei*	6746	42.0
5.	*Oithona plumifera*	3951	45.5
6.	*Oncaea venusta*	3705	48.7
7.	*Parvocalanus latus*	3271	51.6
8.	*Parvocalanus elegans*	3183	54.4
9.	*Oncaea media*	3133	57.1
10.	*Oithona simplex*	3039	59.7

Three species assemblages emerged from cluster analysis: inshore, intermediate and offshore (Othman 1986). The coastal assemblage was consistent in both the wet and dry seasons. During the dry season more stations had assemblages in the offshore cluster, while during the wet season the number of stations in the intermediate assemblage cluster doubled and extended into the southern Gulf.

This warm neritic fauna had at least 88 species in common with adjacent Southeast Asian seas (e.g. Arafura, Timor, Banda and Flores Seas), and 79 species in common with the Coral Sea off the northeastern coast of Australia (Othman et al. 1990). Data files are available through linkages in Davies et al. (2014).

Penaeid Prawn Larval Ecology

The most detailed and systematic studies of the zooplankton were focused on the larval ecology of commercially important penaeid prawns [=shrimp]. Studies of larval ecology involve resolving factors that affect survival and dispersal to determine reproductive dynamics and recruitment success. Worldwide these types of study with penaeids had not been possible because the early larval stages (those close to spawning) could not be identified to the species level. Rothlisberg et al. (1983b) were successful in resolving this dilemma by building a reference collection from known spawners and developing a numerical taxonomic technique using discriminant function analysis. They found that the four commercially important species of *Penaeus*: *P. merguiensis*, *P. esculentus*, *P. semisulcatus* and *P. latisulcatus* had different spawning locations and seasons (Rothlisberg et al. 1987). Further they found the temperatures and salinities where peak larval abundance occurred (Rothlisberg and Jackson 1987).

Preston et al. (1992) studied the role of phytoplankton in the Gulf of Carpentaria as a food source for penaeid prawn larvae. They spawned and cultured larvae at sea and then in novel in situ cage studies showed that diatoms were an important food source but that food availability did not appear to be a key driver of survival. Conversely, a laboratory study with the dominant cyanobacterium, *Trichodesmium*, found that it was a poor food source for penaeid prawn larvae (Preston et al. 1998).

Larval Dispersal Mechanisms

Penaeid prawns in the genus *Penaeus* have a mixed life cycle, with offshore spawning and nearshore or estuarine nursery grounds (type 3 life cycle; Dall et al. 1990). Early studies on the vertical migratory behaviour of the offshore larvae showed that they were lower in the water column by day and higher by night and were exposed to different current strengths and directions depending on their position in the water column (Rothlisberg 1982). More sophisticated sampling and modelling put this behaviour into a wider seasonal and spatial context and demonstrated differences in larval dispersal with different ontogenetic behavioural regimes and seasonal differences in tidally dominated currents (Rothlisberg et al. 1983a). Rothlisberg et al. (1995) also showed the point in the life history and nearshore proximity that postlarvae switched their vertical migratory behaviour from diurnal to a tidal behaviour in a study off the east coast of Australia with the eastern king prawn *P. plebejus*. This study showed that larvae are concentrated onshore and drawn into estuaries on the flood tide once they have settled on the bottom as postlarvae, are in shallow water, and need only respond to a pressure cue. This transition depth, based on nearshore sampling, was estimated to be $\leq$20 m. The study also estimated the spatial extent of the drawing capacity of the estuary based on alongshore current speeds and tidal exchange volumes. The modelling approach was then transplanted back to the Gulf of Carpentaria which estimated the spatial extent (defined as the ‘advective envelope’) of the spawning population (called the ‘effective spawning population’) which gave rise to the postlarvae that reached the obligatory estuarine nursery ground and gave rise to the next generation of spawners and commercial catch (Rothlisberg et al. 1996). The modelled depth of this behavioural switch [= ‘transition depth’] was again 20 m, but was later shown in laboratory studies by Vance and Pendrey (2001) to be 4–8 m depending on species and postlarval size. These very shallow coastal areas lie inshore of most of the commercial fishing grounds.

Summary Points

- While the Gulf lies within tropical latitudes, it is highly seasonal. Seasonal shifts in monsoonal winds and rainfall patterns affect current and mixing regimes; phytoplankton productivity and reproductive seasonality and recruitment of coastal marine species (e.g. penaeid prawns). The extent and offshore mixing of nutrients from continental runoff are limited by a coastal boundary current.
- Phytoplankton productivity is high and dominated by nanoplankton. *Trichodesmium* and diatoms with nitrogen-fixing symbionts dominate the net phytoplankton. Nitrogen inputs from nitrogen fixers are an important nutrient input to deeper waters whilst the shallow waters are seasonally influenced by nutrients from freshwater runoff.
- The Gulf’s zooplankton community is biodiverse with relatively high biomasses and inferred productivity. With only limited water exchange, it shares species with adjacent seas to both the east and west.
- Recruitment of penaeid prawn postlarvae to nearshore and estuarine nursery grounds is controlled by the interaction between ontogenetic changes in larval and postlarval behaviour and both alongshore and tidal currents. This mechanism is controlled by a conservative environmental cue (i.e. pressure) and thereby provides a recruitment mechanism across a wide range of coastal habitats and current regimes. Further, the spatial extent of the recruitment mechanism defines the effective spawning populations of these variable and commercially important coastal species.

These pioneering studies have laid the foundation for future studies including the impact of climate changes, and coastal and catchment development pressures.

Peter C. Rothlisberg and Michele A. Burford

Peter got his Ph.D. in Biological Oceanography in 1975 from Oregon State University, studying the larval ecology of the oceanic pink shrimp *Pandalus jordani* under the supervision of Charlie Miller. Upon graduation he immediately took up a position as a Research Scientist with CSIRO Australia's then Division of Fisheries and Oceanography leading the Larval Ecology Group within the new Tropical Prawn Research Project at the Cleveland Marine Laboratories, near Brisbane. In addition to field sampling in the remote Gulf of Carpentaria, the group was undertaking larval culture work for taxonomic and physiological studies and they needed to boost their phytoplankton skills. Peter put an advertisement in the Brisbane paper and got an application from Michele Burford working in Switzerland! Michele joined CSIRO in 1989, using her skills in algal culturing before moving onto field studies assessing key food sources for penaeid prawn larvae and more generally the controls on primary productivity in this unique environment. Michele had previously graduated with honors from Murdoch University, Perth, and then spent 4 years working in algal biotechnology in Western Australia and Switzerland.

Michele completed a Ph.D. on nutrient cycling in prawn aquaculture ponds at the University of Queensland in Marine Botany under the supervision of Bill Dennison and Pat Glibert (University of Maryland) in 2001. In 2002, Michele and Peter were married. In 2003, Michele moved to Griffith University and broadened her interests into freshwater algal ecology. She is now the Executive Deputy Director of the Australian Rivers Institute at Griffith University. Peter is now a Post-retirement Fellow at CSIRO, still writing and mentoring, and the one who gets to go to conferences as the Accompanying Person.

References

Australian Bureau of Statistics. http://www.abs.gov.au/websitedbs/D3310114.nsf/home/home?opendocument#from-banner=GT

Brown CW, Yoder JA (1994) Coccolithophorid blooms in the global ocean. J Geophys Res 99:7467–7482

Bureau of Meteorology (1988) Climate Atlas of Australia. Department of Administrative Services, Canberra, 67p

Burford MA, Rothlisberg PC (1999) Factors limiting phytoplankton growth in a tropical continental shelf ecosystem. Estuar Coast Shelf Sci 48:541–549

Burford MA, Rothlisberg PC, Wang YG (1995) Spatial and temporal distribution of tropical phytoplankton species and biomass in the Gulf of Carpentaria, Australia. Mar Ecol Prog Ser 118:255–266

Burford MA, Rothlisberg PC, Revill AT (2009) Sources of nutrients driving production in the Gulf of Carpentaria, Australia: a shallow tropical shelf system. Mar Freshw Res 60:1044–1053

Burford MA, Webster IT, Revill AT, Kenyon RA, Whittle M, Curwen G (2012) Controls on phytoplankton productivity in a wet-dry tropical estuary. Estuar Coast Shelf Sci 113:141–151
Chivas AR, Garcia A, van der Kaars S et al (2001) Sea-level and environmental changes since the last interglacial in the Gulf of Carpentaria, Australia: an overview. Quat Int 83–85:19–46
Church JA, Forbes AMG (1983) Circulation in the Gulf of Carpentaria. I. Direct observations of currents in the south-east corner of the Gulf of Carpentaria. Aust J Mar Freshw Res 34:1–10
Condie SA (2011) Modelling seasonal circulation, upwelling and tidal mixing in the Arafura and Timor Seas. Cont Shelf Res 31:1427–1436
Condie SA, Dunn JR (2006) Seasonal characteristics of the surface mixed layer in the Australasian region: implications for primary productivity regimes and biogeography. Mar Freshw Res 57:569–590
Condie SA, Loneragan NR, Die DJ (1999) Modelling of recruitment of tiger prawns (*Penaeus esculentus* and *P. semisulcatus*) to nursery grounds in the Gulf of Carpentaria, northern Australia: implications for assessing the stock-recruitment relationship. Mar Ecol Prog Ser 178:55–68
Dall W, Hill BJ, Rothlisberg PC, Staples DJ (1990) Biology of Penaeidae. Advances in Marine Biology, vol 27. Academic, London, 489p
Davies CH, Armstrong AJ, Baird M, Coman F, Gaughan D, Greenwood J, Gusmão F, Henschke N, Koslow AJ, Leterme SC, McKinnon AD, Pausina S, Palomino JU, Roennfeldt R-L, Rothlisberg PC, Slotwinski A, Strzelecki J, Suthers IM, Swadling KM, Talbot S, Tonks M, Tranter DH, Young JW, Richardson AJ (2014) Over 75 years of zooplankton data from Australia. Ecology 95:3229, http://dx.doi.org/10.1890/14-0697.1
DEWHA (2008) North marine bioregional plan: bioregional profile. Department of the Environment, Water, Heritage and the Arts, Parkes, ACT, 228p., http://www.environment.gov.au/coasts/mbp/publications/north/bioregional-profile.html
Forbes AMG (1984) The contribution of local processes to seasonal hydrology of the Gulf of Carpentaria. Océanogr Trop 19:193–201
Forbes AMG, Church JC (1983) Circulation of the Gulf of Carpentaria. II. Residual currents and mean seal level. Aust J Mar Freshw Res 34:11–22
Furnas MJ, Mitchell AW (1996) Pelagic primary production in the Coral and southern Solomon seas. Mar Freshw Res 47:695–706
Hallegraeff GM, Jeffrey SW (1984) Tropical phytoplankton species and pigments of continental shelf waters of North and North-West Australia. Mar Ecol Prog Ser 20:59–74
Halpern BS, Walbridge S, Selkoe KA, Kappel CV, Micheli F et al (2008) A global map of human impact on marine ecosystems. Science 319:948–952
Motoda S, Kawamura T, Taniguchi A (1978) Differences in productivities between the Great Australian Bight and the Gulf of Carpentaria, Australia, in summer. Mar Biol 46:93–99
Othman BHR (1986) Copepods of the Gulf of Carpentaria. Ph.D. thesis, University of Queensland, St. Lucia, QLD, 392p., https://espace.library.uq.edu.au/view/UQ:366034
Othman BHR, Greenwood JG, Rothlisberg PC (1990) The copepod fauna of the Gulf of Carpentaria, and its Indo-west Pacific affinities. Proceedings of the Snellius II symposium, Jakarta, Indonesia. Netherl J Sea Res 24:561–572
Poiner IR, Staples DJ, Kenyon R (1987) Seagrass communities of the Gulf of Carpentaria, Australia. Aust J Mar Freshw Res 38(1):121–131
Preston NP, Burford MA, Coman FE, Rothlisberg PC (1992) The natural diet and its effect on larval survival of *Penaeus merguiensis* (Decapoda: Penaeidae). Mar Biol 113:181–191
Preston NP, Burford MA, Stenzel D (1998) Effects of *Trichodesmium* spp. blooms on penaeid prawn larvae. Mar Biol 131:671–679
Rothlisberg PC (1982) Vertical migration and its effect on dispersal of penaeid shrimp larvae in the Gulf of Carpentaria, Australia. Fish Bull (US) 80:541–554
Rothlisberg PC, Jackson CJ (1982) Temporal and spatial variation of plankton abundance in the Gulf of Carpentaria, Australia 1975–1977. J Plankton Res 4:19–40
Rothlisberg PC, Jackson CJ (1987) Larval ecology of penaeids in the Gulf of Carpentaria, Australia. II. Hydrographic environment of *Penaeus merguiensis*, *P. esculentus*, *P. semisulcatus* and *P. latisulcatus* zoeae. Aust J Mar Freshw Res 38:19–28

Rothlisberg PC, Church JA, Forbes AMG (1983a) Modelling the advection of vertically migrating shrimp larvae. J Mar Res 41:511–538

Rothlisberg PC, Jackson CJ, Pendrey RC (1983b) Specific identification and assessment of distribution and abundance of early penaeid shrimp larvae. Biol Bull 164:279–298

Rothlisberg PC, Jackson CJ, Pendrey RC (1987) Larval ecology of penaeids in the Gulf of Carpentaria, Australia. I. Assessing the reproductive activity of five species of *Penaeus* from the distribution and abundance of the zoeal stages. Aust J Mar Freshw Res 38:1–17

Rothlisberg PC, Forbes AMG, White N (1989) Hydrographic Atlas of the Gulf of Carpentaria. CSIRO Mar Lab Rep 209, 51p., https://publications.csiro.au/rpr/download?pid=procite:5e08b24e-bd7d-42ff-ab85-f4f9af2acca1&dsid=DS1

Rothlisberg PC, Pollard PC, Nichols PD, Moriarty DJW, Forbes AMG, Jackson CJ, Vaudrey D (1994) Phytoplankton community structure and productivity in relation to the hydrological regime of the Gulf of Carpentaria in summer. Aust J Mar Freshw Res 45:265–282

Rothlisberg PC, Craig PD, Andrewartha JR (1996) Modelling penaeid prawn larval advection in Albatross Bay, Australia: Defining the effective spawning population. Aust J Mar Freshw Res 47:157–168

Rothlisberg P, Condie S, Hayes D, Griffiths B, Edgar S, Dunn J (2005) Collation and analysis of oceanographic datasets for national marine bioregionalisation: the Northern Large Marine Domain. A report to the Australian Government, National Oceans Office. CSIRO Marine Research, 58p., https://publications.csiro.au/rpr/pub?pid=legacy:1047

Vance DJ, Pendrey RC (2001) Vertical migration behaviour of postlarval penaeid prawns: a laboratory study on the effect of tide, water depth and day/night. In: The definition of effective spawning stocks of commercial tiger prawns in the Northern Prawn Fishery and the king prawn in the Eastern King Prawn Fishery – behaviour of postlarval prawns. Fisheries Research and Development Corporation (FRDC) final report (Project 97/108), pp 28–52. http://frdc.com.au/research/Documents/Final_reports/1997-108-DLD.doc

Vance DJ, Staples DJ, Kerr JD (1985) Factors affecting year-to-year variation in the catch of banana prawns (*Penaeus merguiensis*) in the Gulf of Carpentaria, Australia. J Conseil 42:83–97

Wolanski E (1993) Water circulation in the Gulf of Carpentaria. J Mar Syst 4:401–420

Wolanski E, Ridd P (1990) Mixing and trapping in Australian tropical coastal waters. In: Cheng RT (ed) Coastal and Estuarine Studies, vol 38, Residual currents and long-term transport. Springer, New York, NY, pp 165–183

Discerning the Causes of Toxic Cyanobacteria (*Lyngbya majuscula*) Blooms in Moreton Bay, Australia

Judith M. O'Neil and William C. Dennison

Introduction

The waters of Moreton Bay, Queensland, Australia, have experienced large scale blooms of the toxic cyanobacterium *Lyngbya majuscula*, beginning in the late 1990s. *Lyngbya majuscula* had been observed incidentally in the region (Cribb and Cribb 1954), but large scale blooms had not been described previously. The magnitude of the bloom was brought to the attention of scientists in 1997 by the Catchment Coordinator for the Pumicestone Passage in the northern portion of Moreton Bay (Oliver and Dennison 2013). Fishermen from the Deception Bay region of Moreton Bay were experiencing severe skin rashes, which they were attributing to contact with what the fishermen were colloquially calling "fireweed." Samples were subsequently identified as *Lyngbya majuscula* based on cellular characteristics such as wide flattened cells and a prominent sheath (Fig. 1) (Dennison et al. 1999). *Lyngbya majuscula*'s most distinguishing feature, however, was the ability to produce toxins that cause contact dermatitis, which corresponded to fishermen's reports of skin rashes (Osborne et al. 2001, 2008). This prompted the initiation of what would develop into a decade long research project investigating the factors responsible for stimulating the proliferation of these blooms in Moreton Bay.

The initial bloom site was in Deception Bay, an embayment on the northwestern side of Moreton Bay, but the bloom spread to other locations around the Bay, in particular, the Eastern Banks (Fig. 2). Over the years, however, Deception Bay has remained the "hot spot" for these blooms, with very consistent, nearly annual blooms. In an attempt to decipher the specific factors that promoted these consistent blooms, we investigated various chemical, biological, and physical factors that affected *L. majuscula* growth parameters (Watkinson et al. 2005), including nutrient uptake,

J.M. O'Neil (✉) • W.C. Dennison
Horn Point Laboratory, University of Maryland Center for Environmental Science, Cambridge, MD 21613, USA
e-mail: joneil@umces.edu

P.M. Glibert, T.M. Kana (eds.), *Aquatic Microbial Ecology and Biogeochemistry: A Dual Perspective*, DOI 10.1007/978-3-319-30259-1_21

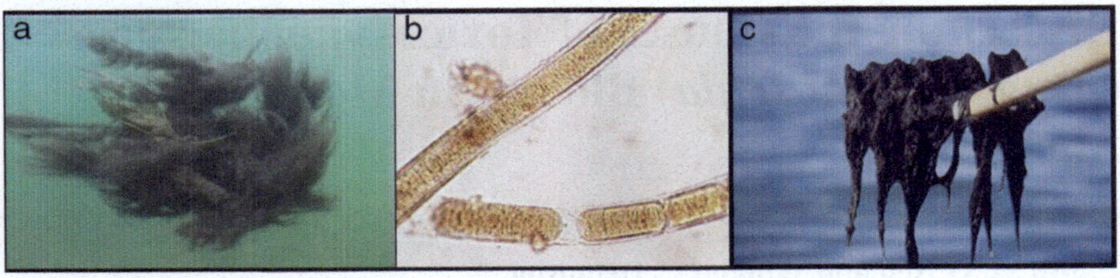

Fig. 1 *Lyngbya majuscula* (**a**) floating up from benthos taking seagrass with it to the surface; (**b**) microscopic view; and (**c**) raked up by fisherman off of seagrasses in Deception Bay, Moreton, Bay Australia

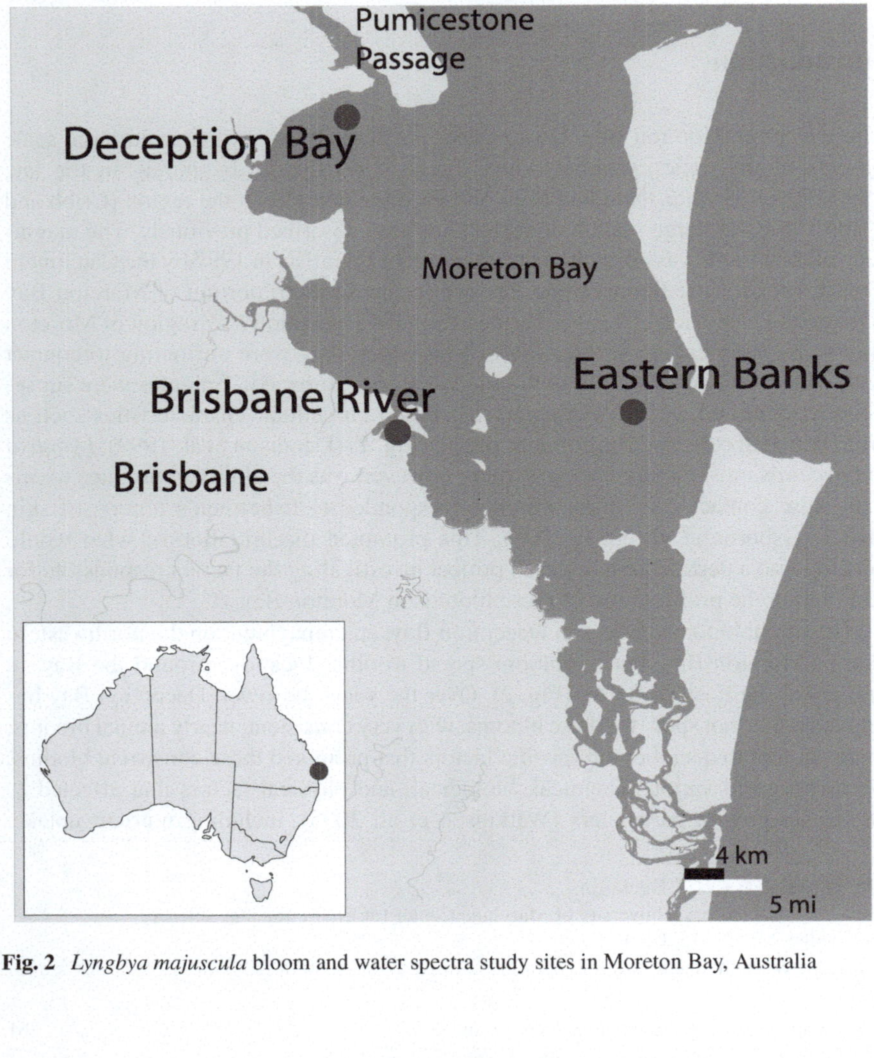

Fig. 2 *Lyngbya majuscula* bloom and water spectra study sites in Moreton Bay, Australia

nitrogen fixation (Dennison et al. 1999), grazing (Capper et al. 2005, 2006a, b), viral lysis (Hewson et al. 2001), toxicity (Osborne et al. 2001) as well as primary productivity and pigment content (Albert 2001; Albert et al. 2005).

As part of our initial studies, we investigated light intensity or "light quantity" effects on *Lyngbya* physiology (Watkinson et al. 2005); however, one aspect we did not focus on originally was how the "light quality" in the humic-rich waters of Deception Bay might enhance *L. majuscula* proliferation. The impetus for investigating light quality at this location was the observation of repeated episodes of dark, stained water that flowed out of the water body directly north of Deception Bay, Pumicestone Passage (Fig. 2), into the region of repeated *L. majuscula* blooms. In addition, the benthic community of marine plants in this region differed from the rest of Moreton Bay at similar water depths (Dennison and Abal 1999). The importance of ultraviolet light penetration into the shallow waters of Deception Bay had been observed as well, in particular, ultraviolet light effects on *Lyngbya* viruses (Hewson et al. 2001) and iron chemistry (Rose and Waite 2003, 2005, 2006; Rose et al. 2005; Salmon et al. 2006).

By integrating light quality results with previous research on nutrient interactions, we have constructed a conceptual model to explain the causal factors that operate in the Deception Bay region of Moreton Bay to promote recurrent prolific blooms of *Lyngbya majuscula*. This conceptual model is presented in a diagrammatic form in this chapter, along with key supporting data. The conceptual diagram is used as a basis to answer the question "What is it about Deception Bay specifically that makes *Lyngbya majuscula* grow so well in this location?". But in a larger sense, this conceptual model provides insight into the initiation of marine cyanobacterial blooms elsewhere in the world.

Nutrient Interactions

Deception Bay has been the "epicenter" of *Lyngbya* blooms in Moreton Bay since the 1990s, as the first site the blooms were noticed by the fisherman, and as the first place the toxic skin rashes were reported (Watkinson et al. 2005; Albert et al. 2005). The most notable difference in Deception Bay, compared to the rest of Moreton Bay, is the drainage of the Pumicestone Passage, and its often "colored" or organic-rich water flows. Streams that flow into the *Lyngbya*-affected portions of Deception Bay were found to have very high levels of iron leaching out from the acid sulfate soils of the region; with some of the streams visibly orange from precipitation of iron oxides. It was hypothesized that sources of both iron and organic compounds in these waters would act as natural chelators, binding the iron, thus keeping the iron from precipitating out of the water column and becoming biologically unavailable (Rose and Waite 2003; 2006; Salmon et al. 2006). Iron leaches out of the acid sulfate soils in the region, especially after rain. The colored water flow contained large amounts of dissolved organic carbon (DOC) that bound iron and transported it to the bloom region (Albert et al. 2005; Rose and Waite 2003; Salmon et al. 2006). Iron is held in solution

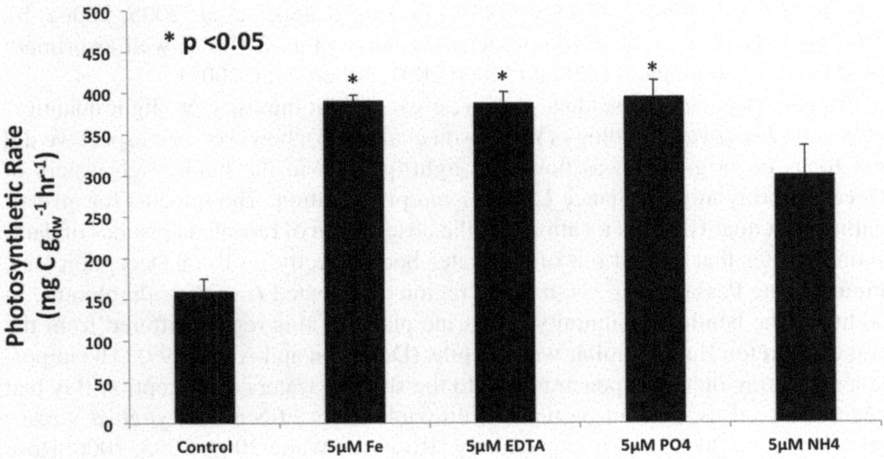

Fig. 3 *Lyngbya majuscula* response to nutrient treatments in bioassay experiments. Photosynthetic rate assessed with ^{14}C uptake in control, 5 μM dissolved Fe^{+2}, 5 μM EDTA (an Fe chelator), 5 μM dissolved PO_4^{-3}, and 5 μM dissolved NH_4^+ treatments. The *asterisk* represents treatments with statistically significant differences. (Adapted from Watkinson et al. 2005)

bound to organic ligands and is transported from the land to the bloom site at Deception Bay. It had been determined that superoxide produced by the *Lyngbya* could reduce the organic bound Fe to Fe^{+2} (Rose et al. 2005; Rose and Waite 2005). The dissolved Fe^{+2} could then be readily absorbed by *Lyngbya* and used for key physiological processes such as photosynthesis and nitrogen fixation, giving the *Lyngbya* an ecological advantage over other primary producers in this location.

In previous studies, the highest rates of nitrogen fixation in cyanobacteria were found with added iron and a chelator (EDTA) (Paerl et al. 1987; Gross and Martin 1996). We conducted a series of bioassay experiments from the bloom area in Deception Bay (Fig. 3), using methods described in Watkinson et al. (2005). In these bioassay experiments, *Lyngbya* photosynthetic rates were significantly stimulated with the additions of iron, a chelator (EDTA), and phosphorus. Since *Lyngbya* can fix atmospheric nitrogen dissolved in the water via nitrogen fixation (Dennison et al. 1999), the availability of dissolved iron and phosphorus are key determinants of bloom initiation, as demonstrated in these bioassay experiments.

Light Interactions

The spectral light quality of Moreton Bay waters was highly variable. Eastern Banks water was largely oceanic water flushed by the 1.5 m tides and was consistently clear. Brisbane River water was highly turbid with high concentrations of suspended sediments. Deception Bay water was periodically colored due to land runoff, occasionally turbid following wind events. Based on samples of these water masses, we observed distinctive wavelength signatures: (1) clear water, (2) turbid water, and (3) colored water, corresponding to the spectra observed at the Eastern Banks, Brisbane River plume, and Deception Bay, respectively. Samples of *Lyngbya majuscula* were collected from the Eastern Banks of Moreton Bay and acclimated in flow-through aquaria exposed to natural light under two layers of neutral density screening, in a rooftop experimental setup at the University of Queensland. A scanning spectrophotometer was used to determine the light wavelength signature of water from Deception Bay, Eastern Banks, and the mouth of the Brisbane River. Spectral filters in combination with neutral density filters were selected that simulated the light wavelength signature of the different water masses. These filters allowed the same quantity of light (25–30 % surface light using a PAR sensor), so that the experiment was testing the differences in light quality rather than quantity. The *L. majuscula* samples were incubated in the four different spectral treatments for 12 days at ~25 °C. At the end of the incubation period, ^{14}C productivity and photosynthetic pigment content, chlorophyll *a* and phycoerythrin, were measured on subsamples for each treatment. Productivity was assessed using ^{14}C-biocarbonate uptake (Parsons et al. 1984), using protocols described previously (Albert et al. 2005; Watkinson et al. 2005). Pigments were extracted from tissue ground with a mortar and pestle using a phosphate buffer (phycoerythrin) (Albert 2001) or acetone (chlorophyll *a*), centrifuged and pigments determined spectrophotometrically using formulas of Rowan (1989) for phycoerythrin and Parsons et al. (1984) for chlorophyll *a*, using protocols described previously (Albert et al. 2005; Watkinson et al. 2005).

Photosynthetic rates of *Lyngbya* were significantly stimulated in the "colored water" treatment simulating Deception Bay light spectra compared with the other treatments (Fig. 4). Clear water and turbid water treatments, simulating the Eastern Banks and Brisbane River, respectively, had the lowest photosynthetic rates, while the treatment with a 50 % mixture of simulated colored (Deception Bay) and turbid (Brisbane River) waters had an intermediate photosynthetic rate. The photosynthetic pigment content of *Lyngbya* also responded to the light treatments (Table 1); however, the differences were not statistically significant. Both phycoerythrin and chlorophyll *a* concentrations were elevated in the colored water treatment, compared with clear, turbid, and turbid + colored water treatments.

Conceptual Model

A conceptual diagram was created to synthesize the results of the nutrient and light interaction studies (Fig. 5). This diagram is based on our understanding that there is not a single factor that causes the stimulation of *Lyngbya majuscula* blooms, but rather a suite of factors that act synergistically to cause *Lyngbya* to grow better in certain sites, under certain circumstances. Even with the same nutrients available in the water column, the light spectra available in the colored water treatment that most closely matched the light spectra of Deception Bay had the highest rates of

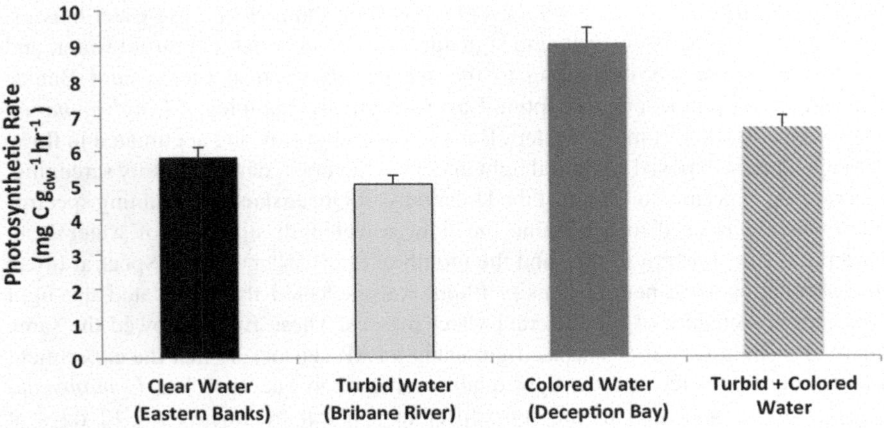

Fig. 4 *Lyngbya majuscula* response to light spectra treatments in bioassay experiments. Photosynthetic rate assessed with ^{14}C uptake in *Lyngbya majuscula* exposed to Eastern Banks (clear), Deception Bay (colored), Brisbane River (turbid), and combination of colored + turbid spectra

Table 1 *Lyngbya majuscula* response to light spectra treatments in bioassay experiments

Sites	Phycoerythin (mg g_{dw}^{-1})	Chlorophyll *a* (mg g_{dw}^{-1})
Clear water (Eastern Banks)	1.17 ± 0.09	1.07 ± 0.11
Turbid water (Brisbane River)	0.94 ± 0.09	0.95 ± 0.06
Colored water (Deception Bay)	1.35 ± 0.15	1.18 ± 0.10
Turbid + colored water	1.12 ± 0.08	0.87 ± 0.06

Phycoerythrin and chlorophyll *a* content with standard errors (S.E.) assessed under different simulated light spectra treatments

primary productivity. Previous studies in Moreton Bay indicated that the concentration of phycoerytherin, a phycobilin accessory photosynthetic pigment found in many cyanobacteria including *Lyngbya majuscula*, is very responsive to nutrient additions (Albert 2001; Ahern et al. 2003). The ability to absorb light at slightly different wavelengths than other phytoplankton and macroalgae that do not contain accessory phycobilins pigments may give these cyanobacteria a photosynthetic advantage over eukaryotic competitors. The results showing that phycoerythrin content was highest in colored water indicate that these conditions allow *L. majuscula* to better absorb the green wavelengths of light, enhancing the photosynthetic capacity of *Lyngbya*. Enhanced photosynthetic capacity was indeed observed in the ^{14}C bicarbonate uptake experiment. Seagrasses, as well as green and brown algae, primarily utilize chlorophyll to absorb the bulk of their photosynthetic light, and chlorophyll effectively absorbs light in the red and blue spectra. In contrast, red algae and some cyanobacteria utilize phycoerythrin and phycocyanin to absorb light in the green portion of the spectrum, which gives them a potential competitive advantage in a light climate in which green light predominates. Water has an

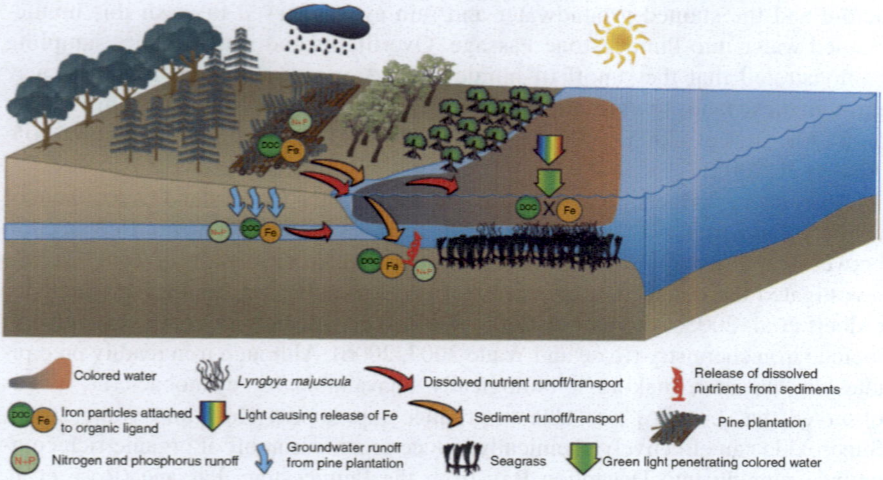

Fig. 5 Conceptual diagram summarizing the interacting light and nutrient factors that act synergistically to make Deception Bay, embayment of Moreton Bay, a "hot spot" for blooms of *Lyngbya majuscula*

intrinsic absorption in the red portion of the spectrum, and turbidity accentuates this absorption in the red spectrum due to the increased optical path lengths created by light bouncing off suspended particles. Thus, underwater plants largely rely on blue light penetration, the most prevalent portion of the light spectrum available. In contrast, humic-stained colored water absorbs blue light preferentially and photosynthetic organisms which can take advantage of the green light that remains available after the intrinsic absorption of red light and humic absorption of blue light are, consequently at a competitive advantage. DOC compounds associated with the humic substances provide both (a) a light spectrum conducive to cyanobacteria pigments and (b) iron complexation substrates that serve to provide bioavailable dissolved iron to *Lyngbya*. These factors work synergistically to create an ideal growing environment for cyanobacteria like *Lyngbya*.

In the catchment adjacent to the bloom site, there had been several land use changes in the previous 2 years leading up to the bloom, including canal estate developments and agricultural expansion. One significant land use change in the year leading up to the bloom was a large proportion of plantation pine forests that had been growing in the region upstream of Deception Bay on Bribie Island adjacent to Pumicestone Passage for 20–30 years had been clearcut. This deforestation was the result of a bushfire that had moved through the region and a change in forest ownership. The result was a massive deforestation, instead of the usual practice of smaller patchworks of forest removal over time. The clearcut plantation pine forest was observed to have standing water in spite of a lack of rainfall, due to the sudden lack of evapotranspiration from the trees, causing the groundwater levels to rise. The pine needles, bark and roots left behind were slightly acidic and the stained groundwater and rain events served to wash this humic-stained water into Pumicestone Passage. Overflights and water quality sampling demonstrated that the runoff of humic stained or colored water found its way through these lands to northern Moreton Bay, extending out to the *Lyngbya* bloom area (Ahern et al. 2003, 2006; Pointon et al. 2003; O'Neil and Dennison 2005; Abal et al. 2005).

An industry partnership with forestry scientists at the Queensland Department of Primary Industries was initiated in 1999 to investigate the potential connection between rapid clearcut of pine trees and downstream expansion of *Lyngbya.* We investigated the connection between soil types, runoff, and *Lyngbya* productivity (Albert et al. 2005; Ahern et al. 2006, 2007). The research focus was extended to include iron chemistry (Rose and Waite 2003, 2006). Although iron readily precipitates in seawater, making it biologically unavailable for cyanobacteria, it was observed that *Lyngbya* was able to produce superoxide (Rose and Waite 2005). Superoxide can effectively chemically reduce organic iron off of organic-rich compounds running into Deception Bay from the Pumicestone Passage (Rose et al. 2005). Several key findings on iron chemistry and the role of superoxide production in scavenging iron bound to organic ligands into a reduced form that could be readily taken up by *L. majuscula* from the water column were elucidated (Rose et al. 2005; Salmon et al. 2006).

One of the reasons *Lyngbya* is able to outcompete other photosynthetic organisms is likely its very flexible metabolism in terms of nutrient acquisition (as is the case with many harmful algal blooms species), making it highly adaptable to fluctuating nutrient sources and pulses in shallow benthic environments where it proliferates. *Lyngbya* needs access to light to support photosynthesis, and the light quality experiments confirm that these cyanobacteria are adapted to the variable water quality, particularly the humic-stained water common in Deception Bay. *Lyngbya* is both a facultative nutrient "sponge" as well as a facultative light quality generalist. *Lyngbya* can take up both inorganic (NO_3^-, NH_4^+) forms of nitrogen as well as organic forms (e.g., urea) (O'Neil et al. 2012; O'Neil and Dennison 2005), but when external sources are limiting, can also fix atmospheric nitrogen. *Lyngbya* has the ability to scavenge iron bound to organic ligands by producing superoxide (Rose and Waite 2005; Rose et al. 2005; Salmon et al. 2006), and to access phosphorus fluxing from the sediments (Watkinson et al 2005). *Lyngbya* also can grow in a variety of light environments and even thrives in colored water conditions.

Broader Significance

Various adaptive strategies allow *Lyngbya* to be very successful in dynamic, shallow coastal environments. Following the blooms in Moreton Bay, we identified several bloom locations along the Queensland, Australia coast (e.g., Fraser Island, Hervey Bay, Shoalwater Bay, and Hardy Reef). Similar to Moreton Bay, *Lyngbya* blooms throughout Queensland had significant ecological impacts (e.g., Shoalwater

Bay; Arthur et al. 2006). Within a few years, blooms were reported from around Australia, including sites in Western Australia and the Northern Territory. In the United States, blooms of *Lyngbya* were observed in Florida. A large study funded by NOAA ECOHAB investigated *Lyngbya* blooms from both the Gulf of Mexico and Atlantic coasts of Florida. Nutrient sources, including bioavailable iron, were identified, as well as humic-stained water draining mangrove areas near Sanibel Island, Florida. The ecophysiology, toxicology, and nutrient interactions were described for the Florida *Lyngbya* blooms (Paerl et al. 2008; Arthur et al. 2009; Capper et al. 2013). These subsequent observations and studies used the Moreton Bay results as a reference point.

In addition to the expanding global research on *Lyngbya*, the initial *Lyngbya* research stimulated a series of ongoing research efforts in Moreton Bay. Much effort went into trying to forecast and predict blooms using various modeling approaches, including Bayesian and neural network decision-making techniques (e.g., Hamilton et al. 2007; Johnson et al. 2010). *Lyngbya* bloom mitigation research included fish (e.g., Capper et al. 2006a) and other grazer interactions (e.g., Pittman and Pittman 2005; Capper et al. 2005, 2006b; Arthur et al. 2007, 2008). Management agencies including the Healthy Waterways program in Southeast Queensland continue to track *Lyngbya* blooms as part of the Ecosystem Health Monitoring Program (e.g., Roelfsema et al. 2006; Ahern et al. 2007). The Queensland Department of Environment and Heritage Protection continues to investigate *Lyngbya* and a *Lyngbya* management strategy was developed. The genus *Lyngbya* is currently undergoing a taxonomic reclassification including new genera designations in different locations including *Moorea* in the South Pacific (Engene et al. 2012) and *Okeania* in Florida (Engene et al. 2013). Preliminary reclassification of Moreton Bay species to *Limnoraphis robusta* has been proposed recently as well (Willis et al. 2015). Further reclassifications are likely as new molecular techniques are brought to bear on these closely related cyanobacteria species, since there are few morphological features that can be readily distinguished microscopically (Engene et al. 2011; O'Neil et al. 2012).

Cyanobacteria have been implicated as beneficiaries of eutrophication and climate change (O'Neil et al. 2012). Developing an understanding of the causal factors of cyanobacteria blooms will become increasingly important. The interactions between various forms of nutrients and light availability (both light quantity and quality) will need to be elucidated to understand the triggers for large scale blooms and the potential toxicity of these blooms. The multidisciplinary approach that we took to discern the causes of Moreton Bay *Lyngbya* blooms will likely be the approach for future investigations of cyanobacteria blooms.

Acknowledgments The coauthors of this chapter are indebted to many colleagues who contributed to the research effort to study light and nutrient interactions of *Lyngbya* in Moreton Bay, particularly B. Longstaff, S. Albert, K. Ahern, C. Ahern, A. Capper, I. Hewson, C. O'Sullivan, A. Rose, T. Salmon, J. Savige, G. Savige, D. Waite, and A. Watkinson. Science communication support was provided by D. Taillie (UMCES). We acknowledge the Healthy Waterways program in Southeast Queensland, especially E. Abal, who was helpful in mobilizing resources, and grateful for the leadership of J. Soorley, A. Sutherland, P. Greenfield, B. Ball, and T. Lloyd. UMCES Contribution # 5191.

Judith M. O'Neil and William C. Dennison

We met in Doug Capone's laboratory at Stony Brook University when Bill was the "Coastal Marine Scholar" postdoc and Judy was working on her master's degree. We collaborated on a seagrass/nutrient interaction project that took us to San Salvador, Bahamas, and the sea, sun, and stars worked their magic and we were married in 1989. Following our stint in New York, we moved to Maryland where Bill started a research faculty position focused on seagrasses and Judy began her Ph.D. on marine cyanobacteria, both at the University of Maryland Center for Environmental Science. We had an opportunity to move to Australia, with positions at the University of Queensland, where we conducted research in Moreton Bay, adjacent to Brisbane, and at Heron Island, Great Barrier Reef. It was during our studies of Moreton Bay that we encountered the cyanobacteria blooms of *Lyngbya*, growing on seagrass—a project that would occupy both of our efforts for many years. After a 10-year stint in Australia which included producing two daughters, we moved back to the University of Maryland Center for Environmental Science, where Bill is Vice President for Science Application and Judy is an Associate Research Professor. We continue to collaborate on research and education programs.

References

Abal EA, Bunn SE, Dennison WC (2005) Healthy waterways healthy catchments: making the connection in South East Queensland, Australia. Moreton Bay Waterways and Catchments Partnership, Brisbane, p 240

Ahern KS, O'Neil JM, Udy JW, Pointon SM, Albert S, Moody P (2003) Assessing the contribution of runoff and groundwater nutrients to blooms of the cyanobacteria *Lyngbya majuscula* in Moreton Bay, SE Queensland: a bioassay technique. Proceedings of the River symposium Conference, Brisbane, http://www.rivversymposium.com/index

Ahern KS, Udy JW, Pointon SM (2006) Investigating the potential for groundwater from different vegetation, soil and landuses to stimulate blooms of the cyanobacterium, Lyngbya majuscula, in coastal waters. Mar Freshw Res 57:177–186

Ahern KS, Ahern CR, Savige GM, Udy JW (2007) Mapping the distribution, biomass and tissue nutrient levels of a marine benthic cyanobacteria bloom (*Lyngbya majuscula*). Mar Freshw Res 58:883–904

Albert SM (2001) The effect of soil extracts on the physiology of *Lyngbya majuscula* (Cyanophyta). Honours thesis, University Queensland

Albert SM, O'Neil JM, Udy JW, Ahern KS, O'Sullivan CM, Dennison WC (2005) Blooms of the cyanobacterium *Lyngbya majuscula* in coastal Queensland, Australia: disparate sites, common factors. Mar Pollut Bull 51(1-4):428–437

Arthur KA, Limpus CJ, Roelfsema CM, Udy JW, Shaw GR (2006) A bloom of *Lyngbya majuscula* in Shoalwater Bay, Queensland, Australia: an important feeding ground for the green turtle (*Chelonia mydas*). Harmful Algae 5:251–256

Arthur KE, O'Neil JM, Limpus CJ, Abernathy K, Marshall GJ (2007) Using animal-borne imaging to assess green turtle (*Chelonia mydas*) foraging ecology in Moreton Bay, Australia. Mar Technol Soc J 41:5–9

Arthur KA, Limpus CJ, Balazs G, Capper A, Udy J, Shaw GR, Keuper-Bennett U, Bennett P (2008) The exposure of green turtles (*Chelonia mydas*) to tumour promoting compounds by the cyanobacterium *Lyngbya majuscula* and their potential role in the aetiology of fibropapillomatosis. Harmful Algae 7:114–125

Arthur KE, Paul VJ, Paerl HW, O'Neil JM, Joyner J, Meickle T (2009) Effects of nutrient enrichment of the cyanobacterium *Lyngbya* sp. on growth, secondary metabolite concentration and feeding by the specialist grazer *Stylocheilus striatus*. Mar Ecol Prog Ser 394:101–110

Capper A, Tibbetts IR, O'Neil JM, Shaw GR (2005) The fate of *Lyngbya majuscula* toxins in three potential consumers. J Chem Ecol 31:1595–1606

Capper A, Tibbetts IR, O'Neil JM, Shaw GR (2006a) Feeding preference and deterrence in rabbitfish *Siganus fuscescens* for the cyanobacterium *Lyngbya majuscula* in Moreton Bay, south-east Queensland, Australia. J Fish Biol 68:1589–1609

Capper A, Tibbetts IR, O'Neil JM, Shaw GR (2006b) Dietary selectivity for the toxic cyanobacterium *Lyngbya majuscula* and resultant growth rates in two species of opistobranch mollusc. J Exp Mar Biol Ecol 331:133–144

Capper A, Flewelling LJ, Arthur KA (2013) Dietary exposure to harmful algal bloom (HAB) toxins in the endangered manatee (*Trichechus manatus latirostris*) and green sea turtle (*Chelonia mydas*) in Florida, USA. Harmful Algae 28:1–9

Cribb AB, Cribb JW (1954) Records of marine algae from south-eastern Queensland. University of Queensland Press, Brisbane, p 1954

Dennison WC, Abal EG (1999) Moreton Bay Study: a scientific basis for the Healthy Waterways Campaign. South East Queensland Regional Water Quality Management Strategy, Brisbane, p 246

Dennison WC, O'Neil JM, Duffy EJ, Oliver PE, Shaw GR (1999) Blooms of the cyanobacterium *Lyngbya majuscula* in coastal waters of Queensland, Australia. Bull Inst Oceanogr Monaco 19:501–506

Engene N, Choi H, Esquenazi E, Rottacker EC, Ellisman MH, Dorrest-ein PC, Gerwick WH (2011) Underestimated biodiversity as a major explanation for the perceived rich secondary metabolite capacity of the cyanobacterial genus *Lyngbya*. Environ Microbiol 13:1601–1610, http://dx.doi.org/10.1111/j.1462-2920.2011.02472.x

Engene N, Rottacker EC, Kaštovský J, Byrum T, Choi H, Ellisman MH, Komárek J, Gerwick WH (2012) *Moorea producens* gen. nov., sp. nov. and *Moorea bouillonii* comb. nov., tropical marine cyanobacteria rich in bio-active secondary metabolites. Int J Syst Evol Microbiol 62:1171–1178, http://dx.doi.org/10.1099/ijs.0.033761-0

Engene N, Paul VJ, Byrum T, Gerwick WH, Thor A, Ellisman MH (2013) Five chemically rich species of tropical marine cyanobacteria of the genus *Okeania* gen. nov. (Oscillatoriales, Cyanoprokaryota). J Phycol 49:1095–1106. doi:10.1111/jpy.12115

Gross ED, Martin DF (1996) Iron dependence of *Lyngbya majuscula*. J Aquat Plant Manag 34:17–20

Hamilton GS, Fielding F, Chiffings AW, Hart BT, Johnstone RW, Mendersen K (2007) Investigating the use of a Bayesian network to model the risk of *Lyngbya majuscula* bloom initiation in Deception Bay, Queensland, Australia. Hum Ecol Risk Assess 13:1271–1287

Hewson I, O'Neil JM, Dennison WC (2001) Virus-like particles associated with *Lyngbya majuscula* (Cyanophyta; Oscillatoriacea) bloom decline in Moreton Bay, Australia. Aquat Microb Ecol 25(3):207–213

Johnson S, Fielding F, Hamilton G, Mengerson K (2010) An integrated Bayesian network approach to *Lyngbya majuscula* bloom initiation. Mar Environ Res 69:27–37

O'Neil JM, Dennison WC (2005) *Lyngbya*. In: Abal EG, Bunn SE, Dennison WC (eds) Healthy waterways healthy catchments: making the connection in South East Queensland, Australia. Moreton Bay Waterways and Catchments Partnership, Brisbane, pp 119–148 (p 240)

O'Neil JM, Davis TM, Burford MA, Gobler CJ (2012) The rise of harmful cyanobacteria blooms: the potential roles of eutrophication and climate change. Harmful Algae 14:313–334

Oliver PE, Dennison WC (2013) Dancing with Dugongs: having fun and developing a practical philosophy for environmental teaching and research. IAN Press, Cambridge, MD, p 250

Osborne NJT, Webb PM, Shaw GR (2001) The toxins of *Lyngbya majuscula* and their human and ecological health effects. Environ Int 27:381–392

Osborne NJT, Seawright A, Shaw GR (2008) Dermal toxicology of *Lyngbya majuscula*, from Moreton Bay, Queensland, Australia. Harmful Algae 7:584–589

Paerl HW, Crocker KM, Prufert LE (1987) Limitation of N_2 fixation in coastal marine waters: Relative importance of molybdenum, iron, phosphorus and organic matter availability. Limnol Oceanogr 32:525–536

Paerl HW, Joyner JJ, Joyner AR, Arthur K, Paul V, O'Neil JM, Heil CA (2008) Co-occurrence of dinoflagellate and cyanobacterial harmful algal blooms in southwest Florida coastal waters: dual nutrient (N and P) input controls. Mar Ecol Prog Ser 371:143–153

Parsons TR, Maita Y, Lalli CM (1984) A manual of chemical and biological methods for seawater analysis. Pergamon, Toronto, ON, p 173

Pittman SJ, Pittman KM (2005) Short-term consequences of a benthic cyanobacterial bloom (*Lyngbya majuscula* Gomont) for fish and penaeid prawns in Moreton Bay (Queensland, Australia). Estuar Coast Shelf Sci 63:619–632

Pointon SM, Moody PW, Preda M, Cox M, Hey KM, Ahern CR, Powell B (2003) Investigation of terrestrial sources of nutrients supporting *Lyngbya majuscula* growth in Moreton Bay http://www.ehmp.org/HealthyWaterways/Resources/ReportsandPublications/PublicationsandReportsArchive.aspx

Roelfsema CM, Phinn SR, Dennison WC, Dekker AG, Brando VE (2006) Monitoring toxic cyanobacteria *Lyngbya majuscula* (Gomont) in Moreton Bay, Australia by integrating satellite image data and field mapping. Harmful Algae 5(1):45–56

Rose AL, Waite TD (2003) Kinetics of iron complexation by dissolved natural organic matter in coastal waters. Mar Chem 84:85–103

Rose AL, Waite TD (2005) Reduction of organically complexed ferric iron by superoxide in a simulated natural water. Environ Sci Tech 39:2645–2650

Rose AL, Waite TD (2006) Kinetic model for Fe (II) oxidation in seawater in the absence and presence of natural organic matter waters. Environ Sci Tech 39(8):2645–2650

Rose AL, Salmon TP, Lukondeh T, Neilan BA, Waite TD (2005) Use of superoxide as an electron shuttle for iron acquisition by the marine cyanobacterium *Lyngbya majuscula*. Environ Sci Tech 39:3708–3715

Rowan KS (1989) Photosynthetic pigments of algae. Cambridge University Press, London, p 427

Salmon TP, Rose AL, Neilan BA, Waite TD (2006) The FeL model of iron acquisition: non-dissociative reduction of ferric complexes in the marine environment. Limnol Oceanogr 51:1744–1754

Watkinson AJ, O'Neil JM, Dennison WC (2005) Ecophysiology of the marine cyanobacterium, *Lyngbya majuscula* (Oscillatoriaceae) in Moreton Bay, Australia. Harmful Algae 4(4):697–715

Willis A, Parks M, Burford MA (2015) Draft genome assembly of filamentous brackish cyanobacterium *Limnoraphis robusta* strain CS-951. Genome Announc 3(5):e00846-15. doi:10.1128/genomeA.00846-15

Copepod, Ctenophore, and Schyphomedusae Control in Structuring the Chesapeake Bay Summer Mesohaline Planktonic Food Web

Kevin G. Sellner and Stella G. Sellner

Introduction

As a global phenomenon, eutrophication of many estuaries is well documented (Bricker et al. 2008; Howarth et al. 2011), with resultant elevated phytoplankton production, increases in potentially toxic harmful taxa, hypoxia and anoxia in bottom waters, loss of submersed vegetation, and increases in the importance of ctenophores and schyphomedusa as top consumers. These alterations are often exacerbated by low wild stocks of shellfish, crabs, and fish (e.g., Purcell 2012) due to overharvest or disease. Food webs in these transformed ecosystems shift from more linear energy transfer (phytoplankton to copepods to fish) to highly complex and less efficient microbial-dominated energy processing (phytoplankton to bacteria and microzooplankton [heterotrophic flagellates, ciliates, rotifers] to copepods to jellies). Through high-DOM production, the latter can further enhance bacterial respiration as well as select for a different bacterial flora (Condon et al. 2011), further reducing carbon transfer to commercially important stocks.

Unique characteristics of the eutrophic Chesapeake Bay indicate even more extremes in the plankton food web. First, elevated winter-spring river discharge and nutrient loads yield high diatom bloom biomass that remains largely ungrazed due to seasonal shifts in copepods from winter *Eurytemora carolleeae* to summer *Acartia tonsa* (Verity 1987); the spring bloom occurs between these two copepod maxima. The diatoms sink to the bottom and support summer hypoxia and anoxia in the mesohaline reach of the estuary. Additionally, the high spring river discharge

K.G. Sellner (✉)
Center for Coastal and Watershed Studies, Hood College,
401 Rosemont Avenue, Frederick, MD 21701, USA
e-mail: kgsellner@gmail.com

S.G. Sellner
Patuxent Environmental and Aquatic Research Laboratory, Morgan State University,
10545 Mackall Road, St. Leonard, MD 20685, USA

P.M. Glibert, T.M. Kana (eds.), *Aquatic Microbial Ecology and Biogeochemistry: A Dual Perspective*, DOI 10.1007/978-3-319-30259-1_22

stratifies the bay supporting summer development of small phytoplankton cells (i.e., picoplankton, <10 μm diatoms) as well as migratory taxa (dinoflagellates, cyanobacteria) that can ascend for light during the day and descend at night for nutrients at or below the pycnocline (Seliger et al. 1975; Malone et al. 1991). Finally, low subpycnocline dissolved oxygen (DO) and high temperatures in surface waters restrict vertical distributions of many shellfish and macrobenthos (e.g., Sturdivant et al. 2014) and fish (Niklitschek and Secor 2005; Brandt et al. 2009), but allow small individuals of the low DO resistant ctenophore *Mnemiopsis leidyi* and the jellyfish *Chrysaora quinquecirrha* to effectively capture prey and grow (Kolesar et al. 2010; Purcell 2012).

The role of predation, specifically grazer abundances that must be present to exert top-down control in the summer, mesohaline Chesapeake Bay, is the primary focus of this manuscript. Questions to be addressed include: (1) Is there a minimum number of adult copepods that appear to control densities of rotifers and ciliates, their preferred microzooplankton prey (Stoecker and Egloff 1987; Gifford and Dagg 1988; White and Roman 1992; Löder et al. 2011; Dhanker et al. 2012; York et al. 2013)? (2) Similarly, as ctenophores are efficient grazers of copepods in the bay and elsewhere (e.g., Breitburg et al. 1997; Purcell and Decker 2005; Condon and Steinberg 2008; Kimmel et al. 2012, 2015; McNamara et al. 2013; Sullivan 2014), can a ctenophore density be identified that appears to limit numbers of *Acartia* present? And (3) do large schyphomedusae (Purcell and Decker 2005; Condon and Steinberg 2008) have a threshold abundance that limits numbers of the ctenophores in the region? The threshold densities (predation control) of each of these three predators on their respective primary prey should be possible to quantify as summer phytoplankton biomass and productivity are high and very likely non-limiting, removing bottom up control of the consumers; average summer biomass and gross primary productivity are 14.5 μg chl L^{-1} and 2.76 gC m^{-2} day^{-1}, respectively, for this bay reach (Harding et al. 2015).

Methods

Water quality, phytoplankton, and zooplankton monitoring was conducted in the Maryland, USA portion of the Chesapeake Bay and its tributaries from 1984 to 2002. Summer mesohaline data (July–September) from 1987 to 2002 were examined in three central bay stations (CB3.3C, 4.3C, 5.2) and the mouth of the Potomac River estuary (LE2.2, Fig. 1) through the Chesapeake Bay Program's Chesapeake Information Management Systems (CIMS, http://www.chesapeakebay.net/data/downloads/baywide_cbp_plankton_database); archived primary productivity at the Morgan State University Patuxent Environmental and Aquatic Research Laboratory were also accessed. Analyses focused on abundances and biomass of key plankton groups and volumes for mixed ctenophores (the Chesapeake Bay taxa are primarily *Mnemiopsis leidyi* but possibly rarer *Beroë* sp.) and jellyfish (primarily *Chrysaora quinquecirrha* and rarer *Aurelia* sp. and *Cyanea* sp.; note that 0.5 m, 202 μm Bongo

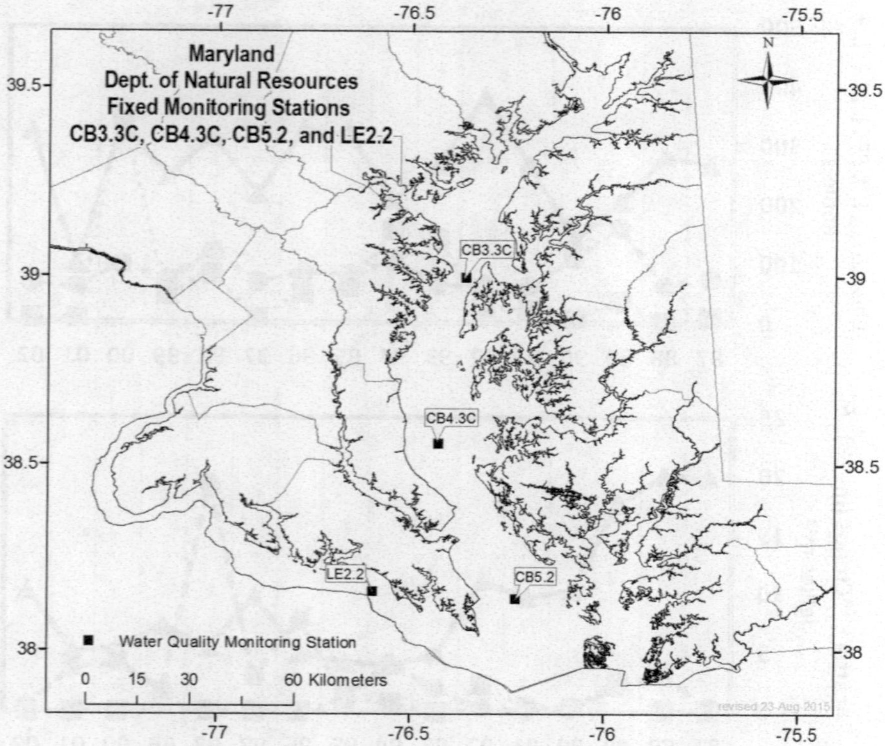

Fig. 1 Sampling stations in mesohaline Chesapeake Bay and the mouth of the Potomac River estuary

nets were employed for copepods and gelatinous zooplankton, likely inappropriate for the latter). Bi-weekly phytoplankton and monthly zooplankton data were compiled and in most cases, regression analyses were conducted to identify possible linkages between plankton groups as well as visual inspection of summer data distributions to identify feeding thresholds for a plankton group on its prey. Mean summer data for each year were computed from the data sets for depiction in Fig. 2.

Results and Discussion

General Patterns in the Summer Mesohaline Chesapeake Bay

Summer mesohaline plankton distributions are highly variable in Chesapeake Bay, from phytoplankton (as chlorophyll *a* and phytoplankton C) through the consumers (rotifers, *Acartia*, ctenophores, and medusae). Year-to-year variability is apparent

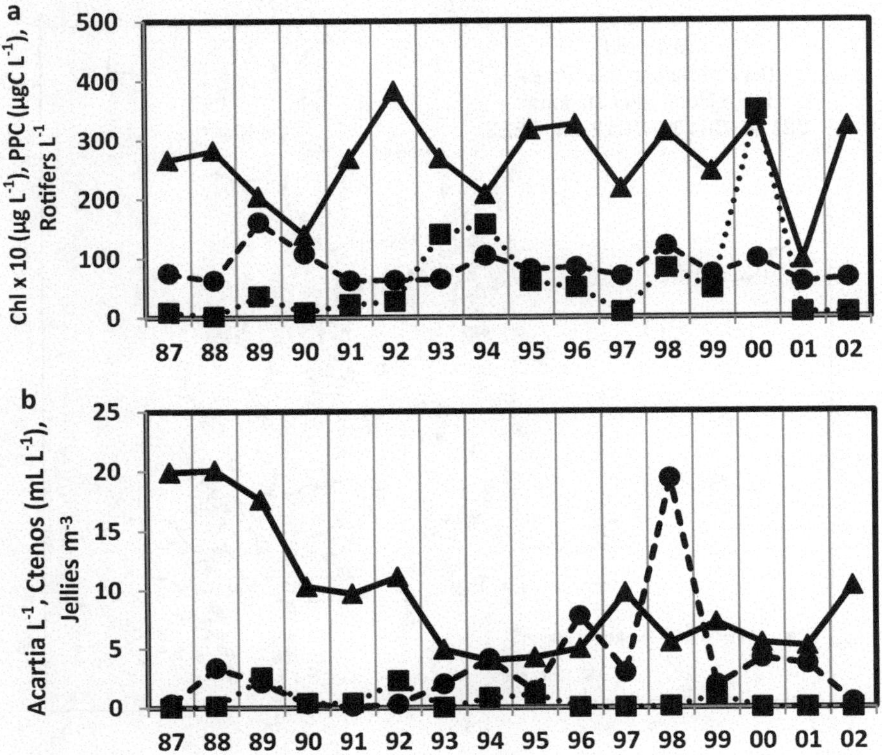

Fig. 2 Mean summer (July, August, September) plankton abundances in the mid-mesohaline Chesapeake Bay station for the period 1987–2002. The *top panel* (**a**) shows surface chl *a* (µg $L^{-1} \times 10$, *filled triangle*), phytoplankton carbon (µgC L^{-1}, *filled circle*), and rotifers (# L^{-1}, *filled square*) while the *bottom panel* (**b**) depicts *Acartia* (L^{-1}, *filled triangle*), ctenophores (mL m^{-3}, *filled circle*), and jellyfish (# m^{-3}, *filled square*). Error bars per year are excluded for clarity. The *vertical bars* allow comparisons within the same time periods for the two data sets

within individual stations (Fig. 2) as well as across the mesohaline area of the bay and its nearby mesohaline station at the mouth of the Potomac River (Table 1).

Top-Down Controls and Thresholds

The zooplankton of the Chesapeake Bay and its tributaries have been characterized for more than a century (Cowles 1930) with a diverse community of microzooplankton (ciliates, rotifers, nauplii), mesozooplankton (copepods, meroplanktonic larvae), and macrozooplankton as jellies (ctenophores and schyphomedusae). The summer dominant copepod, *Acartia tonsa*, has been long believed as the primary herbivore in the region but work in the past 2–3 decades suggests it can prefer

Table 1 A comparison of plankton abundances and DIN and DIP concentrations (mn ± se) across the four mesohaline stations for the period 1985–2002

Stn	Surface Chl *a* (μg L^{-1})	AP Dino C (μg L^{-1})	AP Rotifers (# L^{-1})	*Acartia tonsa* (# m^{-3})	AP DIN (μM)	AP DIP (μM)	*N*/*P*
CB3.3C	23.1 ± 2.6^{a}	283.0 ± 75.6^{a,b}	216.2 ± 45.5^{a}	10,277 ± 1,789^{a}	8.46 ± 1.86^{a}	0.29 ± 0.03^{ab}	30
LE2.2	14.3 ± 2.4^{b}	433.3 ± 108.8^{a}	240.1 ± 40.4^{a}	6050 ± 873^{b}	5.66 ± 0.96^{ab}	0.32 ± 0.03^{a}	18
CB4.3C	8.2 ± 0.7^{c}	126.5 ± 18.4^{c}	58.6 ± 21.3^{b}	10,839 ± 1,672^{a}	4.82 ± 1.03^{b}	0.21 ± 0.03^{bc}	22
CB5.2	8.9 ± 0.9^{c}	127.6 ± 33.4^{bc}	54.0 ± 17.9^{b}	11,519 ± 2,242^{a}	2.52 ± 0.47^{c}	0.15 ± 0.02^{c}	17

AP corresponds to surface mixed layer. Note the high chlorophyll *a*, dinoflagellate, and rotifer contributions in the upper mesohaline Bay station (CB3.3C) and the mouth of the Potomac River (LE2.2) relative to the two other open bay mesohaline stations (CB4.3C and CB5.2)

^{a,b,c}Significantly different, *F* statistic, $p < 0.05$

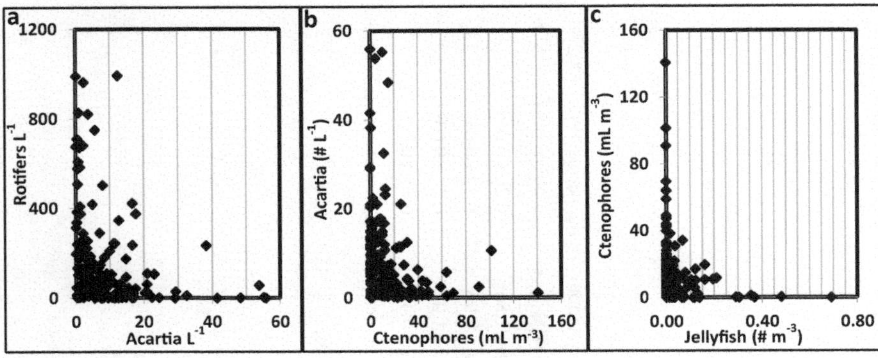

Fig. 3 Summer top-down control thresholds for (**a**) *Acartia* on rotifers, (**b**) ctenophores on *Acartia*, and (**c**) jellyfish on ctenophores for the four mesohaline stations for the period 1987–2002

animal prey, such as rotifers and ciliates (Stoecker and Egloff 1987; Gifford and Dagg 1988; Dolan 1991; White and Roman 1992). The strong copepod control of rotifers by summer *A. tonsa* abundances is seen in Fig. 3a: Adult *A. tonsa* abundances exceeding 5–8 L^{-1} maintain rotifers at low levels. A similar pattern was noted for ciliates for the period 1998–2001 (data not shown). Summer copepod control of rotifers, however, can be modified by unusual events, such as elevated spring-early summer flows in the northern bay where high rotifer abundances of the upper Bay's lower salinity areas were transported into the mid-region of the mesohaline reach (1993, 1994, 2000), temporarily decoupling *A. tonsa* predation on the advected populations.

Ctenophore abundance appears to govern densities of summer *Acartia* populations (Fig. 3b). At volumes exceeding 10–20 mL m^{-3}, *A. tonsa* populations were maintained at low levels, similar to *M. leidyi* densities thought to limit the copepod in the York River (Fig. 4b in Condon and Steinberg 2008). The removal of *Acartia* would presumably free its largely microzooplankton prey from top-down control and instead, limits on ciliates and rotifers would be governed by bottom up drivers such as food availability, local physics, or other parameters (see below).

The largest member of the zooplankton community, jellyfish, also exerted substantial control on its primary prey, ctenophores (Fig. 3c), previously noted by many researchers in the Chesapeake and elsewhere (see references in Breitburg and Burrell 2014). At jellyfish densities >0.05 m^{-3}, ctenophores were only minimal contributors to the summer plankton food web. Breitburg and Burrell (2014) suggest that *Chrysaora* abundances of 0.18 ind m^{-3} are coincident with minimal or no *Mnemiopsis* in the Patuxent River estuary and Mackall Cove (a small embayment off the river) in 2004.

Bottom Up Controls in Summer Mesohaline Stations

The results above indicate that in the absence of top-down control due to low predator densities, accumulations of these prey might be controlled instead by bottom up reservoirs (nutrient concentrations or ratios, food levels, DO). To examine possible linkages of rotifers when *A. tonsa* predation was minimal (<5–8 copepods L^{-1}), summer mesohaline rotifer abundances under low *A. tonsa* densities (<5 L^{-1}) for the four stations were compared to a spectrum of possible food sources for the 1987–2002 period: surface chlorophyll *a* (chl a) concentrations, biomass in cryptophytes, diatoms, and dinoflagellates, and primary productivity (as an indicator of new cell production). Temperature was also examined as a possible limit for rotifers. In no case was there any relationship between these indices of food availability and rotifer numbers (r^2=<0.001–0.18, p=0.13–0.98) and on only two occasions did temperature partially explain rotifer abundances, 25 % (p=0.035) and 30 % (p=0.004), respectively. Hence, it appears that food in the planktonic autotrophs was always sufficiently abundant (nonlimiting) to support this summer mesohaline metazoan group.

Another note of interest is the significantly lower *A. tonsa* abundances in the lower Potomac River estuary (6,050 m^{-3}) than in the open bay (10,277–11,519 m^{-3}, Table 1), not explained by any differences in predatory ctenophores between the two regions (p=0.77). The lower *A. tonsa* densities, however, did overlap with highest dinoflagellate biomass levels (Table 1). Might there be food quality or toxic properties of the elevated dinoflagellates that could depress fecundity in the copepod? Several investigators have identified low *A. tonsa* fecundity and egg hatching success to missing fatty acids within dinoflagellates (Jónasdóttir 1994; Kleppel et al. 1998) while others have suggested feeding deterrents in some dinoflagellates observed in the Chesapeake Bay and its tributaries (*Karlodinium veneficum*) may limit ingestion (Waggett et al. 2008), potentially reducing egg production. Dam and Colin (2005) reported lower *A. tonsa* egg production when fed another dinoflagellate common to the mesohaline Chesapeake Bay and lower Potomac River estuary, *Prorocentrum minimum*. However, with only one station showing low *A. tonsa* abundances and high summer dinoflagellate biomass, a more comprehensive examination of dinoflagellate species and copepod egg production and hatching success from *field exposures* is needed to further assess this possible linkage.

Implications

Identification of the important role of jellyfish and ctenophores in the planktonic food web of the Chesapeake Bay was not surprising considering previous work by several research groups (see references by Purcell, Breitburg, and Condon), but setting thresholds for the abundances of these predators that would limit prey is a unique aspect of a large data set covering the period 1987–2002. The large data set provided sufficient breadth to identify the strong top-down control of rotifers (and ciliates) by *A. tonsa* in the field, strongly suggested by laboratory (Richman et al. 1977; Stoecker and Egloff 1987) and egg production (White and Roman 1992) studies of previous years. However, most disappointing was an inability to identify any bottom up control of rotifer populations from the mesohaline bay when copepod predation was minimal, implying ample food availability. Is summer phytoplankton, then, never limiting to these small metazoans? This might be feasible if the micrograzers are ingesting phytoplankton, bacteria, and non-chlorophyll bearing cells such as heterotrophic nanoflagellates, several not measured in the monitoring program collections and hence not in the database. Alternatively, as noted for *Brachionus calyciflorus* (Golz et al. 2015), ingestion may be independent of total available *C*, *N*, and *P* but instead dependent on maintenance of a constant intracellular *N*/*P* ratio across all available food qualities. The absence of any detectable bottom up control of this simple metazoan in the field remains a paradigm.

In a similar manner to the identified copepod abundance (>5–8 L^{-1}) that would regulate rotifer densities, the ctenophore and jellyfish volumes and numbers that control copepods and ctenophores, respectively, provide general levels (thresholds) for these predators that could shift planktonic food web structure and resulting trophic cascades. For example, schyphomedusae at densities >0.05–0.18 m^{-3} (density noted herein and Breitburg and Burrell 2014) appear to reduce *Mnemiopsis* abundances thereby freeing copepods from ctenophore predation with subsequent increases in copepods, optimal prey for piscivore larvae and forage fish. Copepod prey (ciliates, rotifers) would decline, leading to increases in heterotrophic flagellates and reduced bacteria. Alternatively, for ctenophores at volumes exceeding 10–20 mL m^{-3}, summer copepod abundances should be reduced, allowing accumulations of rotifers and ciliates, lower numbers of small autotrophic and heterotrophic flagellates, and elevated bacteria densities. Unfortunately the routine monitoring program (1984–2002) did not include all of these plankton groups preventing any ability to detect changes across the food web attributable to these predatory meso- and macrozooplankton. Further, with the regional focus on reducing nutrient loads to the bay and its tributaries moving to fiscal commitments to implementation of best management practices on the land rather than detecting food web responses, routine zooplankton monitoring was terminated after 2002, eliminating estimation of top-down impacts at the base of the food web. Documenting energy flow to or away from valued fishery resources is equally important to the on-going water quality improvements and hence, multiple trophic level monitoring should again become a priority for the region.

Acknowledgments The authors are indebted to long-term cooperation from members of the MD Biomonitoring Team, particularly, R. Lacouture and M. Olson, as well as the plankton database manager J. Johnson and her colleague C. Buchanan. M. Trice and B. Cole provided additional data and map, respectively, deeply appreciated. Monitoring funding and ship time were provided by U.S. EPA and MD's Department of the Environment.

Kevin G. Sellner and Stella G. Sellner

The Sellner's met in the early 1980s after Kevin hired Stella to expand microzooplankton taxonomy for the ongoing plankton monitoring program he had been invited to implement for Maryland's portion of Chesapeake Bay in 1984. Following receipt of his M.S. from the Baruch Institute of the University of South Carolina and his Ph.D. from Dalhousie, Kevin had joined the Academy of Natural Sciences in Philadelphia in 1978 and subsequently transferred to its Benedict Estuarine Research Laboratory on the Patuxent River, Chesapeake Bay, in 1981. Stella joined the plankton group from the USDA's Carbohydrate Analysis Laboratory and the two began a three decade collaboration, including marriage in 1995, on phytoplankton-zooplankton interactions for the Chesapeake Bay and Gulf of Finland, focusing on describing the seasonal succession of taxa in tidal fresh to polyhaline reaches of the former as well as phytoplankton-zooplankton relationships in algal blooms of the two systems. Kevin shifted from the lab to administering the initial US multi-agency Federal program ECOHAB (Ecology and Oceanography of Harmful Algal Blooms) for several years, overseeing peer and panel review for approximately $50 M of projects on *Alexandrium, Karenia, Pfiesteria, Karlodinium, and Pseudo-nitzschia*. Stella remained at the laboratory while it transitioned to the Morgan State University PEARL (Patuxent Environmental and Aquatic Research Laboratory) where she became the Education Coordinator and directed the NOAA-funded PLANS (PLankton And Nutrient Studies) science program in Calvert County, Maryland high schools. Kevin returned to the Chesapeake to direct the 6-institution Chesapeake Research Consortium in 2001, committed to fostering multi-institution, multi-discipline research and transferring basic research results to the management community of the basin; he was also able to pursue research on mitigating cyanobacteria blooms in the region which he continues today. Kevin and Stella recently retired to pursue other "fun" science opportunities as they enjoy Frederick, MD, USA culture, nature, and history.

References

Brandt SB, Gerken M, Hartman KJ, Demers E (2009) Effects of hypoxia on food consumption and growth of juvenile stribed bass (*Morone saxatilis*). J Exp Mar Biol Ecol 381(S1):S143–S149

Breitburg D, Burrell R (2014) Predator-mediated landscape structure: seasonal patterns of spatial expansion and prey control by *Chrysaora quinquecirrha* and *Mnemiopsis leidyi*. Mar Ecol Prog Ser 510:183–200

Breitburg DL, Loher T, Pacey CA, Gerstein A (1997) Varying effects of low dissolved oxygen on trophic interactions in an estuarine food web. Ecol Monogr 67:489–507

Bricker SB, Longstaff B, Dennison W, Jones A, Boicourt K, Wicks C, Woerner J (2008) Effects of nutrient enrichment in the nation's estuaries: a decade of change. Harmful Algae 8:21–32

Condon RH, Steinberg DK (2008) Development, biological regulation, and fate of ctenophore blooms in the York River estuary, Chesapeake Bay. Mar Ecol Prog Ser 369:153–168

Condon RH, Steinberg DK, del Giorgio PA, Bouvier TC, Bronk DA, Graham WM, Ducklow HW (2011) Jellyfish blooms result in a major microbial respiratory sink of carbon in marine systems. Proc Natl Acad Sci U S A 108:10225–10230

Cowles RP (1930) A biological study of the offshore waters of Chesapeake Bay. Bull US Bur Fish 46:277–381

Dam HG, Colin SP (2005) *Prorocentrum minimum* (clone Exuv) is nutritionally insufficient, but not toxic to the copepod *Acartia tonsa*. Harmful Algae 4:575–584

Dhanker R, Kumar R, Hwang J-S (2012) Predation by *Pseudodiaptomus annandalei* (Copepoda: Calanoida) on rotifer prey: Size selection, egg predation, and effect of algal diet. J Exp Mar Biol Ecol 414–415:44–53

Dolan JR (1991) Microphagous ciliates in mesohaline Chesapeake Bay waters: estimates of growth rates and consumption by copepods. Mar Biol 111:303–309

Gifford DJ, Dagg MJ (1988) Feeding of the estuarine copepod *Acartia tonsa* Dana: Carnivory vs. herbivory in natural microplankton assemblages. Bull Mar Sci 43:458–468

Golz A-L, Burian A, Winder M (2015) Stoichiometric regulation in micro- and mesozooplankton. J Plankton Res 37:293–305

Harding LW Jr, Adolf JE, Mallonee ME, Miller WD, Gallegos CL, Perry ES, Johnson JM, Sellner KG, Paerl HW (2015) Climate effects on phytoplankton floral composition in Chesapeake Bay. Estuar Coast Shelf Sci 162:53–68

Howarth R, Chan F, Conley DJ, Garnier J, Doney SC, Marino R, Billen G (2011) Coupled biogeochemical cycles: Eutrophication and hypoxia in temperate estuaries and coastal marine ecosystems. Front Ecol Environ 9:18–26

Jónasdóttir SH (1994) Effects of food quality on the reproductive success of *Acartia tonsa* and *Acartia hudsonica*: Laboratory observations. Mar Biol 121:67–81

Kimmel DG, Boynton WR, Roman MR (2012) Long-term decline in the calanoid copepod *Acartia tonsa* in central Chesapeake Bay, USA: an indirect effect of eutrophication? Estuar Coast Shelf Sci 101:76–85

Kimmel DG, McGlaughon BD, Leonard J, Paerl HW, Taylor JC, Cira EK, Wetz MS (2015) Mesozooplankton abundance in relation to the chlorophyll maximum in the Neuse River estuary, North Carolina, USA: implications for trophic dynamics. Estuar Coast Shelf Sci 157:59–68

Kleppel GS, Burkart CA, Houchin L (1998) Nutrition and regulation of egg production in the calanoid copepod *Acartia tonsa*. Limnol Oceanogr 43:1000–1007

Kolesar SE, Breitburg DL, Purcell JE, Decker MB (2010) Effects of hypoxia on *Mnemiopsis leidyi*, ichthyoplankton, and copepods: clearance rates and vertical habitat overlap. Mar Ecol Prog Ser 411:173–188

Löder MGJ, Meunier C, Wiltshire KH, Boersma M, Aberle N (2011) The role of ciliates, heterotrophic dinoflagellates, and copepods in structuring spring plankton communities at Helgoland Roads, North Sea. Mar Biol 158:1551–1559. doi:10.1007/s00227-011-1670-2

Malone TC, Ducklow HW, Peele ER, Pike SE (1991) Picoplankton carbon flux in Chesapeake Bay. Mar Ecol Prog Ser 78:11–22

McNamara ME, Lonsdale DJ, Cerrato RM (2013) Top-down control of mesozooplankton by adult *Mnemiopsis leidyi* influences microplankton abundance and composition enhancing prey conditions for larval ctenophores. Estuar Coast Shelf Sci 133:2–10

Niklitschek EJ, Secor DH (2005) Modeling spatial and temporal variation of suitable nursery habitats for Atlantic sturgeon in the Chesapeake Bay. Estuar Coast Shelf Sci 64:135–148

Purcell JE (2012) Jellyfish and ctenophore blooms coincide with human proliferations and environmental perturbations. Ann Rev Mar Sci 4:209–235

Purcell JE, Decker MB (2005) Effects of climate on relative predation by scyphomedusae and ctenophores on copepods in Chesapeake Bay during 1987–2000. Limnol Oceanogr 50:376–387

Richman S, Heinle DR, Huff R (1977) Grazing by adult estuarine calanoid copepods of the Chesapeake Bay. Mar Biol 42:69–84

Seliger HH, Loftus ME, Subba Rao DV (1975) Dinoflagellate accumulations in Chesapeake Bay. In: LoCicero VR (ed) Proceedings of the 1st international conference on toxic dinoflagellate blooms. MA Sci Technol Fdn, Wakefield, MA

Stoecker DK, Egloff DA (1987) Predation by *Acartia tonsa* Dana on planktonic ciliates and rotifers. J Exp Mar Biol Ecol 110:53–68

Sturdivant SK, Diaz RJ, Llansó R, Dauer DM (2014) Relationship between hypoxia and macrobenthic production in Chesapeake Bay. Estuar Coast Shelf Sci 37:1219–1232

Sullivan LJ (2014) Apparent diel feeding by the ctenophore *Mnemiopsis leidyi* A. Agassiz 1865 (Ctenophora, Lobata). J. Plankton Res 37:1–12

Verity P (1987) Factors driving changes in the pelagic structure of estuaries, with implications for the Chesapeake Bay. In: USEPA (ed) Perspectives on the Chesapeake Bay: recent advances in estuarine sciences. CRC Publications, Edgewater, MD, pp 35–56

Waggett RJ, Tester PA, Place AR (2008) Anti-grazing properties of the toxic dinoflagellate *Karlodinium veneficum* during predator–prey interactions with the copepod *Acartia tonsa*. Mar Ecol Prog Ser 366:31–42

White JR, Roman MR (1992) Egg production by the calanoid copepod *Acartia tonsa* in the mesohaline Chesapeake Bay: the importance of food resources and temperature. Mar Ecol Prog Ser 86:239–249

York JK, McManus GB, Kimmerer WJ, Slaughter AM, Ignoffo TR (2013) Trophic links in the plankton in the low salinity zone of a large temperate estuary: top-down effects of introduced copepods. Estuar Coast Shelf Sci. doi:10.1007/s12237-013-9698-9

Microbiogeochemical Ecophysiology of Freshwater Hydrothermal Vents in Mary Bay Canyon, Yellowstone Lake, Yellowstone National Park WY

Carmen Aguilar and Russell Cuhel

Introduction

Geothermal activity has been a subject of practical use as well as awe since the beginning of human cognizance, no doubt. Hydrothermal systems such as hot springs have been respected practically, as a source of social and medical benefit in thermal baths worldwide. Academically, the latter twentieth century included discovery of TAQ polymerase, a cornerstone of modern molecular biology. *Thermus aquaticus* was isolated from Yellowstone hot springs (Brock and Freeze 1969); an excellent history and significance description was provided by Brock (1997).

Underwater hydrothermal activity is a more recent subject—in marine studies the real significance of hydrothermal venting did not come into focus until discovery of the Eastern Pacific Ocean hydrothermal ecosystems by geologists John Corliss and Robert Ballard in 1977 (Corliss et al. 1979). Within a few years of study their existence solved one of the great marine geochemistry enigmas: the Magnesium Problem (Drever 1974; summary of resolution in Rona et al. 1983). Everyone now knows about the fantastic chemosynthesis-based communities of the deep sea. Competing with the famous brine pool origin of organic life hypothesis (Miller 1953), deep-sea hydrothermal vents have offered an alternative geochemical milieu for origins of life (Martin et al. 2008).

Freshwater hydrothermal vents and gas-emitting fumaroles have been slower to gain notoriety. In contrast to marine counterparts, freshwater habitats have not been stable long enough for evolution to create endemic populations of animals thriving on chemical energy. Geochemical consequences of freshwater hydrothermal systems also do not have the impact of global marine water-basalt interactions.

C. Aguilar (✉) • R. Cuhel
School of Freshwater Sciences, University of Wisconsin-Milwaukee,
Milwaukee, WI 53204, USA
e-mail: aguilar@uwm.edu; rcuhel@uwm.edu

P.M. Glibert, T.M. Kana (eds.), *Aquatic Microbial Ecology and Biogeochemistry: A Dual Perspective*, DOI 10.1007/978-3-319-30259-1_23

Microorganisms that live in high temperature and chemically inhospitable ecosystems are tightly coupled to their environment. A detailed understanding of the geochemistry of hydrothermal environments can be an important aspect in deciphering critical components for microbial life under these dire circumstances. Some of the organisms are chemolithoautotrophs, using CO_2 as a carbon source and inorganic energy sources. Reduced sulfur compounds, Fe(II) and Mn(II), hydrogen, methane, and ammonia are among the available electron donors, while nitrate, oxygen, iron/manganese oxides, and sulfate are common electron acceptors. Favorable conditions for microbial growth in hydrothermal systems depend on the confluence of these electron donors and acceptors, within a temperature range that is tolerable to microorganisms in this extreme environment. The Yellowstone Plateau Volcanic Field exhibits vigorous hydrothermal activity as well as an array of surface thermal springs with diverse biological and chemical characteristics (reviewed by Hurwitz and Lowenstern 2014). Several different microbial assemblages can develop in thermal effluents, each mediating a distinct suite of redox reactions. Our work is designed to elucidate some of these concepts in freshwater hydrothermal vent environments. The venue is additionally a stellar opportunity to recruit excellent students into biogeochemistry career paths because of the inherently interdisciplinary nature of hydrothermal vent research.

In this chapter, we describe an interdisciplinary and intergenerational team effort to elucidate microbiogeochemical ecophysiological processes occurring in deep, hot hydrothermal vents in Yellowstone Lake, Yellowstone National Park, WY. Specifically, we construct a geophysical, hydrographic, geochemical, and microbiological profile from the surface to the bottom of a sublacustrine caldera canyon, into the effluvia of vents, and below them into geothermally influenced sediments. Surveys of lakewater, ventwater, and sediment pore water gradients established zones of direct and subsurface inputs of geochemically altered fluids. Ventwater intrusion into the surrounding sediments was evident in Mary Bay pore water profiles. In hot sediments, chloride approached reservoir concentrations (15 mM) and the silicate concentration at depth was greater than diagenesis-generated inventories. Principal investigators (PIs), US Park Service personnel, technicians, graduate and undergraduate students, and contractors worked together since 1984 (c.f., Klump et al. 1995) to employ technology, harvest physical and numerical products, and create a model for understanding aspects of the Yellowstone Geothermal Ecosystem under the surface of the lake.

Methods

Many of the practical aspects of research infrastructure and specific methodology have been detailed in a symposium volume on Yellowstone (Aguilar et al. 2002; Anderson and Harmon 2002: George Wright volume) and in several peer-reviewed journal articles describing epiphytic bacterial communities (Konkol et al. 2010); vent chemistry, chemosynthesis, and bacterial communities (Yang et al. 2011).

Water column samples were collected with 5-L Niskin bottles after hydrographic characterization with a Hydrolab Water Quality sonde. Hydrothermal vent samples were collected with 2 L polycarbonate syringes mounted on a Mark IV Minirover submersible (Buchholz et al. 1995) piloted by David Lovalvo (Eastern Oceanics Inc.) using a sipper tube on the articulated arm and a *T*-valve "suck and spit" system to rinse tubing and syringe with sample prior to bulk collection. This sediment core was taken with a hand-deployed torpedo corer from a Boston Whaler with hand winch on the second attempt—the first liner melted and could not be stoppered, losing the core. Core pore water was extracted that afternoon with a Jahnke (1988) "squeezer" using acid washed, inert-gas equilibrated Porex porous rod and rubber-free syringes: a 0.2 μm pore size ion chromatography syringe filter was inserted between port fittings and syringes. Pore water and water column samples were assayed as soon as possible.

Many of the components of interest are labile to rapid chemical and biological change and real-time information on vent and pore water chemical composition is valuable for next-day sampling decisions. It was imperative therefore that numerous aspects of analytical chemistry were conducted on site. Those analytes that could be stabilized for later measurement were returned to the Milwaukee laboratory. On site, the PIs in conjunction with graduate and NSF Research Experience for Undergraduates (REU) students ran as immediately as possible ΣCO_2 by Teflon diffusion (Hall and Aller 1992); sulfur species by HPLC on DTNP-derivatized samples (Vairavamurthy and Mopper 1990); most nutrients including $NO_3^- + NO_2^-$ and SiO_2 by FIA (APHA 1996); ammonium by phenol-hypochlorite spectrophotometry (APHA 1996); and chloride–sulfate by ion chromatography (APHA 1996).

Chemo- and photosynthesis were measured as acid-stable $^{14}CO_2$ fixation in dark and light incubations, respectively, using water-cooled aluminum "photosynthetron" blocks with circulator baths underneath or aluminum dry blocks for elevated temperature measurements (Yang et al. 2011). For both lakewater and ventwater samples, subsamples were spiked with $^{14}CO_2$ to ca. 1 μCi/mL and incubated at 10–14 °C or sealed at 50 °C depending on in situ conditions. Measurements were terminated by addition of H_2SO_4. Chemosynthesis measurements by this technique are aerobic, as is the bottom receiving water in Yellowstone Lake, and hence represent in-lake rather than in-vent activity potential.

Big Picture Outcomes

After about 10 years of ad hoc investigations, a hypothetical scenario arose, involving geothermal gas-fluid transport through under-lake conduits and associated biogeochemical processes in receiving waters. In the first NSF Environmental Geochemistry and Biogeochemistry-sponsored expedition of 1998 concerted efforts were made to test this hypothesis, and based on exciting outcomes were further refined in 1999 and subsequent years. Here, we recount a series of sampling activities and analytical outcomes leading to a complete vertical profile from the surface to vent orifices and further into the surrounding sediments. It is a vignette from a synoptic study involving beach, SCUBA, ROV, and small boat sampling; groundwater and tributary inflow analysis; comparisons inside and outside of caldera boundaries; and every source or sink of water we could sample.

The first evidence of impending excitement for lake vents was the appearance of "yellow bubbles" on the surface of the lake. Primitive as it was, the trace from the sonar depth sounder (Fig. 1c) showed a series of peaks and troughs punctuated with tiny vertical signals far too small to be fish, and also oddly arranged for such a source. In reality, they were towers of tiny bubbles emanating from canyon-bottom vents containing entrained gases. Among these was hydrogen sulfide, which gave the region its characteristic "smell of success" (Aguilar, personal remark) (Aguilar et al. 2002). On transit to the distant surface (50+ m) through aerobic water, the H_2S abiotically oxidized to elemental sulfur, an oily yellow substance that formed ephemeral rings on the surface as the bubbles popped.

The next bit of impetus for in-depth study of the canyon came from lowering a Hydrolab hydrographic sonde from the surface to the bottom. Mary Bay has a basin depth in the vicinity of 15–20 m punctuated with deep 40–50 m canyons in the benthiscape (Fig. 1c). An eye-catching component is therefore the basic temperature profile, which shows a standard decrease from the surface with a sharp thermocline at 5–11 m but then remaining above 8 °C all the way to the bottom (Fig. 1b). Typically lakewater at this depth would be 4 °C. This warm basin also demonstrated a weak but decreasing slope of dissolved oxygen with depth where oxygen in oligotrophic lakes generally increases into colder bottom waters. Finally, oxidation–reduction potential (ORP) decreased slightly within canyon (15 m to bottom) but also showed a steep drop very close to the bottom. It is intriguing that all three parameters show sharp inflections in the 2–3 points (30 cm span) just above the bottom (temperature up; oxygen and ORP down)—but only the temperature could not be the result of trace resuspension of bottom sediments. Collectively, these are consistent with a warm bottom containing a mild but persistent dose of readily oxidizable material.

Yellowstone Lake water and its main outflow at Fishing Bridge are strongly enriched in geochemicals relative to the Yellowstone River inflow in Southeast Arm. Chloride, sulfate, silicate, and total CO_2 (5, 33, 180, and 370 μM, respectively, in inlet samples) are all components or products of hydrothermal fluids and expected to be enhanced near vent emanations. Still, chemical analyses demonstrated few distinctly different water characteristics within the canyon below the sill near 15 m (Fig. 2a; sill zone is stippled). Total CO_2 was the only parameter that became elevated above surface values specifically at or below the sill depth (Fig. 2a). Most were consistent within the basin itself. Total CO_2 may have in fact increased slightly near the very bottom. This in itself is a peculiarity, because nutrients produced by diagenesis of sedimenting organic matter, including phosphate (SRP), silicate, and ammonium or nitrate do not increase near the bottom in this confined zone. Geothermal indicator H_2S was no different in deep samples, and was actually enriched only at the surface, where the source was likely atmospheric from the abundant fumarole gas influence (the "smell of success") in the embayment.

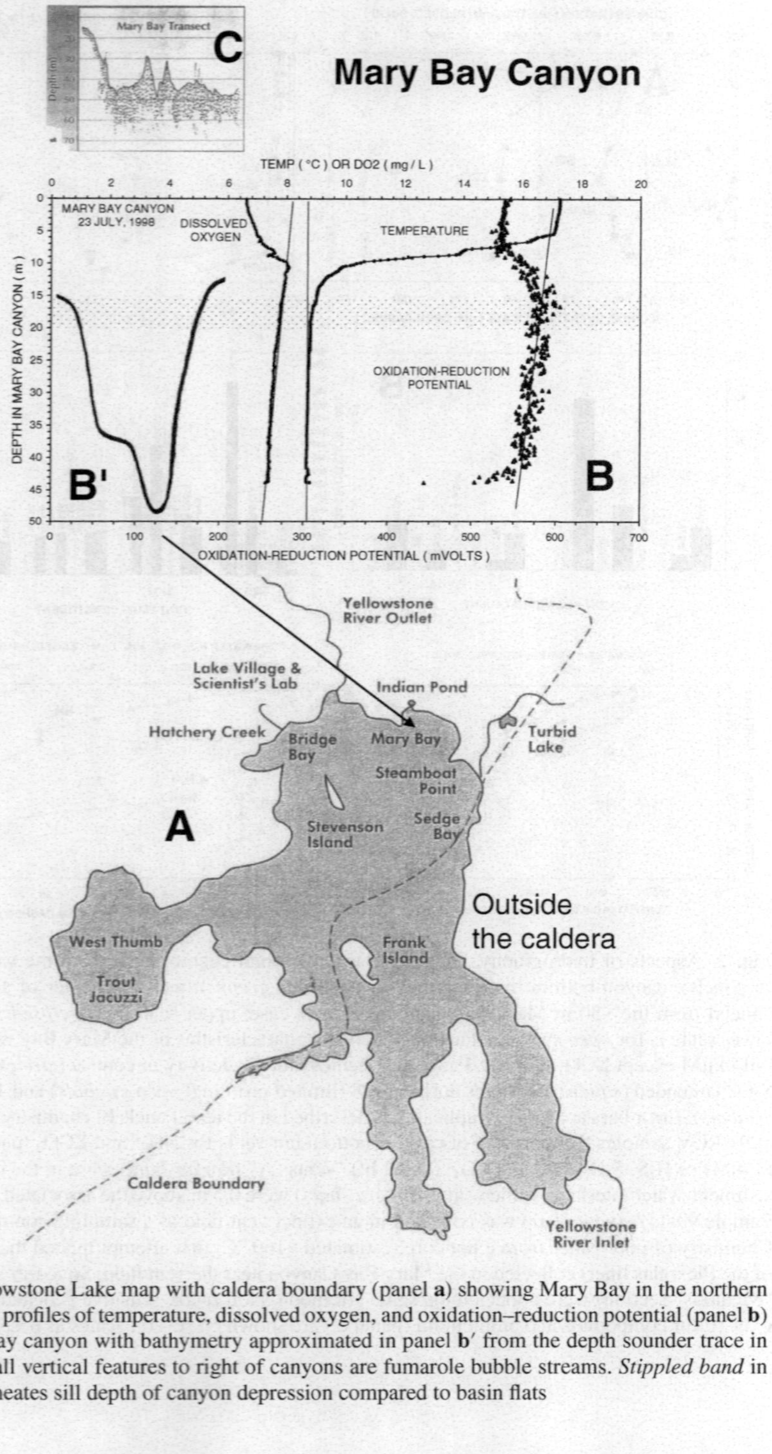

Fig. 1 Yellowstone Lake map with caldera boundary (panel **a**) showing Mary Bay in the northern basin. Depth profiles of temperature, dissolved oxygen, and oxidation–reduction potential (panel **b**) in a Mary Bay canyon with bathymetry approximated in panel **b′** from the depth sounder trace in panel **c**. Small vertical features to right of canyons are fumarole bubble streams. *Stippled band* in panel **b** delineates sill depth of canyon depression compared to basin flats

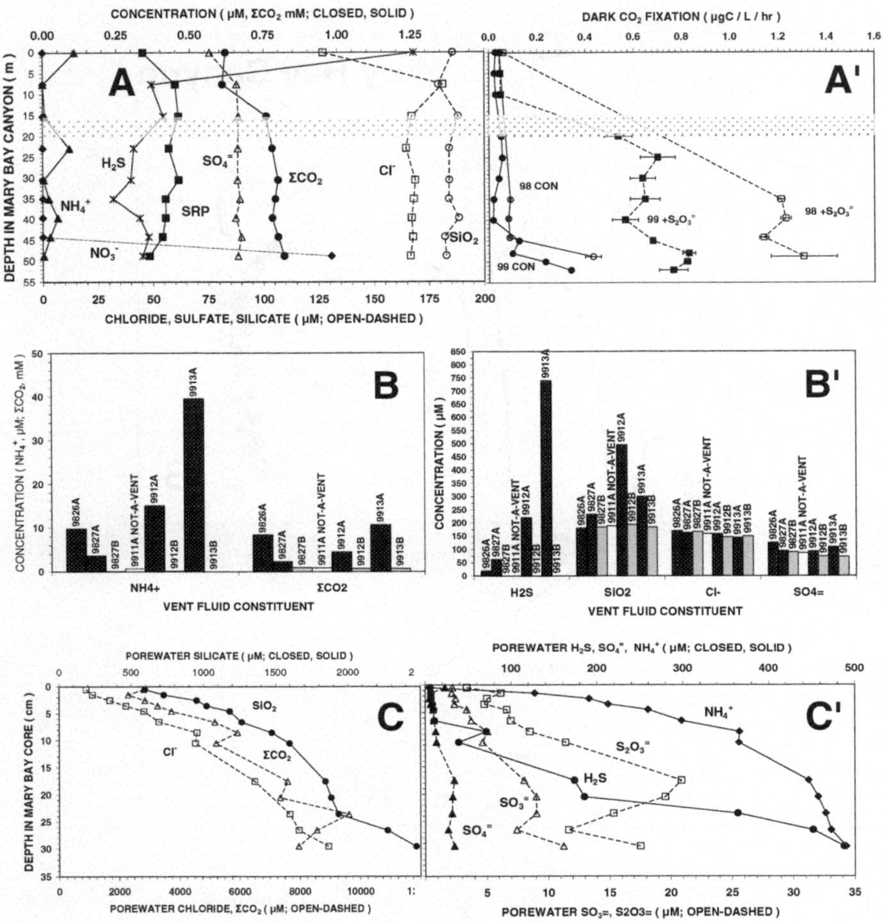

Fig. 2 Aspects of hydrography, chemistry, and microbial chemosynthesis in the water column (**a** panels), canyon-bottom hydrothermal vents (**b** *bar graphs*), and pore water of a hot core (**c** panels) from the ~50 m Mary Bay Canyon. In each case, upper scale is for *closed symbols* and lower scale is for *open symbols*. Panel **a**: Chemical characteristics of the Mary Bay water column (all in µM except ΣCO_2 in mM). Panel **a**′: Chemosynthetic activity in control (*circles*) and 5 mM $S_2O_3^=$-amended (*squares*) samples during 1998 (limited profiling; *open symbols*) and 1999 (*closed symbols*). Error bars are for 3–7 replicates as described in the text. Panels **b**: chemistry of 1998 and 1999 ROV samples from orifices of canyon-bottom hot vents for NH_4^+ and ΣCO_2 (panel **b**; ΣCO_2 in mM) or H_2S, SiO_2, Cl^-, and $SO_4^=$ (panel **b**′). Vents "A" (*darker bars*) were in the orifice at the sediment-water interface, while vents "B" (*gray bars*) were 0.5 m above the associated "A" sample. Sample 99-11A (*white bar*) was collected in an extinct vent field as a sampling control. Panels **c**: Chemistry of pore water from a hot core (estimated ~160 °C; first attempt melted the bottom end of the Plexiglas liner) collected in the Mary Bay Canyon near the vent field. Squeezer samples (see "Methods" section) were collected the same afternoon. Non-redox-sensitive geochemicals (panel **c**) or sulfur compounds and ammonium (panel **c**′) are shown on several scales as described

Intermediate reduced sulfur compounds sulfite and thiosulfate were below detection (<0.1 μM; data not shown). These results indicate that either (efflux plus diagenesis) and biological uptake are in balance, or there is little flux from either source.

For chemosynthesis, the hypothesis is that reduced substrates (e.g., reduced sulfur, hydrogen, ammonium, iron, and methane) and the bacteria that use them for chemosynthetic energy are trapped and hence concentrated in these steep deep canyons. Dark incubations for CO_2 uptake yielded very low levels of autotrophy at any point in the canyon except for a systematic increase of two- to three-fold in the very bottom samples (Fig. 2a′). Incubations enriched with 5 mM thiosulfate ($S_2O_3^=$) were consistently elevated in all canyon samples below the ca. 15 m sill (Fig. 2a′), revealing strong chemosynthesis *potential* in the bottom of the canyon. The replicate error was much higher in the very bottom sample, likely due to chemosynthetic bacteria attached to suspended sediment particles.

Hydrothermal vents at the bottom of Mary Bay Canyon were indeed a source of certain geochemical enrichment based on two ROV sipper samples from 1998 (55, 66 °C) and three from 1999 (thermistors failed). H_2S (19–739 μM) and related reduced sulfur compounds $SO_3^=$ and $S_2O_3^=$ (not shown) were present in vent orifice samples (darkest bars; level of the sediment-water interface) and total CO_2 was enriched up to 12 times (1.04–10.654 mM) over bottom water values (0.8 mM; white bars in Figure). Vent orifices were among rare positive samples for ammonium (3.6–39.5 μM), which was not seen in near-vent or bottom samples. Also enriched were SRP (2–10×), silicate (2×), and a few percent systematic enhancement in sulfate. Conservative tracer chloride, however, was not enriched in any of these vent samples, and nitrate was rarely detected anywhere. Mixing with bottom water was documented with a second sipper taped 0.5 m above the orifice tip on the ROV arm ("$V+0.5$ m"; light gray bars). In hot vent collections, we were indeed surprised to find that the $V+0.5$ m samples were barely distinguishable from the chemical composition of bottom water, indicating rapid advective mixing, rapid transformation of labile species, and slow flow rates (though the shimmering heated water was readily visible in the ROV camera view).

Chemosynthesis was active in the sampled vents at 10–14 °C water temperatures, with an expectedly high degree of variability among them (Fig. 3). In 1998, cooler vent bacteria in 98-26A (55 °C, 18.7 μM H_2S) fixed CO_2 at 1.21 ± 0.09 μgC/L/h ($n=7$) with reduction by 5 mM thiosulfate. Chemical reaction of added $S_2O_3^=$ with an unrelated electron donor (e.g., Fe(II)) could result in lower chemosynthetic CO_2 fixation without actually causing inhibition per se. Microbes in the warmer vent 98-27A (66 °C, 63 μM H_2S) fixed CO_2 at 0.68 ± 0.03 μgC/L/h ($n=7$). Sample 98-27B, 0.5 m above the vent, was identical to its orifice partner in chemosynthetic behavior even though the chemistry was more like bottom water (Fig. 2b). Ammonium never stimulated chemosynthesis, and the ribosomal protein synthesis inhibitor chloramphenicol (CAP) reduced fixation about 85 % in each case.

Incubations at elevated temperature (50 °C) delineated the presence of thermophilic chemolithoautotrophic bacteria in vents sampled in 1999 (Fig. 3). In vent sample 99-12A (220 μM H_2S), dark CO_2 fixation occurred at 4.15 ± 0.67 μgC/L/h (n = 3) with 20 % reduction by thiosulfate, but increased threefold to

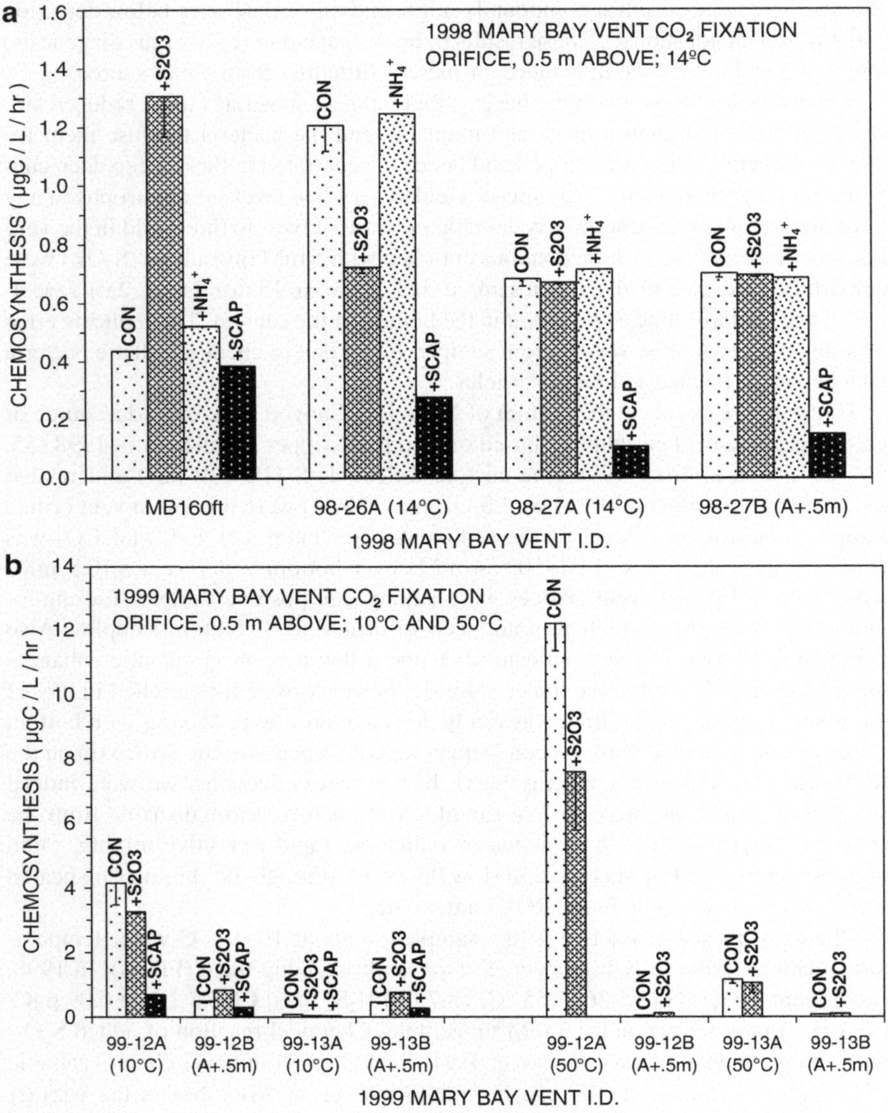

Fig. 3 Chemosynthesis in Mary Bay Canyon hydrothermal vent fluids during 1998 (panel **a**) and 1999 (panel **b**) field collections. In 1998, 14 °C incubations included three controls, three each enrichments with 5 mM thiosulfate ($S_2O_3^=$) or ammonium (NH_4^+), and a negative control with $S_2O_3^=$ plus bacterial protein synthesis inhibitor chloramphenicol (CAP). "A" and "B" designate vents and above-vents as in Fig. 2. Addition of 50 °C incubations in 1999 (panel **b**, *right side*) enabled detection of thermophilic chemosynthesis. No NH_4^+ or CAP additions were made at 50 °C

12.21 ± 0.87 μgC/L/h at 50 °C. The second Mary Bay Canyon sample was extremely high in H_2S (739 μM) and had very low rates of chemosynthesis at 10 °C (0.079 ± 0.003 μgC/L/h, $n = 3$) but responded with a 15-fold increase at 50 °C. Based on both H_2S and thermal characteristics, the second sample 99-13A appeared to be less admixed with bottom lakewater and was populated with obligate thermophiles. In both 1999 cases, the associated half-meter-above samples (B) closely resembled water column samples with low ambient rates (0.49, 0.46 μgC/L/h) responding strongly to thiosulfate stimulation (0.84, 0.76 μgC/L/h) and losing 90 % of activity at 50 °C.

Light-saturated photosynthesis in surface survey samples was 1–4 μgC/L/h (data not shown). This range is similar on a volumetric basis to unammended, 10–14 °C chemosynthesis rates of 0.5–4 μgC/L/h in the five vent orifice samples described here. The *magnitude* of total vent contribution is unquantified, but widespread within the caldera portion of the lake.

Influence of the geothermal system on sediment pore water chemistry is also profound in the vent fields of oligotrophic Yellowstone Lake. In our first attempt to core Mary Bay Canyon, the polycarbonate liner was melted and we were lucky to save the torpedo device itself. Very rapid retrieval after impact enabled us to successfully collect a very hot core on the second attempt. The pore water, squeezed that afternoon, provided clear evidence of hydrothermal intrusion into sediments (Fig. 2c, c′). Every parameter shown in the profiles was highly enriched over lake bottom water (parenthetical values below), indicating flux into the lake. In the 30 cm span of our collections, ΣCO_2 (850 μM) increased from 2 to 10 mM downcore, far greater than control locations outside the caldera that reached up to only 3 mM in deeper horizons. Silicate (180 μM) rose from 600 μM to near 2500 μM, and H_2S (<1 μM) suddenly increased below the 7 cm horizon to attain deep values of near 500 μM. Organic diagenesis would not reach these levels. Conservative chloride (160 μM) displayed a nearly linear gradient reaching 9 mM downcore, further supporting the conclusion that hydrothermally enriched fluids were contributing to pore water composition. Only ammonium is not potentially a geothermal product but rather a leachate of transit (Holloway et al. 2011) and its contrasting steep gradient in the upper core is more consistent with a diagenetic profile.

Closing Remarks

Geothermally influenced water flows under Yellowstone Lake within the caldera boundary and promotes geochemical and microbiological processes that then alter the chemistry of the lake itself. Water emanating from vent orifices in an enclosed canyon location mixes into lake bottom water with rapid dissipation of geochemical signals but supports overlying chemosynthetic bacterial populations within the geologically constrained system. Hot vents reaching over 65 °C sometimes contain obligately thermophilic chemolithoautotrophic bacteria that attain dark CO_2 fixation rates much higher than overlying surface photosynthetic production. Mesophilic chemoautotrophic bacteria are also found in admixtures of ventwater and receiving water, and are reliably stimulated by a model reduced sulfur energy source, thiosulfate. Pore water gradients of a hot core collected near these vents indicate efflux of geothermally sourced chemicals into the lake through bottom substrates as well as through direct venting from specific orifices. Elevated concentrations of intermediate reduced sulfur compounds sulfite and thiosulfate, along with a pore water sulfide gradient sink or barrier at 5–7 cm below the sediment-water interface, points to vigorous biogeochemical oxidation in upper sediments near hydrothermal venting regions. Widespread active or potential chemosynthetic activity by microbial populations indicates significant continuous point source and diffuse input of reduced energy substrates that promotes carbon dioxide fixation by bacterial assemblages. Both biotic and abiotic geochemical processes, arising through interaction with geothermal solid and fluid phases, systematically change lakewater chemistry relative to the major inflow of the Yellowstone River outside of the caldera boundary.

One of the many interesting applications of this 10-year project has been the idea of a model site for testing NASA mission vehicles intended for use on possibly watery moons Enceladus and Titan of Saturn or Europa of Jupiter. A winter expedition to bore through Mary Bay ice, sniff geochemical gradients, navigate to sources, and measure a suite of relevant compounds and potential processes (e.g., add oxygen, seek oxidation chemistry) would be a stunning exercise.

Acknowledgements The research described in this article grew out of years of work by Klump et al. (1995). During the 1998–1999 field seasons, research costs and student participation were funded by National Science Foundation Grants EAR-EGB-9708501 (Environmental Geochemistry and Biogeochemistry to RLC et al.; research) and OCE 9732316 (Research Experience for Undergraduates to R.L.C. and C.A.; education). Any opinions, findings, conclusions, or recommendations expressed in this report are those of the participants and do not necessarily represent the official views, opinions, or policy of the National Science Foundation. Development and operation of ROV resources by contractor David Lovalvo of Eastern Oceanics was far beyond expert level and is tremendously appreciated. The participants also thank the National Park Service in Yellowstone National Park, particularly the Fisheries and Aquatic Sciences Section, Center for Resources and at Bridge Bay and Lake Ranger Stations who have made it possible for us to have conducted research in the Park.

Carmen Aguilar and Russell Cuhel

Russell met Carmen while serving as a visiting scholar for Ken Nealson's Shaw Distinguished Professor program—Ken was Carmen's dissertation advisor. They clearly could party together, and after a couple of visits to Carmen's research site on Oneida Lake it was clear (except for the Cyanobacterial blooms) that they could complement each other in collaborative scientific research as well. When Carmen was a Postdoc at the Carnegie Institution of Washington and the University of North Carolina at Chapel Hill, she obtained space for Russell on research cruises in the Sargasso Sea under the pretext of his providing high sensitivity nitrate analyses for her Acid Precipitation program with Drs. Marilyn Fogel and Hans Paerl. They worked together during the Hurricane Gordon cruise, where Russell saved Christmas by reconstructing a working Niskin bottle from spare parts after the rosette sampler was lost in heavy weather. This led Carmen to propose using blue topaz stud earrings, and they were married several cruises later on the RV Cape Hatteras at Station 5 during the Hurricane Marilyn cruise of September 1995, one of three couples to become hitched on a research vessel. Since that time they have immersed themselves wholeheartedly in a continuum of research, research education, scientific and personal travel, and fabulous cuisine preparation continuing their individual traditions to the max. During the subject years of this paper, they had the following projects running simultaneously during the summer of 1998 and 1999: NSF Research Experience for Undergraduates Site in Aquatic Sciences directors; NSF REU 6-day expedition covering most of Lake Michigan; Elkhart Lake WI invasive species Sea Grant project; NOAA-CISNet Lake Michigan time series program revitalization; and this NSF Yellowstone Lake Environmental Geochemistry and Biogeochemistry Program expedition. They are still like that….

References

Aguilar C, Cuhel RL, Klump JV (2002) Porewater and hydrothermal vent water inputs to Yellowstone Lake, Wyoming. In: Anderson RJ, D. Harmon (eds) Yellowstone Lake: Hotbed of chaos or reservoir of resilience? Proceedings of the 6th biennial conference on the greater Yellowstone ecosystem, 8–10 Oct 2001. Mammoth Hot Springs Hotel, Yellowstone Center for Resources and The George Wright Society, Yellowstone National Park, WY, Hancock MI, p 1–18

American Public Health Association (APHA) (1996) Standard methods for the examination of water and wastewater, 20th edn. APHA, Washington, DC

Anderson RJ, Harmon D (eds) (2002) Yellowstone Lake: hotbed of chaos or reservoir of resilience? Proceedings of the 6th biennial conference on the greater Yellowstone ecosystem, 8–10 Oct 2001. Mammoth Hot Springs Hotel, Yellowstone Center for Resources and The George Wright Society, Yellowstone National Park, WY, Hancock, MI

Brock TD (1997) The value of basic research: discovery of *Thermus aquaticus* and other extreme thermophiles. Genetics 146:1207–1210

Brock TD, Freeze H (1969) *Thermus aquaticus* gen. n. and sp. n., a nonsporulating extreme thermophile. J Bacteriol 98(1):289–297

Buchholz LA, Anderson PD, Cuhel RL, Klump JV, Kostka JE, Paddock RW, Remsen CC, Maki JS, Lovalvo D (1995) Employment of ROV techniques and Scuba in Yellowstone Lake. In: Harper DEJ (ed) Diving for science. American Academy of Underwater Science, Nahant, MA, pp 1–7

Corliss JB, Dymond J, Gordon LI, Edmond JM, Von Herzen RP, Ballard RD, Green K, Williams D, Bainbridge A, Crane K, van Andel T (1979) Submarine thermal springs on the Galapagos Rift. Science 203(4385):1073–1083

Drever JI (1974) The magnesium problem. In: Goldberg EJ (ed) The sea, vol 5. Wiley, New York, NY, pp 337–350

Hall OJ, Aller RC (1992) Rapid, small-volume, flow injection analysis for ΣCO_2 and NH_4^+ in marine and freshwaters. Limnol Oceanogr 37:1113–1119

Holloway JM, Nordstrom DK, Böhlke JK, McCleskey RB, Ball JW (2011) Ammonium in thermal waters of Yellowstone National Park: processes affecting speciation and isotope fractionation. Geochim Cosmochim Acta 75:4611–4636

Hurwitz S, Lowenstern JB (2014) Dynamics of the Yellowstone hydrothermal system. Rev Geophys 52(3):375

Jahnke R (1988) A simple, reliable, and inexpensive pore-water sampler. Limnol Oceanogr 33(3):483–487

Klump JV, Remsen CC, Lovalvo D, Anderson P, Cuhel RL, Kaplinski M, Kaster J, Maki J, Paddock R (1995) 20,000 leagues under Yellowstone Lake. Yellowstone Sci 3(4):10–14

Konkol NR, Bruckner JC, Aguilar C, Maki JS (2010) Dominance of epiphytic filamentous *Thiothrix* spp. on an aquatic macrophyte in a hydrothermal vent plume in Sedge Bay, Yellowstone Lake, Wyoming. Microb Ecol 60(3):528–538

Martin W, Baross J, Kelley D, Russell M (2008) Hydrothermal vents and the origin of life. Nat Rev Microbiol 6:805–814

Miller SL (1953) A production of amino acids under possible primitive Earth conditions. Science 117:528–529

Rona PA, Bostrom K, Laubier L, Smith KL Jr (eds) (1983) Hydrothermal processes at seafloor spreading centers. Springer, Springer, p 796

Vairavamurthy A, Mopper K (1990) Determination of sulfite and thiosulfate in aqueous samples including anoxic seawater by liquid chromatography after derivatization with 2, 2′-dithiobis (5-nitropyridine). Environ Sci Tech 24:333–337

Yang T, Lyons S, Aguilar C, Cuhel RL, Teske A (2011) Microbial communities and chemosynthesis in Yellowstone Lake sublacustrine hydrothermal springs. Front Microbiol 2(130):1–17. doi:10.3389/fmicb.2011.00130

Index

P.M. Glibert, T.M. Kana (eds.), *Aquatic Microbial Ecology and Biogeochemistry: A Dual Perspective*, DOI 10.1007/978-3-319-30259-1

D

E

F

G

H

K

L

M

The manufacturer's authorised representative in the EU is Springer Nature Customer Service Centre GmbH, Europaplatz 3, 69115 Heidelberg, Germany. If you have any concerns regarding our products, please contact ProductSafety@springernature.com

Printed and bound by CPI Group (UK) Ltd, Croydon, CR0 4YY
25/09/2026
02222836-0012